UNIQUENESS AND NONUNIQUENESS CRITERIA FOR ORDINARY DIFFERENTIAL EQUATIONS

SERIES IN REAL ANALYSIS

Published

Vol. 1: Lectures on the Theory of Integration
R Henstock

Vol. 2: Lanzhou Lectures on Henstock Integration
Lee Peng Yee

Vol. 3: The Theory of the Denjoy Integral & Some Applications
V G Celidze & A G Dzvarseisvili
translated by P S Bullen

Vol. 4: Linear Functional Analysis
W Orlicz

Vol. 5: Generalized ODE
S Schwabik

Vol. 6: Uniqueness and Nonuniqueness Criteria for Ordinary Differential Equations
R P Agarwal & V Lakshmikantham

Vol. 7: Henstock–Kurzweil Integration: Its Relation to Topological Vector Spaces
Jaroslav Kurzweil

Vol. 8: Integration between the Lebesgue Integral and the Henstock–Kurzweil Integral: Its Relation to Local Convex Vector Spaces
Jaroslav Kurzweil

Vol. 9: Theories of Integration: The Integrals of Riemann, Lebesgue, Henstock–Kurzweil, and McShane
Douglas S Kurtz & Charles W Swartz

Vol. 10: Topics in Banach Space Integration
Štefan Schwabik & Ye Guoju

Vol. 11: Generalized Ordinary Differential Equations: Not Absolutely Continuous Solutions
Jaroslav Kurzweil

Vol. 12: Henstock–Kurzweil Integration on Euclidean Spaces
Tuo Yeong Lee

Vol. 13: Theories of Integration: The Integrals of Riemann, Lebesgue, Henstock–Kurzweil, and McShane (2nd Edition)
Douglas S Kurtz & Charles W Swartz

Series in Real Analysis Volume 6

UNIQUENESS AND NONUNIQUENESS CRITERIA FOR ORDINARY DIFFERENTIAL EQUATIONS

R P Agarwal

Department of Mathematics, National University of Singapore

V Lakshmikantham

Department of Applied Mathematics, Florida Institute of Technology

World Scientific

Singapore • New Jersey • London • Hong Kong

Published by

World Scientific Publishing Co. Pte. Ltd.
5 Toh Tuck Link, Singapore 596224
USA office: 27 Warren Street, Suite 401-402, Hackensack, NJ 07601
UK office: 57 Shelton Street, Covent Garden, London WC2H 9HE

British Library Cataloguing-in-Publication Data
A catalogue record for this book is available from the British Library.

Series in Real Analysis — Vol. 6
UNIQUENESS AND NONUNIQUENESS CRITERIA FOR ORDINARY DIFFERENTIAL EQUATIONS

ISBN-13 978-981-02-1357-2
ISBN-10 981-02-1357-3

PREFACE

The theory of differential equations, the methods used in their solutions, and their wide applications have all advanced to the extent that aspects in each of these areas have demanded individual attention, resulting in a number of excellent monographs of special interest.

The question of the existence and uniqueness of solutions of differential equations is an age - old problem of great importance. Research has continued to find, refine and generalize uniqueness conditions, resulting in an enormous amount of literature offering various sufficient conditions for uniqueness problems, with implied conditions for the existence of solutions. Surprisingly, though, no attempt appears to have been made to bring all the available uniqueness criteria into one volume, which would enable one to compare and contrast these criteria, and then, perhaps, go on to discover new conditions, without traversing already known ground. Furthermore, it has been recently brought to light that there are still existence and uniqueness problems which are of merit and yet for which existing results or ideas cannot offer any worthwhile solutions.

This monograph aims to fill a void by making available a source book which first systematically describes the existing literature, and compares and contrasts the merits of various known criteria, and second, discusses open problems and offers some directions towards possible solutions.

The authors gratefully acknowledge the valuable help of Drs S.L.Huang, Y.K. Leong, M.P.Navarro, R.Schulz and Q.Sheng for translating several papers from Japanese, Russian, French, German and Chinese. Special mention needs also to be made of the help given by Dr Q.Sheng in unravelling the mysteries of LaTeX. In addition, we are indebted to Prof. Lee Peng Yee for accepting our monograph in his Series in Real Analysis of the World Scientific.

R.P.Agarwal
V.Lakshmikantham

CONTENTS

CHAPTER 2

FIRST ORDER DIFFERENTIAL EQUATIONS (CONTD.)

CHAPTER 3

FIRST ORDER DIFFERENTIAL SYSTEMS

CHAPTER 7

FUNCTIONAL DIFFERENTIAL EQUATIONS

CHAPTER 8

IMPULSIVE DIFFERENTIAL EQUATIONS

CHAPTER 9

DIFFERENTIAL EQUATIONS WITH HYSTERESIS

CHAPTER 10

GENERALIZED DIFFERENTIAL EQUATIONS

CHAPTER 1

FIRST ORDER DIFFERENTIAL EQUATIONS

1.1 INTRODUCTION

Consider the initial value problem

(1.1.1) $$y' = f(x, y), \ y(x_0) = y_0$$

where the function $f(x, y)$ is at least continuous in a domain $D \subseteq R^2$, and $(x_0, y_0) \in D$. By a **solution** of (1.1.1) in an interval J containing x_0, we mean a function $y(x)$ satisfying :

(i) $y(x_0) = y_0$,

(ii) for all $x \in J$, the points $(x, y(x)) \in D$,

(iii) $y'(x)$ exists and continuous for all $x \in J$; and

(iv) $y'(x) = f(x, y(x))$.

If J is closed then at the end points of J only one - sided existence of $y'(x)$ is assumed.

It is well known that the continuity of $f(x,y)$ in a closed rectangle

$$\bar{S} : |x - x_0| \leq a, |y - y_0| \leq b$$

is sufficient for the existence of at least one solution of (1.1.1) in the interval $J_h : |x - x_0| \leq h = \min(a, b/M)$, where $M = \sup_{\bar{S}} |f(x,y)|$. However, if $f(x,y)$ is discontinuous then the nature of the solutions is quite arbitrary. For example, the initial value problem $y' = 2(y-1)/x$, $y(0) = 0$ has no solution; while the problem $y' = 2(y-1)/x$, $y(0) = 1$ has an infinite number of solutions $y(x) = 1 + cx^2$, where c is an arbitrary constant.

Although, continuity of $f(x,y)$ is sufficient for the existence of a solution of (1.1.1), it does not imply the uniqueness of the solutions. For example, the function $f(x,y) = y^{2/3}$ is continuous in R^2, but the problem $y' = y^{2/3}$, $y(0) = 0$ has an infinite number of solutions $y(x) \equiv 0$, and

$$y(x) = \begin{cases} 0, & 0 \leq x \leq c \\ (x-c)^3/27, & x \geq c \end{cases}$$

where c is an arbitrary constant. In fact, the situation is rather bad, in the year 1925 Lavrentieff [96] gave an example of a continuous function $f(x,y)$ in the open rectangle

$$S : |x - x_0| < a, \quad |y - y_0| < b$$

with the property that for every choice of the initial point (x_0, y_0) interior to S, the initial value problem (1.1.1) has more than one solution in every interval $[x_0, x_0 + \epsilon]$ and $[x_0 - \epsilon, x_0]$. Further, in the year 1963 Hartman [58] used methods somewhat similar to that of van Kampen [159] and constructed another continuous function $f(x,y)$ in R^2 so that the problem (1.1.1) has the same nonuniqueness property. Hence, to ensure the uniqueness also, besides continuity we need to impose some additional conditions on $f(x,y)$.

1.2 LIPSCHITZ UNIQUENESS THEOREM

To ensure the uniqueness of the solutions of (1.1.1) we begin with the assumption that the variation of the function $f(x,y)$ relative to y remains bounded. We state this condition in

Definition 1.2.1. The function $f(x,y)$ is said to satisfy the **uniform Lipschitz condition** in a domain D if

$$|f(x,y) - f(x,\bar{y})| \le L|y-\bar{y}| \tag{1.2.1}$$

for all $(x,y),(x,\bar{y})$ in D having the same x. The nonnegative constant L is called the **Lipschitz constant.**

The Lipschitz condition lies in between the continuity and the differentiability of f with respect to y. Indeed, from (1.2.1) the continuity of f with respect to y is obvious, and the continuous function $f(x,y) = y^{2/3}$ violates the Lipschitz condition in any domain containing $y = 0$. However, the nondifferentiable function $f(x,y) = |y|, (x,y) \in R^2$ satisfies (1.2.1) with $L = 1$. If the function f is differentiable with respect to y, then it is easy to compute the Lipschitz constant L. In fact, we shall prove

Lemma 1.2.1. Let the domain D be convex and the function $f(x,y)$ be differentiable with respect to y in D. Then, for the Lipschitz condition (1.2.1) to be satisfied, it is necessary and sufficient that

$$\sup_D \left|\frac{\partial f(x,y)}{\partial y}\right| \le L. \tag{1.2.2}$$

Proof. Since $f(x,y)$ is differentiable with respect to y and the domain D is convex, for all $(x,y),(x,\bar{y}) \in D$ the mean value theorem provides

$$f(x,y) - f(x,\bar{y}) = \frac{\partial f(x,y^*)}{\partial y}(y-\bar{y}), \tag{1.2.3}$$

where $\min\{y,\bar{y}\} < y^* < \max\{y,\bar{y}\}$. Thus, on using (1.2.2) in (1.2.3), we find (1.2.1). Conversely, from (1.2.1) we have

$$\left|\frac{\partial f(x,y)}{\partial y}\right| = \lim_{\bar{y}\to y}\left|\frac{f(x,y)-f(x,\bar{y})}{y-\bar{y}}\right| \le L. \quad \blacksquare$$

To prove the first main result of this section we need the following two lemmas.

Lemma 1.2.2. Let $f(x,y)$ be continuous in the domain D, then any solution of (1.1.1) is also a solution of the integral equation

$$y(x) = y_0 + \int_{x_0}^{x} f(t, y(t))dt \tag{1.2.4}$$

and conversely.

Proof. Any solution of the differential equation $y' = f(x,y)$ converts it into an identity in x, i.e., $y'(x) = f(x, y(x))$. An integration of this equality with $y(x_0) = y_0$ gives (1.2.4). Conversely, if $y(x)$ is any solution of (1.2.4), then the substitution $x = x_0$ in (1.2.4) gives $y(x_0) = y_0$. Further, since $f(x,y)$ is continuous, on differentiating (1.2.4), we find that $y'(x) = f(x, y(x))$. ∎

Lemma 1.2.3. Let $\phi(x)$ and $q(x)$ be nonnegative continuous functions in $|x - x_0| \leq a$, and satisfy

$$\phi(x) \leq \left| \int_{x_0}^{x} q(t)\phi(t)dt \right|. \tag{1.2.5}$$

Then, $\phi(x) = 0$ in $|x - x_0| \leq a$.

Proof. For $x \in [x_0, x_0 + a]$, we define $r(x) = \int_{x_0}^{x} q(t)\phi(t)dt$ so that $r(x_0) = 0$, and $r'(x) = q(x)\phi(x)$. Thus, from (1.2.5) we have $r'(x) \leq q(x)r(x)$. Multiplying both sides of this inequality by $\exp\left(-\int_{x_0}^{x} q(t)dt\right)$, we get

$$\frac{d}{dx}\left(\exp\left(-\int_{x_0}^{x} q(s)ds\right) r(x)\right) \leq 0.$$

Thus, $\exp\left(-\int_{x_0}^{x} q(s)ds\right) r(x)$ is nonincreasing. Since $r(x_0) = 0$, it follows that $r(x) \leq 0$, and hence $\phi(x) \leq r(x) \leq 0$. However, since the function $\phi(x)$ is nonnegative, we find that $\phi(x) = 0$ in $[x_0, x_0 + a]$. The proof is similar in $[x_0 - a, x_0]$. ∎

Theorem 1.2.4 (Lipschitz Uniqueness Theorem). Let $f(x,y)$ be continuous and satisfy the uniform Lipschitz condition (1.2.1) in $\bar{S}$. Then, the initial value problem (1.1.1) has at most one solution in $|x - x_0| \leq a$.

First Proof. Suppose that $y(x)$ and $\bar{y}(x)$ are two solutions of (1.1.1) in

$|x - x_0| \leq a$, then from (1.2.4) and (1.2.1) it follows that

$$\begin{aligned} |y(x) - \bar{y}(x)| &\leq \left| \int_{x_0}^{x} |f(t, y(t)) - f(t, \bar{y}(t))| dt \right| \\ &\leq L \left| \int_{x_0}^{x} |y(t) - \bar{y}(t)| dt \right|. \end{aligned}$$

Now as an application of Lemma 1.2.3, we find that $|y(x) - \bar{y}(x)| \equiv 0$, and hence $y(x) = \bar{y}(x)$ in $|x - x_0| \leq a$. ∎

It is interesting to note that Theorem 1.2.4 can also be proved by using mean value theorem, and the alternative proof does not require the integral equation representation (1.2.4). Further, the novelty in the proof is in its simplicity.

Second Proof. We shall consider the interval $[x_0, x_0 + a]$, whereas the proof in the interval $[x_0 - a, x_0]$ is similar. Suppose $y(x)$ and $\bar{y}(x)$ are two solutions of (1.1.1). Then, $y'(x) = f(x, y(x))$ and $\bar{y}'(x) = f(x, \bar{y}(x))$ in $[x_0, x_0 + a]$ and $y(x_0) = \bar{y}(x_0) = y_0$. Suppose to the contrary that the function $y(x) - \bar{y}(x)$ (which is known to be continuously differentiable in $[x_0, x_0 + a]$ and vanishes at $x = x_0$) does not vanish for all x in $[x_0, x_0 + a]$. Then, there exists a number ε with $x_0 \leq \varepsilon < x_0 + a$ such that $|y(x) - \bar{y}(x)| = 0$ for $x_0 \leq x \leq \varepsilon$ and $|y(x) - \bar{y}(x)|$ is not identically zero in an interval $\varepsilon \leq x \leq \varepsilon + l \leq x_0 + a$, where $l > 0$. Choose $l > 0$ so small that $lL < 1$, where L is the Lipschitz constant of $f(x, y)$. Since the function $|y(x) - \bar{y}(x)|$ is continuous and not identically zero in the closed interval $\varepsilon \leq x \leq \varepsilon + l$, it has a positive maximum at some point $\hat{\varepsilon} > \varepsilon$ in this interval. By the mean value theorem for $\varepsilon \leq x \leq \hat{\varepsilon}$, there exists $\varepsilon < \varepsilon^* < \hat{\varepsilon}$ such that

$$y'(\varepsilon^*) - \bar{y}'(\varepsilon^*) = \frac{[y(\hat{\varepsilon}) - \bar{y}(\hat{\varepsilon})] - [y(\varepsilon) - \bar{y}(\varepsilon)]}{\hat{\varepsilon} - \varepsilon}.$$

Thus, it follows that

$$\begin{aligned} 0 &< |y(\hat{\varepsilon}) - \bar{y}(\hat{\varepsilon})| = |\hat{\varepsilon} - \varepsilon||y'(\varepsilon^*) - \bar{y}'(\varepsilon^*)| \\ &= (\hat{\varepsilon} - \varepsilon)|f(\varepsilon^*, y(\varepsilon^*)) - f(\varepsilon^*, \bar{y}(\varepsilon^*))| \\ &\leq lL|y(\varepsilon^*) - \bar{y}(\varepsilon^*)| \\ &< |y(\varepsilon^*) - \bar{y}(\varepsilon^*)|. \end{aligned}$$

Hence, the positive number $|y(\hat{\varepsilon}) - \bar{y}(\hat{\varepsilon})|$ is not the maximum value of the continuous function $|y(x) - \bar{y}(x)|$ in the interval $\varepsilon \le x \le \varepsilon + l$, which is a contradiction. ∎

We shall provide one more proof of Theorem 1.2.4 which does not require even the mean value theorem.

Third Proof. Once again, we shall consider only the interval $[x_0, x_0 + a]$. Suppose $y(x)$ and $\bar{y}(x)$ are two solutions of (1.1.1), then the function $v(x) = (y(x) - \bar{y}(x))^2$ satisfies

$$\begin{aligned} v'(x) &= 2(y(x) - \bar{y}(x))(y'(x) - \bar{y}'(x)) \\ &= 2(y(x) - \bar{y}(x))(f(x, y(x)) - f(x, \bar{y}(x))) \\ &\le 2Lv(x), \end{aligned}$$

which is the same as $(\exp(-2L(x - x_0))v(x))' \le 0$. Thus, the function $\exp(-2L(x - x_0))v(x)$ is nonincreasing, and hence the condition $v(x_0) = 0$ implies that $v(x) \le 0$. Therefore, $v(x) \equiv 0$, i.e., $y(x) = \bar{y}(x)$ in $[x_0, x_0 + a]$. ∎

We shall illustrate Theorem 1.2.4 with an example.

Example 1.2.1. Consider the initial value problem

$$y' = 1 + y^2, \ y(0) = 1. \tag{1.2.6}$$

The function $f(x, y) = 1 + y^2$ is continuously differentiable in R^2. Further, from Lemma 1.2.1, we find that it satisfies the Lipschitz condition (1.2.1) in $\bar{S}$ with $L = \sup_{\bar{S}} |2y| \le 2(1 + b)$. Thus, from Theorem 1.2.4, the problem (1.2.6) has at most one solution in $|x| \le a$. ∎

The solution $y(x) = \tan(x + \pi/4)$ of (1.2.6) exists only in the interval $(-3\pi/4, \pi/4)$. However, it does not contradict the conclusion of Theorem 1.2.4 because 'at most' does include nonexistence of the solutions.

If the function $f(x, y)$ satisfies one - sided Lipschitz condition, then the uniqueness of the solutions of (1.1.1) is also one - sided. This is the content of our next result.

Theorem 1.2.5. Let $f(x,y)$ be continuous in

$$\bar{S}_+ : x_0 \le x \le x_0 + a, |y - y_0| \le b$$

and for all $(x,y), (x,\bar{y})$ in $\bar{S}_+$ with $\bar{y} \ge y$ it satisfies **one - sided Lipschitz condition**

$$f(x,\bar{y}) - f(x,y) \le L(\bar{y} - y). \tag{1.2.7}$$

Then, the initial value problem (1.1.1) has at most one solution in $[x_0, x_0 + a]$.

Proof. Suppose $y(x)$ and $\bar{y}(x)$ are two solutions of (1.1.1) which differ somewhere in $[x_0, x_0 + a]$. We assume that $\bar{y}(x) > y(x)$ for $x_1 < x \le x_1 + \varepsilon \le x_0 + a$, and $y(x) = \bar{y}(x)$ for $x_0 \le x \le x_1$, i.e., x_1 is the greatest lower bound (glb) of the set A of values of x for which $\bar{y}(x) > y(x)$. This glb exists because the set A is bounded below by x_0 at least. Thus, for all $x \in (x_1, x_1 + \varepsilon]$, we have

$$\bar{y}(x) - y(x) = \int_{x_1}^{x} (f(t,\bar{y}(t)) - f(t,y(t)))dt \le L \int_{x_1}^{x} (\bar{y}(t) - y(t))dt.$$

Applying Lemma 1.2.3 to the above inequality, we get $\bar{y}(x) - y(x) = 0$ in $[x_1, x_1 + \varepsilon]$. This contradiction shows that $y(x) = \bar{y}(x)$ in $[x_0, x_0 + a]$. ∎

The above proof of Theorem 1.2.5 is similar to that of the first proof of Theorem 1.2.4. Its two other proofs based on the second and the third proof of Theorem 1.2.4 can be rather easily extended.

The Lipschitz condition is only a sufficient but not a necessary condition to prove the uniqueness of the solutions of (1.1.1). For this, we present the following two interesting examples.

Example 1.2.2. Consider the initial value problem

$$y' = 1 + y^{2/3}, \; y(0) = 0. \tag{1.2.8}$$

The function $f(x,y) = 1 + y^{2/3}$ is continuous in R^2. However, it does not satisfy the Lipschitz condition in any domain containing $y = 0$.

Separating the variables in the above differential equation and using the substitution $y = z^3$ immediately lead to the unique solution of (1.2.8) as

$$3(y^{1/3} - \tan^{-1} y^{1/3}) = x. \quad ∎$$

Example 1.2.3. Consider the initial value problem

$$(1.2.9)\qquad y' = f(x,y) = \begin{cases} y\ln(1/y), & 0 < y < 1 \\ 0, & y = 0 \end{cases}$$
$$y(0) = \alpha, \quad 0 \le \alpha < 1.$$

Since $\lim_{y\to 0^+} y\ln(1/y) = \lim_{y\to 0^+} \frac{\ln(1/y)}{1/y} = \lim_{y\to 0^+} y\,\frac{-1/y^2}{-1/y^2} = \lim_{y\to 0^+} y = 0$, this function $f(x,y)$ is continuous in the strip $|x| < \infty, 0 \le y < 1$. However, it does not satisfy the Lipschitz condition in any domain containing $y = 0$. The problem (1.2.9) has a unique solution which can be written as

$$x = \ln\left[\ln(\frac{1}{\alpha})\Big/\ln(\frac{1}{y})\right]. \qquad \blacksquare$$

The conclusion of the Example 1.2.2 can be deduced from the following :

Theorem 1.2.6. Let the function $f(x,y) = f(x)g(y)$ be continuous in the domain D, and $g(y) \ne 0$. Then, the initial value problem (1.1.1) has a unique solution $y(x) = G^{-1}(F(x))$, where $F(x) = \int_{x_0}^{x} f(t)dt$ and $G(y) = \int_{y_0}^{y} \frac{dz}{g(z)}$, as long as $(x, F(x)) \in D$.

Proof. Since $g(y) \ne 0$, $\frac{y'}{g(y)} = f(x)$, and hence it follows that

$$G(y) = \int_{y_0}^{y} \frac{dz}{g(z)} = \int_{x_0}^{x} f(t)dt = F(x).$$

Obviously, the function $G(y)$ is strictly monotonic, and hence the conclusion follows by the inverse function theorem. $\blacksquare$

In Section 1.1 we have seen that the initial value problem $y' = g(y) = y^{2/3}$, $y(0) = 0$ has an infinite number of solutions. This function $g(y) = y^{2/3}$ vanishes at $y = 0$. In our next result we shall show that this property is in fact true for all such initial value problems.

Theorem 1.2.7. Let the function $g(y)$ be continuous in the interval $|y-y_0| < b$. Further, let the initial value problem

$$(1.2.10)\qquad y' = g(y), \; y(x_0) = y_0$$

has at least two solutions in every neighborhood of x_0. Then, $g(y_0) = 0$.

Proof. Suppose $y(x)$ and $\bar{y}(x)$ are two solutions of (1.2.10) in $U : |x-x_0| < \varepsilon$. Let $g(y_0) \neq 0$, then there is a neighborhood V of y_0 on which $g(y)$ is bounded away from zero. Let $W : |x - x_0| < \varepsilon_1 \leq \varepsilon$ be such that $y(W) \subset V$ and $\bar{y}(W) \subset V$. For $x \in W$, we define $G(x) = \int_{y(x)}^{\bar{y}(x)} \frac{dz}{g(z)}$. Obviously, if $y(x) \neq \bar{y}(x)$ for some $x \in W$, then $G(x) \neq 0$. But, for all $x \in W$

$$G'(x) = \frac{\bar{y}'(x)}{g(\bar{y}(x))} - \frac{y'(x)}{g(y(x))} = 0.$$

Therefore, $G(x)$ is a constant on W. However, since $G(x_0) = 0$ it follows that $G(x) = 0, x \in W$. This contradiction completes the proof. ∎

From Example 1.2.3 it is clear that the converse of Theorem 1.2.7 is not true. However, the following holds.

Theorem 1.2.8. Let the function $g(y)$ be continuous in the interval $|y-y_0| \leq b$, and $g(y_0) = 0$. Further, let $g(y) \neq 0$ for $y_0 < y \leq y_0 + b$ $(y_0 - b \leq y < y_0)$ and the integral

$$(1.2.11) \qquad \int_{y_0}^{y_0+b} \frac{dz}{g(z)} \qquad \left(\int_{y_0-b}^{y_0} \frac{dz}{g(z)}\right)$$

is divergent. Then, there is no solution of the initial value problem (1.2.10) which will from above (below) join $y = y_0$.

Proof. Assume that there exists a solution $y(x) \not\equiv y_0$ of the initial value problem (1.2.10). Then, let there be (to consider one of the four cases) right to x_0 a point x_1 with $y_0 < y(x_1) = y_1 < y_0 + b$. Thus, by the differential equation it follows that

$$(1.2.12) \qquad \int_{y_1}^{y(x)} \frac{dz}{g(z)} = (x - x_1)$$

at least as long as $y(x)$ satisfies $y_0 < y(x) \leq y_0 + b$. Assume that $\bar{x}$ is the first point left to x_1 with $y(\bar{x}) = y_0$. Then, (1.2.12) leads to a contradiction, because as $x \to \bar{x}$ the integral in the left of (1.2.12) tends to ∞ whereas the right side remains bounded. ∎

From the conditions of Theorem 1.2.8 it is clear that if a solution $y(x) > y_0 (< y_0)$ at some point, then this inequality holds everywhere. Further, if both the integrals in (1.2.11) diverge then $y(x) \equiv y_0$ is the only solution of (1.2.10). As an example the initial value problem $y' = 2y, y(0) = 0$ has a unique solution $y(x) \equiv 0$, and the integral $\int_0^{\pm\alpha} \frac{dz}{z}, (\alpha > 0)$ diverges. Whereas, in Section 1.1 we have seen that the problem $y' = y^{2/3}, y(0) = 0$ has an infinite number of solutions and the integral $\int_0^{\pm\alpha} \frac{dz}{z^{2/3}}$, $(\alpha > 0)$ converges.

1.3 PEANO'S UNIQUENESS THEOREM

In the previous section, we have seen that the initial value problem (1.1.1) may have a unique solution even when $f(x, y)$ does not satisfy the Lipschitz condition. In this section we shall discuss Peano's uniqueness theorem which assumes apparently different conditions on the function $f(x, y)$, such as increasing or decreasing with respect to y.

Theorem 1.3.1 (Peano's Uniqueness Theorem). Let $f(x, y)$ be continuous in $\bar{S}_+$ and nonincreasing in y for each fixed x in $[x_0, x_0 + a]$. Then, the initial value problem (1.1.1) has at most one solution in $[x_0, x_0 + a]$.

Proof. Suppose $y(x)$ and $\bar{y}(x)$ are two solutions of (1.1.1) which differ somewhere in $[x_0, x_0 + a]$. We assume that $\bar{y}(x) > y(x)$ for $x_1 < x \leq x_1 + \varepsilon \leq x_0 + a$, and $y(x) = \bar{y}(x)$ for $x_0 \leq x \leq x_1$. Thus, for all $x \in (x_1, x_1 + \varepsilon]$, we have $f(x, y(x)) \geq f(x, \bar{y}(x))$, and hence $y'(x) \geq \bar{y}'(x)$. This implies that the function $\phi(x) = \bar{y}(x) - y(x)$ is nonincreasing. Further, since $\phi(x_1) = 0$, we have $\phi(x) \leq 0$ in $[x_1, x_1 + \varepsilon]$. This contradiction proves that $y(x) = \bar{y}(x)$ in $[x_0, x_0 + a]$. ∎

The above result is actually a particular case of Theorem 1.2.5, indeed if $L = 0$, then condition (1.2.7) reduces to the hypothesis of Theorem 1.3.1.

The 'nonincreasing' nature in Theorem 1.3.1 cannot be replaced by 'nondecreasing'. For this, we have

Example 1.3.1. Consider the initial value problem

$$y' = |y|^{1/2} \text{sgn } y, \; y(0) = 0 \tag{1.3.1}$$

where $\text{sgn } y = \begin{cases} 1 & \text{if } y \geq 0 \\ -1 & \text{if } y < 0 \end{cases}$. The function $f(x,y) = |y|^{1/2}\text{sgn } y$ is continuous for $0 \leq x < \infty, |y| < \infty$ and it is nondecreasing in y for all $x \in [0,\infty)$. The problem (1.3.1) has two solutions $y(x) \equiv 0$ and $y(x) = x^2/4$ in $[0,\infty)$. ∎

We shall now illustrate an example where $f(x,y)$ does not satisfy the Lipschitz condition (1.2.1) but it satisfies the conditions of Theorem 1.3.1.

Example 1.3.2. Consider the initial value problem

$$y' = -|y|^{1/2}\text{sgn } y, \; y(0) = 0. \tag{1.3.2}$$

The function $f(x,y) = -|y|^{1/2}\text{sgn } y$ does not satisfy the Lipschitz condition (1.2.1) in any domain containing $y = 0$. Indeed, for $y > 0$ we have

$$\frac{|f(x,y) - f(x,0)|}{|y - 0|} = \frac{|y|^{1/2}}{|y|} = \frac{1}{|y|^{1/2}},$$

which is unbounded as $y \to 0$. However, conditions of Theorem 1.3.1 are satisfied since $-|y|^{1/2}\text{sgn } y$ is continuous for $0 \leq x < \infty, |y| < \infty$ and nonincreasing in y for all $x \in [0,\infty)$. Thus, from Theorem 1.3.1 it follows that $y(x) \equiv 0$ in $[0,\infty)$ is the only solution of the problem (1.3.2). ∎

Theorem 1.3.2. Suppose that $f(x,y)$ is continuous in

$$\bar{S}_- : x_0 - a \leq x \leq x_0, |y - y_0| \leq b$$

and nondecreasing in y for each fixed x in $[x_0 - a, x_0]$. Then, the initial value problem (1.1.1) has at most one solution in $[x_0 - a, x_0]$.

Proof. The proof is similar to that of Theorem 1.3.1. ∎

On combining the above results, we obtain

Theorem 1.3.3. If $f(x,y)$ is continuous in $\bar{S}$ and nondecreasing in y for each fixed x in $[x_0 - a, x_0]$, and nonincreasing in y for each fixed x in $[x_0, x_0 + a]$, then the initial value problem (1.1.1) has at most one solution in $|x - x_0| \leq a$. ∎

The conditions of Peano's uniqueness theorem are only sufficient. This is clear from the initial value problem (1.2.8). The function $f(x,y) = 1 + y^{2/3}$ is nondecreasing in y for each x in $0 \le x < \infty$. However, in Example 1.2.2 we have seen that (1.2.8) has a unique solution.

1.4 OSGOOD'S UNIQUENESS THEOREM

In this section we shall study a generalization of the Lipschitz uniqueness theorem which is due to Osgood . For this, we require the following :

Lemma 1.4.1. Let $g(z)$ be a continuous and nondecreasing function in the interval $[0, \infty)$ and $g(0) = 0,\ g(z) > 0$ for $z > 0$. Also,

$$\lim_{\varepsilon \to 0^+} \int_\varepsilon \frac{dz}{g(z)} = \infty. \tag{1.4.1}$$

Let $\phi(x)$ be a nonnegative continuous function in $[0, a]$. Then,

$$\phi(x) \le \int_0^x g(\phi(t))dt,\ 0 < x \le a \tag{1.4.2}$$

implies that $\phi(x) = 0$ in $[0, a]$.

Proof. Define $\Phi(x) = \max_{0 \le t \le x} \phi(t)$ and assume that $\Phi(x) > 0$ for $0 < x \le a$. Then, $\phi(x) \le \Phi(x)$ and for each x there is an $x_1 \le x$ such that $\phi(x_1) = \Phi(x)$. From this, we have

$$\Phi(x) = \phi(x_1) \le \int_0^{x_1} g(\phi(t))dt \le \int_0^x g(\Phi(t))dt,$$

i.e., the increasing function $\Phi(x)$ satisfies the same inequality as $\phi(x)$ does. Let us set $\bar{\Phi}(x) = \int_0^x g(\Phi(t))dt$, then $\bar{\Phi}(0) = 0, \Phi(x) \le \bar{\Phi}(x)$ and $\bar{\Phi}'(x) = g(\Phi(x)) \le g(\bar{\Phi}(x))$. Hence, for $0 < \delta < a$, we have

$$\int_\delta^a \frac{\bar{\Phi}'(x)}{g(\bar{\Phi}(x))}dx \le a - \delta < a.$$

However, from (1.4.1), it follows that

$$\int_\delta^a \frac{\bar{\Phi}'(x)}{g(\bar{\Phi}(x))}dx = \int_\varepsilon^\alpha \frac{dz}{g(z)},\quad \bar{\Phi}(\delta) = \varepsilon, \bar{\Phi}(a) = \alpha$$

becomes infinite when $\varepsilon \to 0^+ (\delta \to 0)$. This contradiction shows that $\Phi(x)$ cannot be positive and so $\Phi(x) \equiv 0$, and hence $\phi(x) = 0$ in $[0, a]$. ∎

Theorem 1.4.2 (Osgood's Uniqueness Theorem). Let $f(x,y)$ be continuous in $\bar{S}$ and for all $(x,y), (x,\bar{y}) \in \bar{S}$ it satisfies **Osgood's condition**

$$(1.4.3) \qquad |f(x,y) - f(x,\bar{y})| \le g(|y - \bar{y}|),$$

where $g(z)$ is the same as in Lemma 1.4.1. Then, the initial value problem (1.1.1) has at most one solution in $|x - x_0| \le a$.

Proof. Suppose $y(x)$ and $\bar{y}(x)$ are two solutions of (1.1.1) in $|x - x_0| \le a$. We shall show that $y(x) = \bar{y}(x)$ in $[x_0, x_0 + a]$. From (1.4.3) it follows that

$$\begin{aligned} |y(x_0 + x) - \bar{y}(x_0 + x)| &\le \int_{x_0}^{x_0+x} |f(t, y(t)) - f(t, \bar{y}(t))| dt \\ &\le \int_{x_0}^{x_0+x} g(|y(t) - \bar{y}(t)|) dt \\ &= \int_0^x g(|y(z + x_0) - \bar{y}(z + x_0)|) dz. \end{aligned}$$

For x in $[0, a]$, we set $\phi(x) = |y(x + x_0) - \bar{y}(x + x_0)|$. Then, the nonnegative continuous function $\phi(x)$ satisfies the inequality (1.4.2), and therefore, Lemma 1.4.1 implies that $\phi(x) = 0$ in $[0, a]$, i.e., $y(x) = \bar{y}(x)$ in $[x_0, x_0 + a]$. If x is in $[x_0 - a, x_0]$, then the proof remains the same except that we need to define the function $\phi(x) = |y(x_0 - x) - \bar{y}(x_0 - x)|$ in $[0, a]$. ∎

Example 1.4.1. Consider the initial value problem

$$(1.4.4) \qquad y' = Ly, \ y(0) = 0$$

where $L > 0$. For this problem, we choose $g(z) = Lz$, which is clearly continuous and nondecreasing in the interval $[0, \infty)$. Further, since $g(0) = 0,\ g(z) > 0$ for $z > 0$, and $\lim_{\varepsilon \to 0^+} \int_\varepsilon [g(z)]^{-1} dz = \frac{1}{L} \lim_{\varepsilon \to 0^+} \ln \frac{1}{\varepsilon} = \infty$. This function $g(z)$ satisfies the conditions of Lemma 1.4.1. Next, for any y and $\bar{y}$ we have

$$|f(x,y) - f(x,\bar{y})| = |Ly - L\bar{y}| = g(|y - \bar{y}|)$$

and hence Osgood's condition (1.4.3) is also satisfied. Therefore, from Theorem 1.4.2 the problem (1.4.4) has a unique solution, namely, $y(x) \equiv 0$. ∎

The importance of Example 1.4.1 is in the fact that Osgood's condition (1.4.3) with $g(z) = Lz$ reduces to the Lipschitz condition (1.2.1).

Example 1.4.2. Consider the initial value problem

$$y' = f(x,y) = \begin{cases} -y\ln y, & 0 < y \le 1/e \\ 0, & y = 0 \end{cases} \tag{1.4.5}$$
$$y(0) = 0.$$

From Example 1.2.3 it is clear that for this problem Lipschitz uniqueness theorem is not applicable. Further, since this function $f(x,y)$ is increasing with respect to y, Peano's uniqueness theorem also cannot be used. For this problem we choose $g(z) = \begin{cases} 0, & z = 0 \\ -z\ln z, & 0 < z \le 1/e \\ 1/e, & z > 1/e \end{cases}$. Clearly, $g(z)$ is continuous and nondecreasing in $[0,\infty)$ with $g(0) = 0,\ g(z) > 0$ for $z > 0$. Moreover, we have

$$\lim_{\varepsilon\to 0+}\int_\varepsilon \frac{dz}{g(z)} = \lim_{\varepsilon\to 0+}\int_\varepsilon \frac{dz}{-z\ln z} = \lim_{\varepsilon\to 0+}\ln(\ln(1/\varepsilon)) = \infty.$$

Thus, this function $g(z)$ satisfies the conditions of Lemma 1.4.1. Now, we shall verify the condition (1.4.3). For this, we note that $\frac{d^2f}{dy^2} = -\frac{1}{y} < 0, 0 < y \le 1/e$ and therefore f is concave. Let $0 < y - \bar{y} < \bar{y} < y \le 1/e$, then it follows that

$$\frac{f(x,y-\bar{y}) - f(x,0)}{y-\bar{y}} \ge \frac{f(x,y) - f(x,\bar{y})}{y-\bar{y}},$$

which is the same as

$$-(y-\bar{y})\ln(y-\bar{y}) \ge -y\ln y + \bar{y}\ln\bar{y}$$

and hence

$$|f(x,y) - f(x,\bar{y})| = |(-y\ln y) - (-\bar{y}\ln\bar{y})| \le -|y-\bar{y}|\ln|y-\bar{y}| = g(|y-\bar{y}|).$$

Thus, for the problem (1.4.5) Osgood's uniqueness theorem is applicable. ∎

In Theorem 1.4.2 the condition that the function $g(z)$ is nondecreasing is in fact superfluous. To show this we need the following result which is in a sense complementary to Theorem 1.2.8.

Theorem 1.4.3. Let the function $g(y) > 0$ be continuous for $y_0 < y \le y_0 + \beta$ for some $\beta > 0$ and $g(y_0) = 0$. Further, let the integral $\int_{y_0+\varepsilon}\frac{dz}{g(z)}$ diverges as $\varepsilon \to 0^+$. Then, the initial value problem $y' = g(y)$, $y(x_1) = y_1$ where $y_0 < y_1 \leq y_0 + \beta$ has a unique solution. This solution $y(x)$ is such that $y_0 < y(x) \leq y_0 + \beta$ and tends to y_0 as $x \to -\infty$. Further, if $\int_{y_0+\varepsilon} \frac{dz}{g(z)}$ converges as $\varepsilon \to 0^+$, then there are infinite number of solutions of (1.2.10).

Proof. If $y(x)$ is a solution of $y' = g(y)$, $y(x_1) = y_1$, then it follows that

$$\int_{y_1}^{y(x)} \frac{dz}{g(z)} = (x - x_1). \tag{1.4.6}$$

Since $g(y)$ is defined only for $y_0 \leq y \leq y_0 + \beta$, it is necessary that $y_0 \leq y(x) \leq y_0 + \beta$. From (1.4.6) and the fact that g is continuous and $g(y) > 0$ for $y_0 < y \leq y_0 + \beta$, we find that $y(x)$ exists for as long as $y_0 < y(x) \leq y_0 + \beta$. Since equation (1.4.6) defines only candidate for a solution (implicitly), and since by direct substitution it gives a solution, it follows that for $y' = g(y)$, $y(x_1) = y_1$, (1.4.6) defines the unique solution.

To show that $\lim_{x \to -\infty} y(x) = y_0$, when

$$\lim_{\varepsilon \to 0^+} \int_{y_0+\varepsilon}^{y_1} \frac{dz}{g(z)} = \infty \tag{1.4.7}$$

we proceed as follows. The left side of (1.4.6) is finite provided $y_0 < y(x) \leq y_0 + \beta$. Since the integral in (1.4.7) diverges, this solution $y(x)$ cannot become y_0 for a finite value of x and at the same time satisfy (1.4.6). Thus, $y_0 < y(x) \leq y_0 + \beta$ for all $x < x_1$. As $x \to -\infty$ the right side of (1.4.6) tends to $-\infty$, and therefore $\lim_{x \to -\infty} y(x) = y_0$.

Since $g(y_0) = 0$, clearly $y(x) \equiv y_0$ is a solution of (1.2.10). If there is another solution $\phi(x)$ of (1.2.10) such that $y_0 < \phi(x) \leq y_0 + \beta$ for $x \neq x_0$, then it follows that

$$\lim_{\varepsilon \to 0^+} \int_{y_0+\varepsilon}^{\phi(x)} \frac{dz}{g(z)} = (x - x_0). \tag{1.4.8}$$

If the integral in (1.4.7) diverges, then (1.4.8) does not define a function. If the integral in (1.4.7) converges, then the equation (1.4.8) defines a function; in fact, then there is a family of solutions $\phi_c(x)$ of (1.2.10) defined by

$$\begin{cases} \phi_c(x) = y_0, & x \leq c \\ \displaystyle\int_{y_0}^{\phi_c(x)} \frac{dz}{g(z)} = (x - c), & x > c \end{cases}$$

for every $c \geq x_0$. Here, it is understood that

$$\int_{y_0}^{\phi_c(x)} \frac{dz}{g(z)} = \lim_{\varepsilon \to 0+} \int_{y_0+\varepsilon}^{\phi_c(x)} \frac{dz}{g(z)},$$

where this integral converges. ∎

As an example once again consider the initial value problem $y' = y^{2/3}, y(0) = 0$ discussed in Sections 1.1 and 1.2.

Theorem 1.4.4. Let $f(x,y)$ be continuous in $\bar{S}$ and for all $(x,y),(x,\bar{y}) \in \bar{S}$ it satisfies Osgood's condition (1.4.3), where $g(z) > 0$ is a continuous function for $0 < z \leq 2b$, $g(0) = 0$ and satisfies the condition (1.4.1). Then, the initial value problem (1.1.1) has at most one solution in $|x - x_0| \leq a$.

First Proof. Suppose $y(x)$ and $\bar{y}(x)$ are two solutions of (1.1.1) in $|x-x_0| \leq a$. Once again we shall show that $y(x) = \bar{y}(x)$ in $[x_0, x_0+a]$. Let $\phi(x) = \bar{y}(x)-y(x)$ and assume that there is a point $x_1 \in (x_0, x_0 + a]$ such that $\phi(x_1) > 0$. Then, we have

$$\begin{aligned} |\phi'(x)| &= |f(x,\bar{y}(x)) - f(x,y(x))| \leq g(|\bar{y}(x) - y(x)|) \\ &= g(|\phi(x)|) < 2g(|\phi(x)|). \end{aligned}$$

By Theorem 1.4.3 the initial value problem $z' = 2g(z)$, $z(x_1) = \phi(x_1)$ has a unique solution $z(x)$ existing for $-\infty < x \leq x_1$ and $z(x) > 0$. Since initially $z(x_1) = \phi(x_1)$, the solution $z(x)$ of $z' = 2g(z)$ may follow $\phi(x)$ for some $x \leq x_1$. Let $\bar{x} = \text{glb}\{x : \phi(x) = z(x)\}$. Then, $x_0 < \bar{x} \leq x_1$, because $\phi(x_0) = 0$ and $z(x_0) > 0$. Since $z(x) > 0$ for $-\infty < x \leq x_1, \phi(\bar{x}) = z(\bar{x}) > 0$. But,

$$\phi'(\bar{x}) \leq |\phi'(\bar{x})| < 2g(|\phi(\bar{x})|) = 2g(\phi(\bar{x})) = 2g(z(\bar{x})) = z'(\bar{x});$$

thus $\phi'(\bar{x}) < z'(\bar{x})$.

By continuity this inequality must hold in some interval $\beta \leq x \leq \bar{x}$. Then,

$$\int_x^{\bar{x}} \phi'(t)dt < \int_x^{\bar{x}} z'(t)dt$$

implies that $\phi(\bar{x}) - \phi(x) < z(\bar{x}) - z(x)$ which yields the inequality $\phi(x) > z(x)$ in $[\beta, \bar{x}]$. But, $\bar{x}$ is the smallest value of x for which $\phi(x) = z(x)$ and so the graphs $y = \phi(x)$, $y = z(x)$ cannot cross again. Hence, we have $\phi(x) > z(x)$ in

$[x_0, \bar{x}]$. But, $\phi(x_0) = 0 < z(x_0)$, which is a contradiction. Thus, $\phi(x) \equiv 0$, and hence $y(x) = \bar{y}(x)$ in $[x_0, x_0 + a]$. ∎

Like Lipschitz uniqueness theorem, Theorem 1.4.4 can also be proved by using only the mean value theorem.

Second Proof. We shall consider the interval $[x_0, x_0 + a]$, whereas the proof in the interval $[x_0 - a, x_0]$ is similar. Let $G(z)$ be a function such that $G(z)$ has a continuous first order derivative in $(0, 2b], G(0^+) = -\infty$, and $G'(z) = \frac{1}{g(z)}$ for $0 < z \leq 2b$ (hence $G(z) > -\infty$ for $0 < z \leq 2b$). For example, we may simply choose

$$G(z) = \int_b^z \frac{dt}{g(t)} \quad \text{for } 0 < z \leq 2b.$$

Suppose $y(x)$ and $\bar{y}(x)$ are two solutions of (1.1.1) which differ somewhere in $[x_0, x_0 + a]$. As in Theorem 1.2.5 we assume that $\bar{y}(x) > y(x)$ for $x_1 < x \leq x_1 + \varepsilon \leq x_0 + a$, and $y(x) = \bar{y}(x)$ for $x_0 \leq x \leq x_1$. This means that we also have

$$e^{G(\bar{y}(x_1) - y(x_1))} = 0, \quad \text{while} \quad e^{G(\bar{y}(x) - y(x))} > 0 \text{ for } x_1 < x \leq x_1 + \varepsilon.$$

Then, by the mean value theorem, applied to the function $e^{-x}e^{G(\bar{y}(x)-y(x))}$ in the interval $[x_1, x_1 + \varepsilon]$, there exists a number $\bar{x}$, with $x_1 < \bar{x} < x_1 + \varepsilon$, such that

$$\begin{aligned}
0 &< e^{-(x_1+\varepsilon)}e^{G(\bar{y}(x_1+\varepsilon) - y(x_1+\varepsilon))} \\
&= e^{-(x_1+\varepsilon)}e^{G(\bar{y}(x_1+\varepsilon) - y(x_1+\varepsilon))} - e^{-x_1}e^{G(\bar{y}(x_1) - y(x_1))} \\
&= \varepsilon e^{-\bar{x}}e^{G(\bar{y}(\bar{x}) - y(\bar{x}))}\left[\frac{\bar{y}'(\bar{x}) - y'(\bar{x})}{g(\bar{y}(\bar{x}) - y(\bar{x}))} - 1\right] \\
&= \varepsilon e^{-\bar{x}}e^{G(\bar{y}(\bar{x}) - y(\bar{x}))}\left[\frac{f(\bar{x}, \bar{y}(\bar{x})) - f(\bar{x}, y(\bar{x}))}{g(\bar{y}(\bar{x}) - y(\bar{x}))} - 1\right] \\
&\leq 0,
\end{aligned}$$

where the last inequality follows from Osgood's condition (1.4.3). This contradiction proves the result. ∎

A one - sided analog of Theorem 1.4.4 is the following :

Theorem 1.4.5. Let $f(x,y)$ be continuous in $\bar{S}_+$ and for all $(x,y),(x,\bar{y})$ in $\bar{S}_+$ with $\bar{y} \geq y$ it satisfies **one - sided Osgood condition**

$$f(x,\bar{y}) - f(x,y) \leq g(\bar{y} - y), \tag{1.4.9}$$

where $g(z) > 0$ is a continuous function for $0 < z \leq 2b, g(0) = 0$ and satisfies the condition (1.4.1). Then, the initial value problem (1.1.1) has at most one solution in $[x_0, x_0 + a]$. ∎

Often, the uniqueness of the solutions of (1.1.1) is achieved only in a small interval containing x_0, and not in the entire interval of interest. Such type of results are called **local uniqueness theorems.** In the following we state a local Osgood's uniqueness theorem.

Theorem 1.4.6. Let $f(x,y)$ be continuous in a domain D and for all $(x,y),(x,\bar{y}) \in D$ it satisfies Osgood's condition (1.4.3), where $g(z) > 0$ is a continuous function for $0 < z \leq \alpha$, $g(0) = 0$ for some $\alpha > 0$ and satisfies the condition (1.4.1). Then, the initial value problem (1.1.1) has at most one solution in a small interval containing x_0. ∎

We note that the functions $Lz, Lz|\ln z|, Lz|\ln z|\ln|\ln z|, Lz|\ln z|\ln|\ln z|\cdot \ln\ln|\ln z|, \cdots$, where L is a positive constant, all have the properties required for the function $g(z)$ in Theorem 1.4.6. Further, in progression these functions provide weaker restrictions on the function $f(x,y)$, i.e., provide stronger theorems. However, there is no "strongest" theorem of this type, i.e., if $g(z)$ satisfies the conditions of Theorem 1.4.6, then there always exists another function $g_1(z)$ satisfying the same conditions, but

$$\lim_{z\to 0} \frac{g_1(z)}{g(z)} = \infty.$$

We also remark that if the graph of $g(z)$ is concave, then in Theorem 1.4.6 the condition (1.4.1) is not only sufficient but necessary also.

1.5 MONTEL - TONELLI'S UNIQUENESS THEOREM

A generalization of Osgood's uniqueness theorem is embodied in the following :

Theorem 1.5.1 (Montel - Tonelli's Uniqueness Theorem). Let $f(x,y)$ be continuous in $\bar{S}$ and for all $(x,y),(x,\bar{y}) \in \bar{S}$ it satisfies **Montel - Tonelli's condition**

$$(1.5.1) \qquad |f(x,y) - f(x,\bar{y})| \leq h(x)g(|y-\bar{y}|),$$

where the function $h(x) \geq 0$ is integrable in $|x - x_0| \leq a$, and $g(z) > 0$ is a continuous function for $0 < z \leq 2b$, $g(0) = 0$ and satisfies the condition (1.4.1). Then, the initial value problem (1.1.1) has at most one solution in $|x - x_0| \leq a$.

Proof. Suppose $y(x)$ and $\bar{y}(x)$ are two solutions of (1.1.1) in $|x-x_0| \leq a$. Once again we shall show that $y(x) = \bar{y}(x)$ in $[x_0, x_0 + a]$. Let $\phi(x) = \bar{y}(x) - y(x)$ and assume that there is a point $x_1 \in (x_0, x_0 + a]$ such that $\phi(x) > 0$ for $x_1 < x \leq x_1 + \varepsilon$ and $\phi(x) = 0$ for $x_0 \leq x \leq x_1$. Then, for all $x \in [x_1, x_1 + \varepsilon]$ it follows that

$$\phi'(x) = f(x,\bar{y}(x)) - f(x,y(x)) \leq h(x)g(\phi(x))$$

and hence for $\bar{x} \in (x_1, x_1 + \varepsilon)$, we have

$$\int_{\phi(\bar{x})}^{\phi(x_1+\varepsilon)} \frac{dz}{g(z)} \leq \int_{\bar{x}}^{x_1+\varepsilon} h(t)dt.$$

However, in the above inequality as $\bar{x}$ tends to x_1 the right side remains finite, whereas the left side in view of (1.4.1) becomes unbounded. This contradiction completes the proof. ∎

1.6 NAGUMO'S UNIQUENESS THEOREM

The main result in this section assumes that the variation of the function

$f(x,y)$ relative to y is bounded by a certain function of x. We state this condition in

Definition 1.6.1. The function $f(x,y)$ is said to satisfy **Nagumo's condition** in the domain D if

$$|f(x,y) - f(x,\bar{y})| \le k|x - x_0|^{-1}|y - \bar{y}|; \; x \ne x_0, k \le 1 \tag{1.6.1}$$

for all $(x,y),(x,\bar{y}) \in D$.

Lemma 1.6.1. Let $\phi(x)$ be a nonnegative continuous function in $|x - x_0| \le a$, and $\phi(x_0) = 0$, and let $\phi(x)$ be differentiable at $x = x_0$ with $\phi'(x_0) = 0$. Then, the inequality

$$\phi(x) \le \left| \int_{x_0}^{x} \frac{\phi(t)}{t - x_0} dt \right| \tag{1.6.2}$$

implies that $\phi(x) = 0$ in $|x - x_0| \le a$.

Proof. For $x \in [x_0, x_0 + a]$, we define $\Phi(x) = \int_{x_0}^{x} \frac{\phi(t)}{t - x_0} dt$. This integral exists since $\lim_{x \to x_0} \frac{\phi(x)}{x - x_0} = \phi'(x_0) = 0$. Further, we have

$$\Phi'(x) = \frac{\phi(x)}{x - x_0} \le \frac{\Phi(x)}{x - x_0}$$

and hence

$$\frac{d}{dx}\left(\frac{\Phi(x)}{x - x_0}\right) = \frac{(x - x_0)\Phi'(x) - \Phi(x)}{(x - x_0)^2} \le 0,$$

which implies that $\frac{\Phi(x)}{x - x_0}$ is nonincreasing. Since $\Phi(x_0) = 0$, this gives $\Phi(x) \le 0$, which is a contradiction to $\Phi(x) \ge 0$. So, $\Phi(x) \equiv 0$, and hence $\phi(x) = 0$ in $[x_0, x_0 + a]$. The proof is similar in $[x_0 - a, x_0]$. ∎

We shall use Lemma 1.6.1 to prove

Theorem 1.6.2 (Nagumo's Uniqueness Theorem). Let $f(x,y)$ be continuous and satisfy Nagumo's condition (1.6.1) in $\bar{S}$. Then, the initial value problem (1.1.1) has at most one solution in $|x - x_0| \le a$.

First Proof. Suppose $y(x)$ and $\bar{y}(x)$ are two solutions of (1.1.1) in $|x - x_0| \le a$.

Then, from (1.6.1) it follows that

$$\begin{aligned} |y(x)-\bar{y}(x)| &\leq \left|\int_{x_0}^{x} |f(t,y(t))-f(t,\bar{y}(t))|dt\right| \\ &\leq \left|\int_{x_0}^{x} |t-x_0|^{-1}|y(t)-\bar{y}(t)|dt\right|. \end{aligned}$$

We set $\phi(x)=|y(x)-\bar{y}(x)|$, then the nonnegative continuous function $\phi(x)$ in $|x-x_0|\leq a$ satisfies the inequality (1.6.2). Further, since $\phi(x)$ is continuous in $|x-x_0|\leq a$, and $\phi(x_0)=0$, from the mean value theorem, we have

$$\begin{aligned} \phi'(x_0) &= \lim_{h\to 0}\frac{\phi(x_0+h)-\phi(x_0)}{h} = \lim_{h\to 0}\frac{|y(x_0+h)-\bar{y}(x_0+h)|}{h} \\ &= \lim_{h\to 0}\frac{|(y(x_0+h)-y(x_0))-(\bar{y}(x_0+h)-\bar{y}(x_0))|}{h} \\ &= \lim_{h\to 0}\frac{|hy'(x_0+\theta_1 h)-h\bar{y}'(x_0+\theta_2 h)|}{h}, \quad 0<\theta_1,\theta_2<1 \\ &= (\text{sgn } h)\lim_{h\to 0}|y'(x_0+\theta_1 h)-\bar{y}'(x_0+\theta_2 h)| \\ &= 0. \end{aligned}$$

Thus, the conditions of Lemma 1.6.1 are satisfied, and hence $\phi(x)\equiv 0$, i.e., $y(x)=\bar{y}(x)$ in $|x-x_0|\leq a$. ∎

Theorem 1.6.2 can also be proved without using integral equations and the integral inequality. The alternative proof we shall now give requires only the mean value theorem.

Second Proof. We shall consider the interval $[x_0,x_0+a]$, whereas the proof in the interval $[x_0-a,x_0]$ is similar. As before, let $y(x)$ and $\bar{y}(x)$ be solutions in $[x_0,x_0+a]$. Consider the function $v(x)$ defined by $|y(x)-\bar{y}(x)|/(x-x_0)$ in $(x_0,x_0+a]$, and the value zero at $x=x_0$. It will be first shown that this function $v(x)$ is continuous at $x=x_0$. For this, on using the mean value theorem, and the fact that $y(x)-\bar{y}(x)$ vanishes at $x=x_0$, we find that

$$\begin{aligned} \frac{|y(x)-\bar{y}(x)|}{x-x_0} &= \frac{|(y(x)-\bar{y}(x))-(y(x_0)-\bar{y}(x_0))|}{x-x_0} \\ &= |y'(x*)-\bar{y}'(x*)| \\ &= |f(x*,y(x*))-f(x*,\bar{y}(x*))|, \end{aligned}$$

where $x_0 < x* < x$. But, $f(x,y)$ is continuous at (x_0, y_0), therefore

$$\lim_{x \to x_0} \frac{|y(x) - \bar{y}(x)|}{x - x_0} = |f(x_0, y_0) - f(x_0, y_0)| = 0,$$

as desired.

Now, suppose contrary to what we wish to prove, that the function $y(x) - \bar{y}(x)$ is not identically zero in $[x_0, x_0 + a]$. Then, there exists a number $x_m > x_0$ such that $v(x)$ attains its positive maximum in $[x_0, x_0 + a]$ at $x = x_m$. Furthermore,

$$\frac{|y(x) - \bar{y}(x)|}{x - x_0} < \frac{|y(x_m) - \bar{y}(x_m)|}{x_m - x_0},$$

whenever $x_0 < x < x_m$. However, by the mean value theorem applied to the function $y(x) - \bar{y}(x)$ in the interval $x_0 \le x \le x_m$, and by Nagumo's condition (1.6.1) satisfied by f, we have

$$\begin{aligned} 0 < \frac{|y(x_m) - \bar{y}(x_m)|}{x_m - x_0} &= \frac{|(y(x_m) - \bar{y}(x_m)) - (y(x_0) - \bar{y}(x_0))|}{x_m - x_0} \\ &= |y'(\hat{x}) - \bar{y}'(\hat{x})| \\ &= |f(\hat{x}, y(\hat{x})) - f(\hat{x}, \bar{y}(\hat{x}))| \\ &\le \frac{|y(\hat{x}) - \bar{y}(\hat{x})|}{\hat{x} - x_0}, \end{aligned}$$

where $\hat{x}$ satisfies $x_0 < \hat{x} < x_m$. The last inequality is in direct contradiction to the way in which x_m was chosen, and so the theorem is proved. ∎

In condition (1.6.1), the constant $k = 1$ is the best possible, i.e., it cannot be replaced by $k > 1$. This is illustrated in the following :

Example 1.6.1. Consider the initial value problem

$$(1.6.3) \qquad y' = f(x,y) = \begin{cases} 0, & 0 \le x \le 1, y \le 0 \\ \dfrac{(1+\varepsilon)y}{x}, & 0 \le x \le 1, \ 0 < y < x^\delta, \\ & \qquad \delta = 1 + \varepsilon > 1 \\ (1+\varepsilon)x^{\delta-1}, & 0 \le x \le 1, x^\delta \le y \end{cases}$$

$$y(0) = 0.$$

It is easily seen that this function $f(x,y)$ is continuous at $(0,y)$ for $0 < y < x^\delta$ since

$$\left|\frac{(1+\varepsilon)y}{x}\right| < \frac{(1+\varepsilon)x^\delta}{x} \to 0 \quad \text{as} \quad x \to 0^+.$$

Hence, $f(x,y)$ is continuous in the strip $0 \le x \le 1, |y| < \infty$. Next, $f(x,y)$ satisfies the condition (1.6.1) in this strip except that $k = 1+\varepsilon > 1$. We shall verify this by considering the following cases :

Suppose $0 < y, \bar{y} < x^\delta$, then

$$|f(x,y) - f(x,\bar{y})| = \left|\frac{(1+\varepsilon)y}{x} - \frac{(1+\varepsilon)\bar{y}}{x}\right| = \frac{(1+\varepsilon)}{x}|y - \bar{y}|.$$

Suppose $0 < y < x^\delta \le \bar{y}$, then

$$\begin{aligned}|f(x,y) - f(x,\bar{y})| &= \left|\frac{(1+\varepsilon)y}{x} - (1+\varepsilon)x^{\delta-1}\right| = \frac{(1+\varepsilon)}{x}|y - x^\delta| \\ &= \frac{(1+\varepsilon)}{x}(x^\delta - y) \le \frac{(1+\varepsilon)}{x}|\bar{y} - y|.\end{aligned}$$

Suppose $y \le 0 < \bar{y} < x^\delta$, then

$$|f(x,y) - f(x,\bar{y})| = \left|0 - \frac{(1+\varepsilon)\bar{y}}{x}\right| \le \frac{(1+\varepsilon)}{x}|y - \bar{y}|.$$

Suppose $y \le 0, x^\delta \le \bar{y}$, then

$$\begin{aligned}|f(x,y) - f(x,\bar{y})| &= |0 - (1+\varepsilon)x^{\delta-1}| = \frac{(1+\varepsilon)}{x}x^\delta \\ &\le \frac{(1+\varepsilon)}{x}\bar{y} \le \frac{(1+\varepsilon)}{x}|y - \bar{y}|.\end{aligned}$$

The initial value problem (1.6.3) has an infinite number of solutions $y(x) = cx^\delta$ in the interval $0 \le x \le 1$, where c is an arbitrary constant such that $0 < c < 1$. ∎

The continuity requirement of the function $f(x,y)$ particularly at the point (x_0, y_0) is also essential for the validity of the uniqueness theorem. For this, we have

Example 1.6.2. Consider the initial value problem

$$(1.6.4) \qquad y' = f(x,y) = \begin{cases} 1, & 0 \le x \le 1, y > x \\ \dfrac{y}{x}, & 0 \le x \le 1, 0 < y \le x \\ 0, & 0 \le x \le 1, y \le 0 \end{cases}$$
$$y(0) = 0.$$

This function $f(x,y)$ satisfies Nagumo's condition (1.6.1) in the strip $0 \le x \le 1, |y| < \infty$. For this, we look at the following cases :

Suppose $0 < y, \bar{y} \le x$, then

$$|f(x,y) - f(x,\bar{y})| = \left|\frac{y}{x} - \frac{\bar{y}}{x}\right| = \frac{1}{x}|y - \bar{y}|.$$

Suppose $y > x \ge \bar{y} > 0$, then

$$|f(x,y) - f(x,\bar{y})| = \left|1 - \frac{\bar{y}}{x}\right| = \frac{1}{x}(x - \bar{y}) < \frac{1}{x}(y - \bar{y}) = \frac{1}{x}|y - \bar{y}|.$$

Suppose $y > x, \bar{y} \le 0$, then

$$|f(x,y) - f(x,\bar{y})| = |1 - 0| = \frac{x}{x} < \frac{1}{x}y \le \frac{1}{x}|y - \bar{y}|.$$

Suppose $0 < y \le x, \bar{y} \le 0$, then

$$|f(x,y) - f(x,\bar{y})| = \left|\frac{y}{x} - 0\right| \le \frac{1}{x}|y - \bar{y}|.$$

Since for $y > x$, $\lim_{(x,y)\to(0^+,0)} f(x,y) = 1$; and for $0 < y \le x$, $\lim_{\substack{y=x \\ x\to 0^+}} f(x,y) = \lim_{\substack{y=x \\ x\to 0^+}} \frac{y}{x} = 1$, but $f(0,0) = 0$ this function $f(x,y)$ is not continuous at $(0,0)$.

The initial value problem (1.6.4) has an infinite number of solutions $y(x) = cx$ in the interval $0 \le x \le 1$, where c is an arbitrary constant such that $0 \le c \le 1$. ∎

Once again, Nagumo's condition is only a sufficient condition to prove the uniqueness of the solutions of the initial value problems. To show this, we have

Example 1.6.3. Consider the initial value problem

$$(1.6.5)\qquad y' = f(x,y) = \begin{cases} 0, & -\infty < x \le 0, |y| < \infty \\ 2x, & 0 < x \le 1, -\infty < y < 0 \\ 2x - 4\dfrac{y}{x}, & 0 < x \le 1, 0 \le y \le x^2 \\ -2x, & 0 < x \le 1, x^2 < y < \infty \end{cases}$$

$$y(0) = 0.$$

This function $f(x,y)$ is continuous at $(0,y)$ for $0 \le y \le x^2$ since

$$|2x - 4y/x| \le 2x + 4x^2/x = 6x \to 0 \quad \text{as} \quad x \to 0.$$

Hence, $f(x,y)$ is continuous in the region $-\infty < x \le 1, |y| < \infty$.

Next, we shall verify that $f(x,y)$ satisfies condition (1.6.1) except that $k = 4$. Once again, it suffices to consider the following cases :

Suppose $0 \le y, \bar{y} \le x^2$, then

$$|f(x,y) - f(x,\bar{y})| = \left| -\frac{4y}{x} + \frac{4\bar{y}}{x} \right| = \frac{4}{x}|y - \bar{y}|.$$

Suppose $-\infty < y < 0 \le \bar{y} \le x^2$, then

$$|f(x,y) - f(x,\bar{y})| = \left| 2x - 2x + \frac{4\bar{y}}{x} \right| < \frac{4}{x}|y - \bar{y}|.$$

Suppose $-\infty < y < 0, x^2 < \bar{y} < \infty$, then

$$|f(x,y) - f(x,\bar{y})| = 4x < \frac{4}{x}\bar{y} \le \frac{4}{x}|y - \bar{y}|.$$

Suppose $0 \le y \le x^2 < \bar{y} < \infty$, then

$$|f(x,y) - f(x,\bar{y})| = \left| 2x + 2x - \frac{4y}{x} \right| = \frac{4}{x}(x^2 - y) < \frac{4}{x}|y - \bar{y}|.$$

The initial value problem (1.6.5) has a unique solution

$$y(x) = \begin{cases} 0, & -\infty \le x \le 0 \\ x^2/3, & 0 < x \le 1. \end{cases} \qquad \blacksquare$$

Another example where the function $f(x,y)$ satisfies the conditions of Theorem 1.6.2 except that $k = 1 + \varepsilon$, but has a unique solution is the following :

Example 1.6.4. The initial value problem

$$(1.6.6)\qquad y' = f(x,y) = \begin{cases} 0, & x = 0, |y| < \infty \\ x^{\varepsilon}, & 0 < x \le 1, -\infty < y < 0 \\ x^{\varepsilon} - (1+\varepsilon)\dfrac{y}{x}, & 0 < x \le 1, 0 \le y \le x^{1+\varepsilon} \\ -\varepsilon x^{\varepsilon}, & 0 < x \le 1, x^{1+\varepsilon} < y < \infty \end{cases}$$

$$y(0) = 0$$

has a unique solution $y(x) = \frac{1}{2(1+\varepsilon)}x^{1+\varepsilon}, 0 \le x \le 1.$ ∎

In our final example here the function $f(x,y)$ satisfies all the conditions of Theorem 1.6.2, and hence has at most one solution.

Example 1.6.5. For the initial value problem

$$(1.6.7)\qquad y' = f(x,y) = \begin{cases} 0, & x = 0, |y| < \infty \\ \frac{1}{2}x^{\varepsilon}, & 0 < x \le 1, -\infty < y < 0 \\ \frac{1}{2}x^{\varepsilon} - \dfrac{y}{x}, & 0 < x \le 1, 0 \le y \le x^{1+\varepsilon} \\ -\frac{1}{2}x^{\varepsilon}, & 0 < x \le 1, x^{1+\varepsilon} < y < \infty \end{cases}$$

$$y(0) = 0,$$

$y(x) = \frac{1}{2(2+\varepsilon)}x^{1+\varepsilon}, 0 \le x \le 1$ is the only solution. ∎

A one - sided analog of Nagumo's uniqueness theorem is the following :

Theorem 1.6.3. Let $f(x,y)$ be continuous in $\bar{S}_+$ and for all $(x,y), (x,\bar{y})$ in $\bar{S}_+$ with $\bar{y} \ge y$ it satisfies **one - sided Nagumo condition**

$$(1.6.8)\qquad f(x,\bar{y}) - f(x,y) \le k(x-x_0)^{-1}(\bar{y}-y);\ x \ne x_0,\ k \le 1.$$

Then, the initial value problem (1.1.1) has at most one solution in $[x_0, x_0 + a]$. ∎

1.7 KRASNOSEL'SKII - KREIN'S UNIQUENESS THEOREM

In the previous section we have seen that in condition (1.6.1) the Nagumo constant k must not be greater than one is a necessary condition. Thus, if k

is allowed to be any constant greater then zero, then an additional condition must be imposed on the function $f(x,y)$ to achieve the uniqueness of the solutions of the initial value problem (1.1.1). One such condition is added in the following :

Theorem 1.7.1 (Krasnosel'skii - Krein's Uniqueness Theorem). Let $f(x,y)$ be continuous in $\bar{S}$ and for all $(x,y),(x,\bar{y}) \in \bar{S}$ it satisfies

$$(1.7.1) \qquad |f(x,y)-f(x,\bar{y})| \leq k|x-x_0|^{-1}|y-\bar{y}|, \ x \neq x_0, k>0$$

and the Hölder condition

$$(1.7.2) \qquad |f(x,y)-f(x,\bar{y})| \leq c|y-\bar{y}|^{\alpha}, \ c>0, 0<\alpha<1, k(1-\alpha)<1.$$

($k(1-\alpha)<1$ is no restriction at all, when $k \leq 1$.)

Then, the initial value problem (1.1.1) has at most one solution in $|x-x_0| \leq a$.

Proof. Suppose $y(x)$ and $\bar{y}(x)$ are two solutions of (1.1.1) in $|x-x_0| \leq a$. We shall show that $y(x)=\bar{y}(x)$ in the interval $[x_0, x_0+a]$. We let $\phi(x) = |y(x)-\bar{y}(x)|$. For this, from (1.7.2), we have

$$\begin{aligned}\phi(x) &\leq \int_{x_0}^{x} |f(t,y(t))-f(t,\bar{y}(t))|dt \\ &\leq \int_{x_0}^{x} c|y(t)-\bar{y}(t)|^{\alpha}dt = \int_{x_0}^{x} c\phi^{\alpha}(t)dt.\end{aligned}$$

Now, we let $R(x) = \int_{x_0}^{x} c\phi^{\alpha}(t)dt$, then $R(x_0)=0$, and $R'(x) = c\phi^{\alpha}(x) \leq cR^{\alpha}(x)$, which gives $R'(x)-cR^{\alpha}(x) \leq 0$. Since $R(x)>0$ for $x>x_0$, on multiplying this inequality by $(1-\alpha)R^{-\alpha}(x)$, we obtain $(R^{1-\alpha}(x))' \leq c(1-\alpha) \leq c$, and hence $R^{1-\alpha}(x) \leq c(x-x_0)$. Thus, it follows that $\phi(x) \leq (c(x-x_0))^{(1-\alpha)^{-1}}$. Hence, the function $\psi(x) = \frac{\phi(x)}{(x-x_0)^k}$ satisfies the inequality

$$0 \leq \psi(x) \leq c^{(1-\alpha)^{-1}}(x-x_0)^{(1-\alpha)^{-1}-k}.$$

Since $k(1-\alpha)<1$, it is immediate that $\lim_{x\to x_0^+} \psi(x) = 0$. Therefore, if we define $\psi(x_0)=0$, then the function $\psi(x)$ is continuous in $[x_0, x_0+a]$. We wish to show that $\psi(x)=0$ in $[x_0, x_0+a]$. If $\psi(x)>0$ at any point in $[x_0, x_0+a]$, then there exists a point $x_1 > x_0$ such that $0 < m = \psi(x_1) = \max_{x_0 \le x \le x_0+a} \psi(x)$. However, from (1.7.1) we have

$$\begin{aligned} m = \psi(x_1) = \frac{\phi(x_1)}{(x_1 - x_0)^k} &\le (x_1 - x_0)^{-k} \int_{x_0}^{x_1} k(t - x_0)^{-1} \phi(t) dt \\ &\le (x_1 - x_0)^{-k} \int_{x_0}^{x_1} k(t - x_0)^{k-1} \psi(t) dt \\ &< m(x_1 - x_0)^{-k} \int_{x_0}^{x_1} k(t - x_0)^{k-1} dt \\ &= m, \end{aligned}$$

which is the desired contradiction. So, $\psi(x) \equiv 0$, and hence $\phi(x) = 0$ in $[x_0, x_0 + a]$. The proof is similar in the interval $[x_0 - a, x_0]$. ∎

We shall illustrate the above theorem with an example.

Example 1.7.1. Consider the initial value problem

$$(1.7.3) \qquad y' = f(x, y) = \begin{cases} 0, & 0 \le x \le 1, x^{(1-\alpha)^{-1}} < y < \infty \\ kx^{\alpha(1-\alpha)^{-1}} - k\dfrac{y}{x}, & 0 \le x \le 1, 0 \le y \le x^{(1-\alpha)^{-1}} \\ kx^{\alpha(1-\alpha)^{-1}}, & 0 \le x \le 1, -\infty < y < 0 \end{cases}$$

$$y(0) = 0,$$

where $0 < \alpha < 1, k > 0$ and $k(1 - \alpha) < 1$.

This function $f(x, y)$ is continuous at $(0, y)$ for $0 \le y \le x^{(1-\alpha)^{-1}}$ since

$$\left| kx^{\alpha(1-\alpha)^{-1}} - k\frac{y}{x} \right| \le kx^{\alpha(1-\alpha)^{-1}} + \frac{k}{x} x^{(1-\alpha)^{-1}} = 2kx^{\alpha(1-\alpha)^{-1}} \to 0 \;\text{ as } x \to 0^+.$$

Thus, it is clear that $f(x, y)$ is continuous in the strip $0 \le x \le 1, |y| < \infty$.

Next, we shall verify the conditions (1.7.1) and (1.7.2) by considering the following cases :

Suppose $0 \le y, \bar{y} \le x^{(1-\alpha)^{-1}}$, then

$$
\begin{aligned}
|f(x,y) - f(x,\bar{y})| &= \left|-k\frac{y}{x} + k\frac{\bar{y}}{x}\right| \\
&= \frac{k}{x}|y-\bar{y}|; \text{ and} \\
|f(x,y) - f(x,\bar{y})| &= \frac{k}{x}|y-\bar{y}|^{1-\alpha}|y-\bar{y}|^{\alpha} \\
&\le \frac{k}{x}(|y|+|\bar{y}|)^{1-\alpha}|y-\bar{y}|^{\alpha} \le \frac{k}{x}\, 2^{1-\alpha}x|y-\bar{y}|^{\alpha} \\
&= 2^{1-\alpha}k|y-\bar{y}|^{\alpha}.
\end{aligned}
$$

Suppose $x^{(1-\alpha)^{-1}} < y < \infty, -\infty < \bar{y} < 0$, then

$$
\begin{aligned}
|f(x,y) - f(x,\bar{y})| &= |-kx^{\alpha(1-\alpha)^{-1}}| < \frac{k}{x}y \\
&< \frac{k}{x}|y-\bar{y}|; \text{ and} \\
|f(x,y) - f(x,\bar{y})| &= kx^{\alpha(1-\alpha)^{-1}} < ky^{\alpha} < k|y-\bar{y}|^{\alpha} \\
&< 2^{1-\alpha}k|y-\bar{y}|^{\alpha}.
\end{aligned}
$$

Suppose $x^{(1-\alpha)^{-1}} < y < \infty, 0 \le \bar{y} \le x^{(1-\alpha)^{-1}}$, then

$$
\begin{aligned}
|f(x,y) - f(x,\bar{y})| &= \left|-kx^{\alpha(1-\alpha)^{-1}} + k\frac{\bar{y}}{x}\right| = \frac{k}{x}\left|x^{(1-\alpha)^{-1}} - \bar{y}\right| \\
&< \frac{k}{x}|y-\bar{y}|; \text{ and} \\
|f(x,y) - f(x,\bar{y})| &= k\left[\frac{x^{(1-\alpha)^{-1}} - \bar{y}}{(x^{(1-\alpha)^{-1}})^{(1-\alpha)}}\right] \\
&\le k\left[x^{(1-\alpha)^{-1}} - \bar{y}\right]^{\alpha} < k(y-\bar{y})^{\alpha} \\
&< 2^{1-\alpha}k|y-\bar{y}|^{\alpha}.
\end{aligned}
$$

Suppose $0 \le y \le x^{(1-\alpha)^{-1}}, -\infty < \bar{y} < 0$, then

$$\begin{aligned} |f(x,y)-f(x,\bar{y})| &= \left|kx^{(1-\alpha)^{-1}} - k\frac{y}{x} - kx^{(1-\alpha)^{-1}}\right| = k\frac{y}{x} \\ &< \frac{k}{x}|y-\bar{y}|; \text{ and} \\ |f(x,y)-f(x,\bar{y})| &= k\frac{y}{x} \le ky^{\alpha} \\ &< 2^{1-\alpha}k|y-\bar{y}|^{\alpha}. \end{aligned}$$

Since all the conditions of Theorem 1.7.1 are satisfied, the initial value problem (1.7.3) has a unique solution in $[0,1]$, namely

$$y(x) = \frac{k(1-\alpha)}{1+k(1-\alpha)} x^{(1-\alpha)^{-1}}. \quad \blacksquare$$

For the uniqueness of the solutions of (1.1.1) in Theorem 1.7.1 the condition $k(1-\alpha) < 1$ is not necessary. For this, it suffices to note that the function $f(x,y)$ in (1.7.3) is nonincreasing in y for each x. Thus, by Theorem 1.3.1 the problem (1.7.3) has at most one solution.

1.8 KOOI'S UNIQUENESS THEOREM

A generalization of Krasnosel'skii - Krein's uniqueness theorem is the following :

Theorem 1.8.1 (Kooi's Uniqueness Theorem). Let $f(x,y)$ be continuous in $\bar{S}$ and for all $(x,y),(x,\bar{y}) \in \bar{S}$ it satisfies

(1.8.1) $$|x-x_0||f(x,y)-f(x,\bar{y})| \le k|y-\bar{y}|, \; k>0; \text{ and}$$

(1.8.2) $$|x-x_0|^{\beta}|f(x,y)-f(x,\bar{y})| \le c|y-\bar{y}|^{\alpha}, \; c>0$$

where the constants k, α and β satisfy the inequalities

(1.8.3) $$0<\alpha<1, \; \beta<\alpha \quad \text{and} \quad k(1-\alpha) < 1-\beta.$$

Then, the initial value problem (1.1.1) has at most one solution in $|x-x_0| \le a$.

Proof. The proof is similar to that of Theorem 1.7.1. ∎

Example 1.8.1. Consider the initial value problem

$$(1.8.4)\qquad y' = f(x,y) = \begin{cases} 0, & 0 \le x \le 1, \\ & x^{(1-\beta)(1-\alpha)^{-1}} < y < \infty \\ kx^{(\alpha-\beta)(1-\alpha)^{-1}} - k\dfrac{y}{x}, & 0 \le x \le 1, \\ & 0 \le y \le x^{(1-\beta)(1-\alpha)^{-1}} \\ kx^{(\alpha-\beta)(1-\alpha)^{-1}}, & 0 \le x \le 1, -\infty < y < 0 \end{cases}$$

$$y(0) = 0,$$

where the constants k, α and β satisfy the inequalities (1.8.3). As in Example 1.7.1 it is easily seen that this function $f(x,y)$ is continuous in the strip $0 \le x \le 1, |y| < \infty$ and satisfies the inequalities (1.8.1),(1.8.2). Thus, the initial value problem (1.8.4) has a unique solution $y(x) = \frac{\gamma}{1+\gamma}x^{(1-\beta)(1-\alpha)^{-1}}$, where $\gamma = k(1-\alpha)(1-\beta)^{-1}$. ∎

From Example 1.8.1 it is clear that for the conclusion of Theorem 1.8.1 also the condition $k(1-\alpha) < 1-\beta$ is not necessary.

The next theorem we shall prove is based on Kooi's work.

Theorem 1.8.2. Let $f(x,y)$ be continuous in $\bar{S}$ and

$$(1.8.5)\qquad |f(x,y)| \le A|x-x_0|^p;\ p > -1, A > 0.$$

Further, let for all $(x,y),(x,\bar{y}) \in \bar{S}$ it satisfy

$$(1.8.6)\qquad |f(x,y) - f(x,\bar{y})| \le \frac{c}{|x-x_0|^r}|y-\bar{y}|^q;\ q \ge 1, c > 0$$

where

$$(1.8.7)\qquad q(1+p) - r = p, \quad \text{and} \quad \rho = \frac{c(2A)^{q-1}}{(p+1)^q} < 1.$$

Then, the initial value problem (1.1.1) has at most one solution in $|x-x_0| \le a$.

Proof. Suppose $y(x)$ and $\bar{y}(x)$ are two solutions of (1.1.1) in $|x-x_0| \le a$. We shall show that $y(x) = \bar{y}(x)$ in $[x_0 - a, x_0]$. For this, from (1.8.5) we have

$$\begin{aligned}\phi(x) = |y(x) - \bar{y}(x)| &\le \int_x^{x_0} |f(t, y(t)) - f(t, \bar{y}(t))| dt \\ &\le 2A \int_x^{x_0} (x_0 - t)^p dt \\ &= \frac{2A}{p+1}(x_0 - x)^{p+1}.\end{aligned}$$

Using this estimate and (1.8.6), we obtain

$$\begin{aligned}\phi(x) &\le c \int_x^{x_0} \frac{1}{(x_0 - t)^r} \phi^q(t) dt \\ &\le c \left(\frac{2A}{p+1}\right)^q \int_x^{x_0} (x_0 - t)^{q(p+1)-r} dt \\ &= \rho \left(\frac{2A}{p+1}\right)(x_0 - x)^{p+1}.\end{aligned}$$

Now on using this new estimate and (1.8.6), we get

$$\begin{aligned}\phi(x) &\le c\rho^q \left(\frac{2A}{p+1}\right)^q \int_x^{x_0} (x_0 - t)^{q(p+1)-r} dt \\ &= \rho^{1+q} \left(\frac{2A}{p+1}\right)(x_0 - x)^{p+1}.\end{aligned}$$

Continuing in this way, we find that

$$\phi(x) \le \rho^{1+q+q^2+\cdots+q^m} \left(\frac{2A}{p+1}\right)(x_0 - x)^{p+1}; \quad m = 1, 2, \cdots .$$

Since $q \ge 1$ and $\rho < 1$, it follows that $\phi(x) = 0$ in $[x_0 - a, x_0]$. The same is true for $x \in [x_0, x_0 + a]$. ∎

Example 1.8.2. Consider the initial value problem (1.7.3). The function $f(x, y)$ in (1.7.3) satisfies (1.8.5) and (1.8.6) with $A = 2k, p = \alpha(1-\alpha)^{-1}, c = k, r = q = 1$, and therefore $q(1+p) - r = p$ and $\rho = \frac{c(2A)^{q-1}}{(p+1)^q} = k(1-\alpha) < 1$. Thus, from Theorem 1.8.2 also we can conclude that the problem (1.7.3) has at most one solution. We also note that for this problem the conditions of

Theorem 1.8.2 are also satisfied with $A = 2k, p = \alpha(1-\alpha)^{-1}, c = 2^{1-\alpha}k, q = \alpha$ and $r = 0$. ∎

1.9 ROGER'S UNIQUENESS THEOREM

Nagumo's condition (1.6.1) for $x \geq x_0 = 0$ is the same as

$$(1.9.1) \qquad |f(x,y) - f(x,\bar{y})| \leq \frac{1}{x}|y - \bar{y}|, \; x \neq 0.$$

In this section, we shall establish a uniqueness criterion when $\frac{1}{x}$ is replaced by $\frac{1}{x^2}$ in the above condition. For this, we require the following :

Lemma 1.9.1. If

(i) $\phi(x)$ is continuous and nonnegative in $[0, a]$,

(ii) $\phi(x) \leq \int_0^x \frac{1}{t^2}\phi(t)dt$; and

(iii) $\phi(x) = o(e^{-1/x})$, as $x \to 0$,

then $\phi(x) \equiv 0$.

Proof. Let $\psi(x) = \int_0^x \frac{1}{t^2}\phi(t)dt$. Differentiating $\psi(x)$ with respect to x and using (ii), we obtain for $x > 0, \psi'(x) = \frac{1}{x^2}\phi(x) \leq \frac{1}{x^2}\psi(x)$, which is the same as $(e^{1/x}\psi(x))' \leq 0$, so that $e^{1/x}\psi(x)$ is nonincreasing. If $\varepsilon > 0$, then from (iii), we have for small x, that

$$e^{1/x}\psi(x) = e^{1/x}\int_0^x \frac{1}{t^2}\phi(t)dt \leq e^{1/x}\int_0^x \frac{1}{t^2}\varepsilon e^{-1/t}dt = \varepsilon.$$

Hence, $\lim_{x \to 0^+} e^{1/x}\psi(x) = 0$, and this implies that $e^{1/x}\psi(x) \leq 0$ for $x > 0$. This gives $\psi(x) \leq 0$. However, $\psi(x)$ is nonnegative, and thus $\psi(x) \equiv 0$. The result now follows from (i). ∎

Theorem 1.9.2 (Roger's Uniqueness Theorem). Let $f(x,y)$ be continuous in

$$\bar{\Delta} : 0 \leq x \leq a, \; |y| < \infty$$

and satisfy the condition

$$f(x,y) = o(e^{-1/x}x^{-2}) \tag{1.9.2}$$

uniformly for $0 \le y \le \delta$, $\delta > 0$ arbitrary. Further, let

$$|f(x,y) - f(x,\bar{y})| \le \frac{1}{x^2}|y - \bar{y}|, \ x \ne 0 \tag{1.9.3}$$

for all $(x,y), (x,\bar{y}) \in \bar{\Delta}$. Then, the initial value problem

$$y' = f(x,y), \ y(0) = 0 \tag{1.9.4}$$

has at most one solution in $[0,a]$.

Proof. Suppose $y(x)$ and $\bar{y}(x)$ are two solutions of (1.9.4) in $[0,a]$. Then, from (1.2.4) and (1.9.3), we get

$$|y(x) - \bar{y}(x)| \le \int_0^x \frac{1}{t^2}|y(t) - \bar{y}(t)|dt.$$

Also, if $\varepsilon > 0$, then from (1.2.4) and (1.9.2) for small x, we have

$$|y(x) - \bar{y}(x)| \le 2\varepsilon \int_0^x e^{-1/t}t^{-2}dt = 2\varepsilon e^{-1/x}.$$

Now as an application of Lemma 1.9.1, we find that $|y(x) - \bar{y}(x)| \equiv 0$, and hence $y(x) = \bar{y}(x)$ in $[0,a]$. ∎

We shall illustrate Theorem 1.9.2 with an example.

Example 1.9.1. Consider the initial value problem

$$y' = f(x,y) = \begin{cases} \left(1 + \frac{1}{x}\right) e^{-1/x}, & 0 \le x \le 1, xe^{-1/x} \le y < \infty \\ \frac{y}{x^2} + e^{-1/x}, & 0 \le x \le 1, 0 \le y \le xe^{-1/x} \\ e^{-1/x}, & 0 \le x \le 1, -\infty < y \le 0 \end{cases} \tag{1.9.5}$$

$$y(0) = 0.$$

Since $\lim_{x\to 0^+} e^{-1/x} = 0$, and $\lim_{x\to 0^+} \frac{1}{x}e^{-1/x} = 0$, and for $0 \le y \le xe^{-1/x}$,

$$\left|\frac{y}{x^2} + e^{-1/x}\right| \le \frac{xe^{-1/x}}{x^2} + e^{-1/x} \to 0 \quad \text{as} \quad x \to 0^+$$

it is clear that this function $f(x,y)$ is continuous at $(0,y)$ for $|y| < \infty$, and hence it is continuous in the strip $0 \le x \le 1, |y| < \infty$.

Next, we shall verify that $f(x,y)$ satisfies condition (1.9.2). For this, since

$$\frac{f(x,y)}{e^{-1/x}x^{-2}} = \begin{cases} x+x^2, & 0 < x \le 1, xe^{-1/x} \le y < \infty \\ ye^{1/x}+x^2, & 0 < x \le 1, 0 \le y \le xe^{-1/x} \\ x^2, & 0 < x \le 1, -\infty < y \le 0 \end{cases} \tag{1.9.6}$$

and for $0 \le y \le xe^{-1/x}$,

$$|ye^{1/x}+x^2| \le (xe^{-1/x})e^{1/x}+x^2 = x+x^2 \to 0 \quad \text{as} \quad x \to 0^+,$$

we find that each expression on the right of the equation (1.9.6) tends to zero as x tends to zero. Thus, (1.9.2) holds.

The condition (1.9.3) can be verified by considering the following cases :

Suppose $0 \le y, \bar{y} \le xe^{-1/x}$, then

$$|f(x,y)-f(x,\bar{y})| = \left|\frac{y}{x^2}-\frac{\bar{y}}{x^2}\right| = \frac{1}{x^2}|y-\bar{y}|.$$

Suppose $xe^{-1/x} \le y < \infty, 0 \le \bar{y} \le xe^{-1/x}$, then

$$\begin{aligned} |f(x,y)-f(x,\bar{y})| &= \left|\frac{e^{-1/x}}{x}-\frac{\bar{y}}{x^2}\right| = \frac{1}{x^2}(xe^{-1/x}-\bar{y}) \\ &\le \frac{1}{x^2}(y-\bar{y}) = \frac{1}{x^2}|y-\bar{y}|. \end{aligned}$$

Suppose $xe^{-1/x} \le y < \infty, -\infty < \bar{y} \le 0$, then

$$|f(x,y)-f(x,\bar{y})| = \frac{1}{x^2}xe^{-1/x} \le \frac{1}{x^2}y \le \frac{1}{x^2}|y-\bar{y}|.$$

Suppose $0 \le y \le xe^{-1/x}, -\infty < \bar{y} \le 0$, then

$$|f(x,y)-f(x,\bar{y})| = \frac{1}{x^2}y \le \frac{1}{x^2}|y-\bar{y}|.$$

Thus, all conditions of Theorem 1.9.2 are satisfied, and hence in conclusion this initial value problem has a unique solution in $[0,1]$, namely, $y(x) = xe^{-1/x}$. ∎

In Theorem 1.9.2 the assumption $f(x,y) = o(e^{-1/x}x^2)$ is necessary. For this, we have the following :

Example 1.9.2. Consider the initial value problem

$$(1.9.7) \qquad y' = f(x,y) = \begin{cases} 0, & 0 \le x \le 1, -\infty < y \le 0 \\ \dfrac{y}{x^2}, & 0 \le x \le 1, 0 \le y \le e^{-1/x} \\ \dfrac{1}{x^2}e^{-1/x}, & 0 \le x \le 1, e^{-1/x} \le y < \infty \end{cases}$$
$$y(0) = 0.$$

Since $\lim_{x\to 0^+} \frac{1}{x^2}e^{-1/x} = 0$, it follows that for $0 \le y \le e^{-1/x}$, $\left|\frac{y}{x^2}\right| \le \frac{1}{x^2}e^{-1/x} \to 0$ as $x \to 0^+$. Thus, it is clear that this function $f(x,y)$ is continuous in the strip $0 \le x \le 1, |y| < \infty$.

As usual, we shall verify condition (1.9.3) by considering the following cases :

Suppose $0 \le y, \bar{y} \le e^{-1/x}$, then

$$|f(x,y) - f(x,\bar{y})| = \left|\frac{y}{x^2} - \frac{\bar{y}}{x^2}\right| = \frac{1}{x^2}|y - \bar{y}|.$$

Suppose $-\infty < y \le 0, 0 \le \bar{y} \le e^{-1/x}$, then

$$|f(x,y) - f(x,\bar{y})| = \left|-\frac{\bar{y}}{x^2}\right| = \frac{1}{x^2}|y - \bar{y}|.$$

Suppose $-\infty < y \le 0, e^{-1/x} \le \bar{y} < \infty$, then

$$|f(x,y) - f(x,\bar{y})| = \left|-\frac{1}{x^2}e^{-1/x}\right| \le \frac{1}{x^2}\bar{y} \le \frac{1}{x^2}|y - \bar{y}|.$$

Suppose $0 \le y \le e^{-1/x}, e^{-1/x} \le \bar{y} < \infty$, then

$$\begin{aligned} |f(x,y) - f(x,\bar{y})| &= \left|\frac{y}{x^2} - \frac{e^{-1/x}}{x^2}\right| = \frac{1}{x^2}(e^{-1/x} - y) \\ &\le \frac{1}{x^2}(\bar{y} - y) \le \frac{1}{x^2}|y - \bar{y}|. \end{aligned}$$

This function $f(x,y)$ does not satisfy the condition (1.9.2). Indeed, for $e^{-1/x} \le y < \infty, \lim_{x\to 0^+} \frac{f(x,y)}{e^{-1/x}x^{-2}} \ne 0$.

The initial value problem (1.9.7) has an infinite number of solutions $y(x) = ce^{-1/x}$ in $[0, 1]$, where c is an arbitrary constant such that $0 \le c \le 1$. ∎

1.10 WITTE'S UNIQUENESS THEOREM

In the year 1974 the following result was established.

Theorem 1.10.1 (Witte's Uniqueness Theorem). Let $f(x, y)$ be continuous in

$$\Delta : 0 < x \le a, \ |y| < \infty$$

and satisfy the conditions

$$(1.10.1) \qquad |f(x,y) - f(x,\bar{y})| \le h(x)|y - \bar{y}|, \ (x,y),(x,\bar{y}) \in \Delta; \text{ and}$$

$$(1.10.2) \qquad |f(x,y)| \le \rho(x)h(x)\exp\left(\int_a^x h(t)dt\right) \quad \text{in} \quad \Delta,$$

where $h(x) > 0$ is continuous in $(0, a]$ and $\rho(x)$ is continuous in $[0, a]$ with $\rho(0) = 0$. Then, the initial value problem (1.1.1) with $x_0 = 0$ has at most one solution in $[0, a]$.

We shall generalize this result slightly. For this, we need the following :

Lemma 1.10.2. Let $\phi(x)$ be a nonnegative continuous function in $[0, a]$ and let

(i) $h(x) > 0$ be a continuous function in $(0, a]$,

(ii) there exist a function $H(x)$ in $(0, a]$ such that $H'(x) = h(x)$ for almost all $x \in (0, a]$, and $\lim_{x \to 0^+} H(x)$ exists (finite or infinite),

(iii) $\phi(x) \le \int_0^x h(t)\phi(t)dt, \ x \in [0, a]$; and

(iv) $\phi(x) = o(\exp(H(x)))$ as $x \to 0^+$.

Then, $\phi(x) = 0$ in $[0, a]$.

Proof. Let $\psi(x) = \int_0^x h(t)\phi(t)dt,\ x \in [0, a]$. The existence and the continuity of $\psi(x)$ is clear from the given hypotheses. Thus, it follows that

$$\psi'(x) = h(x)\phi(x) \le h(x)\psi(x), \quad x \in (0, a]$$

and hence

$$\frac{d}{dx}(e^{-H(x)}\psi(x)) \le 0, \text{ for almost all } x \in [0, a].$$

This allows us to deduce that the function $\exp(-H(x))\psi(x)$ is nonincreasing. Hence, choosing $\varepsilon > 0$ and taking x sufficiently small, in view of (iv) we get

$$\begin{aligned} e^{-H(x)}\psi(x) = \;& e^{-H(x)} \int_0^x h(t)\phi(t)dt \\ & \le \varepsilon e^{-H(x)} \int_0^x h(t)e^{H(t)}dt \\ & \le \varepsilon e^{-H(x)} e^{H(x)} \\ & = \varepsilon, \end{aligned}$$

so that $\lim_{x\to 0^+} \exp(-H(x))\psi(x) = 0$. Consequently, $\exp(-H(x))\psi(x) \le 0$ for $x > 0$, which implies that $\int_0^x h(t)\phi(t)dt \le 0$. Therefore, $\phi(x) \equiv 0$. ∎

If $h(x)$ is continuous in $[0, a]$ and $H(x) = \int_0^x h(t)dt$, then Lemma 1.10.2 is the same as Lemma 1.2.3. Taking $h(x) = x^{-1}$ and $H(x) = \ln x$ so that condition (iv) becomes $\phi(x) = o(x)$ as $x \to 0^+$. Thus, in this case Lemma 1.10.2 reduces to Lemma 1.6.1. Further, we note that in the case $h(x) = x^{-1-\alpha}$, $\alpha > 0$ and $H(x) = -\alpha^{-1}x^{-\alpha}$ the condition (iv) has the form $\phi(x) = o(\exp(-\alpha^{-1}x^{-\alpha}))$ as $x \to 0^+$. Thus, for $\alpha = 1$ we obtain Lemma 1.9.1. Finally, consider $h(x) = kx^{-1}$, where $k > 1$ and $H(x) = x^k$ so that condition (iv) has the form $\phi(x) = o(x^k)$. Thus, Lemma 1.10.2 also covers situations where Lemma 1.6.1 is not applicable.

In view of the above remarks the following result not only unifies Theorems 1.2.4, 1.6.2 and 1.9.2 but also provides new uniqueness criterion.

Theorem 1.10.3. Let $f(x, y)$ be continuous in Δ and in addition to (1.10.1) it satisfies the condition

$$(1.10.3)\quad |f(x,y)-f(x,\bar{y})| = o(h(x)\exp(H(x))) \text{ as } x\to 0^+ \text{ uniformly with respect to } y,\bar{y}\in[-\delta,\delta],\ \delta>0 \text{ arbitrary,}$$

where $h(x)$ and $H(x)$ are such as in Lemma 1.10.2. Then, the initial value problem (1.9.4) has at most one solution in $[0,a]$.

Proof. As usual let $y(x)$ and $\bar{y}(x)$ be two solutions of (1.9.4). Then, by (1.10.1), we have

$$|y(x)-\bar{y}(x)| \le \int_0^x h(t)|y(t)-\bar{y}(t)|dt.$$

Further, for an arbitrary $\varepsilon>0$ and x sufficiently small, from (1.10.3) we get

$$|y(x)-\bar{y}(x)| \le \varepsilon\int_0^x h(t)e^{H(t)}dt \le \varepsilon e^{H(x)}.$$

The result is now an immediate consequence of Lemma 1.10.2. ∎

It is clear that (1.10.2) implies (1.10.3), however the converse is not true. For this, we illustrate the following :

Example 1.10.1. Consider the initial value problem

$$(1.10.4)\qquad y' = f(x,y) = \begin{cases} \left(1+\dfrac{1}{x}\right)e^{-1/x}+1, & 0\le x\le 1,\ xe^{-1/x}\le y<\infty \\ \dfrac{y}{x^2}+e^{-1/x}+1, & 0\le x\le 1,\ 0\le y\le xe^{-1/x} \\ e^{-1/x}+1, & 0\le x\le 1, -\infty<y\le 0 \end{cases}$$

$$y(0)=0.$$

From Example 1.9.1 it is clear that this function $f(x,y)$ is continuous in the strip $0\le x\le 1, |y|<\infty$. Further, it satisfies the conditions (1.10.1) and (1.10.3) with $h(x) = x^{-2}$ and $H(x) = -x^{-1}$. The problem (1.10.4) has a unique solution $y(x) = x(1+e^{-1/x})$. It is of interest to note that for this function $f(x,y)$ the condition (1.9.2) of Roger's uniqueness theorem as well as condition (1.10.2) of Witte's uniqueness theorem is not satisfied.

Corollary 1.10.4. In Theorem 1.10.3 conditions (1.10.1) and (1.10.3) can be replaced by the following one - sided conditions

(1.10.5) $(f(x,y)-f(x,\bar{y}))(y-\bar{y}) \le h(x)(y-\bar{y})^2,\ (x,y),(x,\bar{y}) \in \Delta;$ and

(1.10.6) $f(x,y)-f(x,\bar{y}) = o(h(x)\exp(H(x)))$ as $x \to 0^+$ uniformly with respect to $y, \bar{y} \in [-\delta, \delta]$, $\delta > 0$ arbitrary.

Proof. Let $y(x)$ and $\bar{y}(x)$ be two solutions of (1.9.4). We define $v(x) = (y(x)-\bar{y}(x))^2$ so that in view of (1.10.5) it follows that

$$\begin{aligned} v'(x) &= 2(y(x)-\bar{y}(x))(y'(x)-\bar{y}'(x)) \\ &= 2(y(x)-\bar{y}(x))(f(x,y(x))-f(x,\bar{y}(x))) \\ &\le 2h(x)(y(x)-\bar{y}(x))^2 \\ &= 2h(x)v(x). \end{aligned}$$

Thus, we have $v'(x)-2h(x)v(x) \le 0$, and consequently

$$\frac{d}{dx}(\exp(-2H(x))v(x)) \le 0$$

for almost all $x \in [0,a]$. The above inequality implies that the function $e^{-2H(x)}v(x)$ is nonincreasing. On the other hand, taking $\varepsilon > 0$ arbitrary and x sufficiently small, from (1.10.6) we find

$$\begin{aligned} e^{-2H(x)}v(x) &= e^{-2H(x)}(y(x)-\bar{y}(x))^2 \\ &= e^{-2H(x)}\left(\int_0^x (f(t,y(t))-f(t,\bar{y}(t)))dt\right)^2 \\ &\le e^{-2H(x)}\left(\int_0^x \varepsilon h(t)e^{H(t)}dt\right)^2 \\ &\le \varepsilon^2 e^{-2H(x)}e^{2H(x)} \\ &= \varepsilon^2 \end{aligned}$$

and hence $\lim_{x\to 0^+}\exp(-2H(x))v(x) = 0$. Therefore, $v(x) \equiv 0$, i.e., $y(x) = \bar{y}(x)$ in $[0,a]$. ∎

Finally, we remark that in Theorem 1.10.1 the condition (1.10.2) can be replaced by any one of the following :

(i) **Lemmert's Condition.**

The function

$$F(x,y) = f(x,y)\exp\left(\int_x^a h(t)dt\right)\Big/h(x) \tag{1.10.7}$$

is uniformly continuous for $0 < x \le a,\ |y| \le b,\ b > 0$ arbitrary.

(ii) **Došlà - Došlý's Condition.**

$|f(x,y)| = o(H(x))$ as $x \to 0^+$ uniformly with respect to $y \in [-\delta, \delta]$, $\delta > 0$ arbitrary, where $H(x)$ is a nonnegative function defined in $(0, a]$ and satisfies

$$\int_0^a H(t)dt < \infty; \quad \text{and} \tag{1.10.8}$$

$$\liminf_{x\to 0^+}\left[\int_0^x H(t)dt\ \exp\left(\int_x^a h(t)dt\right)\right] < \infty.$$

1.11 PERRON'S UNIQUENESS THEOREM

To prove the main result of this section we need the following :

Definition 1.11.1. A solution $r(x)$ $(\rho(x))$ of the initial value problem (1.1.1) which exists in an interval J is said to be maximal (minimal) solution if for an arbitrary solution $y(x)$ of (1.1.1) existing in J, the inequality $y(x) \le r(x)$ $(\rho(x) \le y(x))$ holds for all $x \in J$.

Clearly, the maximal solution $r(x)$ and the minimal solution $\rho(x)$ if they exist, are unique.

Theorem 1.11.1 [92]. Let $f(x,y)$ be continuous in $\bar{S}_+$, then there exists a maximal solution $r(x)$ and a minimal solution $\rho(x)$ of the initial value problem (1.1.1) in the interval $[x_0, x_0 + \alpha]$ for some positive α. ∎

Let $y(x)$ be a continuous function in an interval J. We shall adopt the following notation for Dini derivatives

$$D^+y(x) = \limsup_{h\to 0+}\left(\frac{y(x+h)-y(x)}{h}\right),$$

$$D_+y(x) = \liminf_{h\to 0+}\left(\frac{y(x+h)-y(x)}{h}\right),$$

$$D^-y(x) = \limsup_{h\to 0-}\left(\frac{y(x+h)-y(x)}{h}\right),$$

$$D_-y(x) = \liminf_{h\to 0-}\left(\frac{y(x+h)-y(x)}{h}\right).$$

When $D^+y(x) = D_+y(x)$, the right derivative of $y(x)$ exists and it will be denoted by $y'_+(x)$. Similarly, $y'_-(x)$ denotes the left derivative.

As an application of the maximal solution $r(x)$, we have

Theorem 1.11.2 [92]. Let $f(x,y)$ be continuous in a domain D, and let $r(x)$ be the maximal solution of (1.1.1) in the interval $J = [x_0, x_0+a)$. Also, let $y(x)$ be a solution of the differential inequality

$$D^+y(x) \le f(x, y(x))$$

in J. Then, $y(x_0) \le y_0$ implies that $y(x) \le r(x)$, for all $x \in J$. ∎

Theorem 1.11.3 [92]. Let $f(x,y)$ be continuous in a domain D, and let $\rho(x)$ be the minimal solution of (1.1.1) in the interval $J = (x_0 - a, x_0]$. Also, let $y(x)$ be a solution of the differential inequality

$$D_-y(x) \le f(x, y(x))$$

in J. Then, $y_0 \le y(x_0)$ implies that $\rho(x) \le y(x)$, for all $x \in J$. ∎

Theorem 1.11.4 (Perron's uniqueness Theorem). Assume that

(i) the function $g(x,z)$ is continuous and nonnegative in $x_0 \le x \le x_0+a$, $0 \le z \le 2b$ and for every x_1, $x_0 < x_1 < x_0 + a$, $z(x) \equiv 0$ is the only differentiable function in $[x_0, x_1)$, which satisfies

$$\begin{aligned} z'(x) &= g(x, z(x)),\ x_0 \le x < x_1 \\ z(x_0) &= 0, \end{aligned} \tag{1.11.1}$$

(ii) the function $f(x,y)$ is continuous in $\bar{S}_+$ and for all $(x,y),(x,\bar{y}) \in \bar{S}_+$ satisfies **Perron's condition**

$$|f(x,y) - f(x,\bar{y})| \le g(x, |y-\bar{y}|). \tag{1.11.2}$$

Then, the initial value problem (1.1.1) has at most one solution in $[x_0, x_0+a]$.

Proof. Suppose there are two solutions $y(x)$ and $\bar{y}(x)$ of (1.1.1) in the interval $[x_0, x_0+a]$. Define $\phi(x) = |y(x) - \bar{y}(x)|$. Clearly, $\phi(x_0) = 0$ and from (1.11.2), we have

$$\begin{aligned} D^+\phi(x) &\le |y'(x) - \bar{y}'(x)| = |f(x,y(x)) - f(x,\bar{y}(x))| \\ &\le g(x, |y(x) - \bar{y}(x)|) \\ &= g(x, \phi(x)). \end{aligned}$$

Now, for any x_1 such that $x_0 < x_1 < x_0 + a$, we obtain from Theorem 1.11.2 the inequality $\phi(x) \le r(x)$, $x_0 \le x < x_1$, where $r(x)$ is the maximal solution of (1.1.1). However, from the hypothesis of the theorem, $r(x) \equiv 0$, and hence $\phi(x) = 0$ in $[x_0, x_1)$, which proves the result. ∎

Corollary 1.11.5. The uniform Lipschitz condition (1.2.1) is a particular case of Perron's condition (1.11.2). ∎

1.12 GYÖRI'S UNIQUENESS THEOREM

We begin with the following generalization of the well - known Viswanatham's lemma [162].

Lemma 1.12.1. Assume that

(i) $\phi(x)$ is a nonnegative and bounded function in $[x_0, x_0+a]$,

(ii) $\psi(x)$ is increasing and continuous in (x_0, x_0+a); and

(iii) $g(x,z) \geq 0$ is continuous for $x_0 \leq x \leq x_0 + a$, $0 \leq z \leq k + \sup_{x_0 \leq x \leq x_0+a} \phi(x) = \phi_0$, where k is an arbitrary positive constant. Further, for fixed x, $g(x,z)$ is a nondecreasing function of z, and $g(x, \phi(x))$ is integrable.

Then, the inequality

$$\phi(x) \leq k + \int_{x_0}^{x} g(t, \phi(t)) d\psi(t), \; x_0 \leq x \leq x_0 + a \tag{1.12.1}$$

implies that for some $a < x_1 \leq x_0 + a$

$$\phi(x) \leq r(x), \; x_0 \leq x \leq x_1 \tag{1.12.2}$$

where $r(x)$ is the maximal solution of the integral equation

$$z(x) = k + \int_{x_0}^{x} g(t, z(t)) d\psi(t) \tag{1.12.3}$$

in the interval $x_0 \leq x \leq x_1$, where this solution exists.

Proof. Since the notion of the maximal solution of an integral equation is analogous to that of in the case of differential equations, it suffices to show that (1.12.3) has a solution satisfying (1.12.2). For this, we define the following successive approximations

$$z_{n+1}(x) = k + \int_{x_0}^{x} g(t, z_n(t)) d\psi(t); \; n = 0, 1, 2, \cdots \tag{1.12.4}$$

where $z_0(x) = \phi(x)$.

These functions $z_n(x)$ are obviously continuous. Let $M > 0$ be the maximum of $g(x,z)$ for $x_0 \leq x \leq x_0 + a$, $0 \leq z \leq \phi_0$. We shall show that for a suitable $x_0 < x_1 \leq x_0 + a$, we have

$$0 \leq z_n(x) \leq \phi_0 \tag{1.12.5}$$

for all positive integers n, whenever $x_0 \leq x \leq x_1$. For this, we note that the case $n = 0$ is obvious. Thus, if (1.12.5) is true for some $n \geq 1$, then by (1.12.4) it follows that

$$0 \leq z_{n+1}(x) \leq k + M(\psi(x) - \psi(x_0)) \leq \phi_0,$$

whenever $x \geq x_0$ is so small that

$$\psi(x) \leq \frac{\phi_0 - k}{M} + \psi(x_0) = \frac{\sup_{x_0 \leq x \leq x_0 + a} \phi(x)}{M} + \psi(x_0).$$

The continuity of $\psi(x)$ implies the existence of such a x. This shows that the sequence $\{z_n(x)\}$ is bounded in the interval $x_0 \leq x \leq x_1$, for some $x_1 > x_0$. Moreover, for every x and n the points $(x, z_n(x))$ belong to the domain of $g(x, z)$.

Now we shall show that $z_n(x) \leq z_{n+1}(x)$ for every $x_0 \leq x \leq x_1$, $n \geq 0$. For this, the case $n = 0$ is obvious. Thus, if $z_{n-1}(x) \leq z_n(x)$, $n \geq 1$ then by the monotonic nature of $g(x, z)$ and $\psi(x)$, we have

$$\begin{aligned} z_n(x) &= k + \int_{x_0}^{x} g(t, z_{n-1}(t))d\psi(t) \\ &\leq k + \int_{x_0}^{x} g(t, z_n(t))d\psi(t) \\ &= z_{n+1}(x), \quad x_0 \leq x \leq x_1. \end{aligned}$$

The definition of the functions $\{z_n(x)\}$ implies that for $\varepsilon > 0$ there exists a $\delta > 0$ such that for every $x_0 \leq \bar{x}, \hat{x} \leq x_1$ and $n \geq 1$ the inequality

$$|z_n(\bar{x}) - z_n(\hat{x})| \leq M|\psi(\bar{x}) - \psi(\hat{x})| < \varepsilon$$

holds, provided $|\bar{x} - \hat{x}| \leq \delta$. Therefore, Arzela's theorem implies that this sequence of functions converges uniformly. Let $\bar{z}(x) = \lim_{n \to \infty} z_n(x)$, $x_0 \leq x \leq x_1$. It is clear that this function is a solution of the equation (1.12.3). Further, in view of

$$\phi(x) = z_0(x) \leq z_1(x) \leq \ldots \leq \bar{z}(x) \leq r(x)$$

the inequality (1.12.2) is satisfied. ∎

Corollary 1.12.2. Assume that

(i) $\phi(x)$ is a nonnegative and bounded function in $[x_0, x_0 + a]$,

(ii) $\psi(x)$ is a continuous nondecreasing function in $[x_0, x_0 + a]$; and

(iii) $g(z) \geq 0$ is a continuous nondecreasing function for $0 \leq z < \infty$ and there exists a $\phi_0 > 0$ such that $g(z) > 0$ for $z \geq \phi_0$, moreover $g(\phi(x))$ is integrable.

Then, for an arbitrary positive constant k the inequality

$$\phi(x) \leq k + \int_{x_0}^{x} g(\phi(t))d\psi(t), \ x_0 \leq x \leq x_0 + a \tag{1.12.6}$$

implies that for some $a < x_1 \leq x_0 + a$

$$\phi(x) \leq G^{-1}(G(k) + \psi(x) - \psi(x_0)), \ x_0 \leq x \leq x_1 \tag{1.12.7}$$

where

$$G(\phi) = \int_{\phi_0}^{\phi} \frac{dz}{g(z)}, \quad \phi \geq 0 \tag{1.12.8}$$

and G^{-1} is the inverse of G; moreover x_1 is chosen so that $G(k)+\psi(x)-\psi(x_0)$ belongs to the domain of G^{-1} whenever $x_0 \leq x \leq x_1$.

Proof. If $r(x)$ is the maximal solution of the integral equation

$$z(x) = k + \int_{x_0}^{x} g(z(t))d\psi(t), \tag{1.12.9}$$

then in view of Lemma 1.12.1 we have $\phi(x) \leq r(x)$. Therefore, it suffices to show that any solution $\bar{z}(x)$ of (1.12.9) satisfies the inequality

$$\bar{z}(x) < G^{-1}(G(k+\varepsilon) + \psi(x) - \psi(x_0)) = z_\varepsilon(x), \tag{1.12.10}$$

where $z_\varepsilon(x)$ is a solution of the integral equation

$$z_\varepsilon(x) = k + \varepsilon + \int_{x_0}^{x} g(z_\varepsilon(t))d\psi(t). \tag{1.12.11}$$

However, since

$$\int_{x_0}^{x} g(G^{-1}(G(k+\varepsilon)+\psi(t)-\psi(x_0)))d\psi(t) = \int_{\psi(x_0)}^{\psi(x)} g(G^{-1}(G(k+\varepsilon)+t-\psi(x_0)))dt$$

and

$$\frac{dG^{-1}(\phi)}{d\phi} = g(G^{-1}(\phi))$$

it follows that

$$\int_{x_0}^{x} g(z_\varepsilon(t))d\psi(t)$$

$$= \int_{x_0}^{x} g(G^{-1}(G(k+\varepsilon)+\psi(t)-\psi(x_0)))d\psi(t)$$

$$= G^{-1}(G(k+\varepsilon)+\psi(x)-\psi(x_0)) - G^{-1}(G(k+\varepsilon)+\psi(x_0)-\psi(x_0))$$

$$= -k-\varepsilon+G^{-1}(G(k+\varepsilon)+\psi(x)-\psi(x_0))$$

$$= -k-\varepsilon+z_\varepsilon(x). \quad \blacksquare$$

Remark 1.12.1. If $q(x) \geq 0$ is continuous and $\psi(x) = \int_{x_0}^{x} q(t)dt$, then Corollary 1.12.2 is the same as LaSalle's lemma (better known as Bihari's lemma). Further, in addition if $g(z) = z$, then the inequality (1.12.6) reduces to

$$\phi(x) \leq k + \int_{x_0}^{x} q(t)\phi(t)dt \tag{1.12.12}$$

for which (1.12.7) becomes

$$\phi(x) \leq k \exp\left(\int_{x_0}^{x} q(t)dt\right), \tag{1.12.13}$$

which is the famous **Gronwall's inequality**. Obviously, if $k = 0$, then (1.12.12) is the same as (1.2.5) and then (1.12.13) also gives $\phi(x) \equiv 0$.

Theorem 1.12.3 (Györi's Uniqueness Theorem). Assume that the function $f(x,y)$ is continuous in $\bar{S}_+$ and

(i) for every sufficiently small δ, $0 < \delta < a$, the function $\psi(x;\delta)$ is continuous and nondecreasing in x, $x_0+\delta \leq x \leq x_0+a$,

(ii) $g(z) \geq 0$ is continuous and nondecreasing for $0 \leq z < \infty$,

(iii) there is no positive constant k such that the inequality

$$\int_{\varepsilon+\delta\mu(\delta)}^{\varepsilon+k} \frac{dz}{g(z)} \leq \psi(x_0+a;\delta) - \psi(x_0+\delta;\delta) \tag{1.12.14}$$

holds for any ε, $\delta > 0$ and for any function $\mu(\delta)$ tending to 0 as $\delta \to 0$; and

(iv) for all $x_0+\delta \le x \le x_0+a$ and $y_0-b \le y, \bar{y} \le y_0+b$ the function $f(x,y)$ satisfies the inequality

$$|f(x,\bar{y}) - f(x,y)| \le \psi'(x;\delta)g(|\bar{y}-y|). \tag{1.12.15}$$

Then, there exists a $0 < \eta \le a$ such that the initial value problem (1.1.1) has at most one solution in the interval $[x_0, x_0+\eta]$.

Proof. Let $y(x)$ and $\bar{y}(x)$ be two solutions of the initial value problem (1.1.1). Let $x > x_0$ and fix $\delta > 0$ so that $x_0 + \delta < x$. Then, in view of (1.12.15) and on omitting the singular part of $\psi(x;\delta)$ from its canonical decomposition as a continuous and monotone function, we find that

$$\begin{aligned}\int_{x_0+\delta}^{x} |f(t,\bar{y}(t)) - f(t,y(t))|dt &\le \int_{x_0+\delta}^{x} g(|\bar{y}(t)-y(t)|)\psi'(t;\delta)dt \\ &\le \int_{x_0+\delta}^{x} g(|\bar{y}(t)-y(t)|)d\psi(t;\delta).\end{aligned}$$

Thus, it follows that

$$|\bar{y}(x) - y(x)| \le \int_{x_0}^{x} |f(t,\bar{y}(t)) - f(t,y(t))|dt$$

$$\le \delta \sup_{x_0 \le x \le x_0+\delta} |f(x,\bar{y}(x)) - f(x,y(x))| + \int_{x_0+\delta}^{x} g(|\bar{y}(t)-y(t)|)d\psi(t;\delta).$$

Let the function $\mu(\delta)$ in (1.12.14) be $\sup_{x_0 \le x \le x_0+\delta} |f(x,\bar{y}(x)) - f(x,y(x))|$ then, from the above inequality we have

$$|\bar{y}(x) - y(x)| \le \delta\mu(\delta) + \int_{x_0+\delta}^{x} g(|\bar{y}(t)-y(t)|)d\psi(t;\delta).$$

Thus, if $\varepsilon > 0$ is an arbitrary but fixed number, then

$$|\bar{y}(x) - y(x)| + \varepsilon \le \delta\mu(\delta) + \varepsilon + \int_{x_0+\delta}^{x} g(|\bar{y}(t)-y(t)| + \varepsilon)d\psi(t;\delta)$$

holds. Now an application of Corollary 1.12.2 gives

$$G(|\bar{y}(x) - y(x)| + \varepsilon) \le G(\delta\mu(\delta) + \varepsilon) + \psi(x;\delta) - \psi(x_0+\delta;\delta),$$

where the function $G(\phi)$ is defined in (1.12.8). However, since for every ε, $\delta > 0$

$$\begin{aligned}\int_{\varepsilon+\delta\mu(\delta)}^{|\bar{y}(x)-y(x)|+\varepsilon} \frac{dz}{g(z)} &\le \psi(x;\delta) - \psi(x_0+\delta;\delta) \\ &\le \psi(x_0+a;\delta) - \psi(x_0+\delta;\delta)\end{aligned}$$

hypothesis (iii) implies that $|\bar{y}(x) - y(x)| \equiv 0$, i.e., $y(x) = \bar{y}(x)$, $x \geq x_0$. Since $x > x_0$ is arbitrary the result follows. ∎

Corollary 1.12.4. Assume that $g(z)$ is as in Lemma 1.4.1 and Osgood's condition (1.4.3) is satisfied. Then, the conclusion of Theorem 1.12.3 holds.

Proof. It suffices to note that the conditions of Theorem 1.12.3 with

$$\psi(x;\delta) - \psi(x_0 + \delta;\delta) = x - x_0 - \delta$$

are satisfied. ∎

Remark 1.12.2. For $g(z) = \sqrt{z}$, and

$$\psi(x;\delta) - \psi(x_0 + \delta;\delta) = (x - x_0)^{\delta} - \delta^{\delta}, \quad x > x_0 + \delta \ (\delta > 0)$$

the inequality (1.12.14) for $\varepsilon,\ \delta > 0$ becomes

$$a^{\delta} - \delta^{\delta} \geq \int_{\varepsilon+\delta\mu(\delta)}^{\varepsilon+k} \frac{dz}{\sqrt{z}} = 2\sqrt{\varepsilon + k} - 2\sqrt{\varepsilon + \delta\mu(\delta)},$$

which implies that $k = 0$. Therefore, the condition

$$|f(x,\bar{y}) - f(x,y)| \leq \delta(x - x_0)^{\delta-1}\sqrt{|\bar{y} - y|}$$

implies the conclusion of Theorem 1.12.3. Hence, in Corollary 1.12.4 it is not necessary that the function $g(z)$ must satisfy the condition (1.4.1).

Corollary 1.12.5. Assume that $g(z) = z$ and Nagumo's condition (1.6.1) is satisfied. Then, the conclusion of Theorem 1.12.3 holds.

Proof. It suffices to note that the conditions of Theorem 1.12.3 with

$$\psi(x;\delta) - \psi(x_0 + \delta;\delta) = \int_{x_0+\delta}^{x} \frac{dt}{t - x_0}, \quad x_0 + \delta < x$$

are satisfied. ∎

Corollary 1.12.6 (Yang's Uniqueness Theorem). Assume that $g(z) = z^{\beta}$ and **Yang's condition**

$$|f(x,\bar{y}) - f(x,y)| \leq \frac{L}{(x - x_0)^{\alpha}}|\bar{y} - y|^{\beta} \tag{1.12.16}$$

is satisfied, where $\beta > 1$, $0 \le \alpha < \beta$, $\alpha \ne 1$ and $L > 0$ are arbitrary constants. Then, the conclusion of Theorem 1.12.3 holds.

Proof. It suffices to note that the conditions of Theorem 1.12.3 with

$$\psi(x;\delta) - \psi(x_0+\delta;\delta) = L\int_{x_0+\delta}^{x} \frac{dt}{(t-x_0)^{\alpha}}, \quad x_0+\delta \le x \le x_0+a$$

are satisfied. For this, (i) and (ii) are obvious, whereas for (iii) we assume the contrary, i.e., that for some positive k the inequality (1.12.14) holds for any ε, $\delta > 0$ and for any function $\mu(\delta)$ tending to 0 as $\delta \to 0$. But, then for $x > x_0 + \delta$, we have

$$\frac{1}{(1-\beta)(\varepsilon+k)^{\beta-1}} - \frac{1}{(1-\beta)(\varepsilon+\delta\mu(\delta))^{\beta-1}} \le \frac{L}{1-\alpha}\left[\frac{1}{a^{\alpha-1}} - \frac{1}{\delta^{\alpha-1}}\right],$$

which is the same as

$$\frac{1}{(1-\beta)(\varepsilon+k)^{\beta-1}} - \frac{L}{(1-\alpha)a^{\alpha-1}} \le \frac{1}{(1-\beta)(\varepsilon+\delta\mu(\delta))^{\beta-1}} - \frac{L}{(1-\alpha)\delta^{\alpha-1}}$$

for all ε, $\delta > 0$. Let $\varepsilon = \delta - \delta\mu(\delta)$, then the above inequality becomes

$$\text{(1.12.17)} \quad \frac{1}{(1-\beta)(\delta-\delta\mu(\delta)+k)^{\beta-1}} - \frac{L}{(1-\alpha)a^{\alpha-1}} \le \frac{1}{(1-\beta)\delta^{\beta-1}} - \frac{L}{(1-\alpha)\delta^{\alpha-1}}.$$

However, from the assumptions on the constants we find that

$$\lim_{\delta\to 0}\left[\frac{1}{(1-\beta)\delta^{\beta-1}} - \frac{L}{(1-\alpha)\delta^{\alpha-1}}\right] = \lim_{\delta\to 0}\frac{(1-\alpha) - L(1-\beta)\delta^{\beta-\alpha}}{(1-\alpha)(1-\beta)\delta^{\beta-1}} = -\infty.$$

Thus, if $\delta \to 0$, then (1.12.17) implies that

$$\frac{1}{(1-\beta)k^{\beta-1}} - \frac{L}{(1-\alpha)a^{\alpha-1}} \le -\infty,$$

which is an obvious contradiction. The case $x < x_0$ can be treated similarly. ∎

1.13 IYANAGA'S UNIQUENESS THEOREM

This is the first general uniqueness criterion from which several earlier results, with strict inequalities only, can be deduced.

Theorem 1.13.1 (Iyanaga's Uniqueness Theorem). Assume that

(i) the function $g(x,z)$ is continuous and nonnegative in $x_0 < x \leq x_0 + a$, $0 \leq z \leq 2b$ and for every x_1, $x_0 < x_1 \leq x_0 + a$ and z_1, $0 < z_1 \leq 2b$ solutions $z(x) = z(x, x_1, z_1)$ of the differential equation

$$z' = g(x,z) \tag{1.13.1}$$

exist for $x_0 < x \leq x_1$, $0 \leq z(x) \leq 2b$, and $\lim_{x \to x_0^+} z(x) > 0$, or $\lim_{x \to x_0^+} z(x) = 0$ and $\lim_{x \to x_0^+} z'(x) > 0$; and

(ii) the function $f(x,y)$ is continuous in $\bar{S}_+$ and for all (x,y), $(x,\bar{y})$ in $\bar{S}_+$ with $\bar{y} > y$ the following inequality is satisfied

$$g(x, \bar{y} - y) > f(x,\bar{y}) - f(x,y). \tag{1.13.2}$$

Then, the initial value problem (1.1.1) has at most one solution in $[x_0, x_0 + a]$.

Proof. Let $y(x)$ and $\bar{y}(x)$ be two solutions of (1.1.1) in the interval $[x_0, x_0 + a]$. Then, the function $\phi(x) = \bar{y}(x) - y(x)$ satisfies $\phi(x_0) = \phi'(x_0) = 0$. We assume that $\phi(x) > 0$ for $x_1 < x \leq x_2 \leq x_0 + a$, and $\phi(x) = 0$ for $x_0 \leq x \leq x_1$. Let $\rho(x)$ be the minimal solution of (1.13.1) satisfying the initial condition $\rho(x_2) = z_2 = \phi(x_2)$. We note that from the given hypotheses $\rho(x_0 + \varepsilon) > 0$ for all sufficiently small $\varepsilon > 0$, and hence we can choose $\varepsilon > 0$ so small that $\rho(x_0 + \varepsilon) > \phi(x_0 + \varepsilon)$. Since the function $g(x,z)$ is nonnegative it is clear that $\rho(x) > 0$ in the interval $[x_0 + \varepsilon, x_2]$. For $x \in (x_1, x_2]$ the inequality (1.13.2) implies that

$$\phi'(x) = f(x, \bar{y}(x)) - f(x, y(x)) < g(x, \bar{y}(x) - y(x)) = g(x, \phi(x)).$$

Thus, in view of Theorem 1.11.3 it follows that $\rho(x) \le \phi(x)$ for all $x \in (x_1, x_2]$. However, since $\phi(x_1) = 0$, from the continuity we find that $\rho(x_1) = 0$. Therefore, if at all x_1 must be the same as x_0. But, then it contradicts $\rho(x_0+\varepsilon) > \phi(x_0+\varepsilon)$ for small ε. ∎

Corollary 1.13.2. For the uniqueness of the solutions of the initial value problem (1.9.4) each of the following conditions is sufficient.

(i) $|f(x,y) - f(x,\bar{y})| < (1+\varepsilon(x))\dfrac{|y-\bar{y}|}{x}$, where $\varepsilon(x) > 0$ and $\lim_{x\to 0^+}\displaystyle\int_\delta^x \frac{\varepsilon(t)}{t}dt > -M$, $0 < M < \infty$, $\delta > 0$;

(ii) $|f(x,y) - f(x,\bar{y})| < \alpha\dfrac{|y-\bar{y}|}{x} + \beta|y-\bar{y}|\ln\dfrac{1}{|y-\bar{y}|}$, $0 \le \alpha < 1$, $0 \le \beta$;

(iii) $|f(x,y) - f(x,\bar{y})| < \dfrac{|y-\bar{y}|}{x} \times \dfrac{\left|\ln\dfrac{1}{|y-\bar{y}|}\right|^l}{\left|\ln\dfrac{1}{x}\right|^k}$, $0 < l < k$, $l \le 1$.

Proof. For the proof of (i) we can apply Theorem 1.13.1 and consider the differential equation

$$z' = g(x,z) = (1+\varepsilon(x))\frac{z}{x} \tag{1.13.3}$$

and put an indefinite integral $\displaystyle\int \frac{\varepsilon(x)}{x}dx = I(x)$.

The general solution of (1.13.3) is $z(x) = xe^{c+I(x)}$, where c is an arbitrary constant. Since $I(x) > -M$ for any $c \ne -\infty$, we have

$$\lim_{x\to 0^+}\frac{dz}{dx} = e^{c+I(x)} + \varepsilon(x)e^{c+I(x)} > 0.$$

For the proof of (ii) we consider the differential equation

$$z' = g(x,z) = \alpha\frac{z}{x} - \beta z\ln z. \tag{1.13.4}$$

The general solution of (1.13.4) is $z(x) = \exp\left(\alpha e^{-\beta x}G(x) + ce^{-\beta x}\right)$, where $G(x)$ is the indefinite integral $\displaystyle\int \frac{e^{\beta x}}{x}dx$, and c is an arbitrary constant. Since

$$\begin{aligned}\ln\frac{dz}{dx} &= \alpha e^{-\beta x}G(x) + ce^{-\beta x} + \ln\left(-\alpha\beta e^{-\beta x}G(x) + \frac{\alpha}{x} - c\beta e^{-\beta x}\right)\\ &= \alpha e^{-\beta x}G(x) - \ln x + O(1)\end{aligned}$$

on putting $\alpha = 1/(1+\gamma)$, $\gamma > 0$ and choosing $\varepsilon < \gamma$ and then $\delta(\varepsilon)$ so that

$$\ln \frac{\delta(\varepsilon)}{x} < G(\delta(\varepsilon)) - G(x) < (1+\varepsilon) \ln \frac{\delta(\varepsilon)}{x},$$

we find

$$\ln \frac{dz}{dx} = \left(1 - \frac{1+\varepsilon}{1+\gamma} e^{-\beta x}\right) \ln \frac{1}{x} + O(1) \quad \longrightarrow \infty.$$

Hence, $\lim_{x \to 0^+} \frac{dz}{dx} = \infty$.

For the proof of (iii), we consider

$$\frac{dz}{dx} = \frac{z}{x} \frac{(-\ln z)^l}{(-\ln x)^k}. \tag{1.13.5}$$

The general solution of (1.13.5) for $0 < l < k < 1$, is

$$z(x) = \exp\left(-\left\{\frac{1-l}{1-k}(-\ln x)^{1-k} + c\right\}^{1/(1-l)}\right),$$

where c is an arbitrary constant. Since,

$$\ln \frac{dz}{dx} = -\left\{\frac{1-l}{1-k}(-\ln x)^{1-k} + c\right\}^{1/(1-l)} + \ln(-\ln x)^{-k}$$

$$+ \ln\left\{\frac{1-l}{1-k}(-\ln x)^{1-k} + c\right\}^{l/(1-l)} + \ln \frac{1}{x} \quad \longrightarrow \infty$$

we find that $\lim_{x \to 0^+} \frac{dz}{dx} = \infty$. A similar situation holds for $l < k = 1$ as well as $1 = l < k$. ∎

Finally, we note that the condition $|f(x,y) - f(x,\bar{y})| \le L(x)|y - \bar{y}|$, where $\lim_{x \to 0^+} x \exp(-\int_\delta^x L(t)dt) < M$, is identical with the condition (i). For this, putting $L(x) = (1+\varepsilon(x))/x$ we have $\int \frac{\varepsilon(x)}{x} dx > -M$, and conversely, putting $(1+\varepsilon(x))/x = L(x)$, we have $\lim_{x \to 0^+} x \exp(-\int_\delta^x L(t)dt) < M$.

1.14 INABA'S UNIQUENESS THEOREM

The motivation of the following result comes from the Corollary 1.13.2.

Theorem 1.14.1 (Inaba's Uniqueness Theorem). Assume that

(i) the function $\varepsilon(x)$ is nonnegative in $[0, a]$ and for every $\delta > 0$ there exists a positive number M such that

$$\lim_{x\to 0+} \int_x^\delta \frac{\varepsilon(t)}{t} dt < M \ , \tag{1.14.1}$$

(ii) the function $h(x)$ is nonnegative in $[0, a]$ and for every $\delta > 0$

$$\lim_{x\to 0+} \int_x^\delta h(t)dt \ , \tag{1.14.2}$$

is a finite positive number N, and there exists a positive number $\bar{N}$ such that

$$\lim_{x\to 0+} \int_x^\delta \frac{\int_0^t h(s)ds}{t} dt < \bar{N}, \tag{1.14.3}$$

(iii) the function $f(x, y)$ is continuous in $\bar{S}_+$ where $(x_0, y_0) = (0, 0)$ and for all (x, y), $(x, \bar{y})$ in $\bar{S}_+$ with $\bar{y} > y$ the following inequality is satisfied

$$f(x, \bar{y}) - f(x, y) \le \frac{1 + \varepsilon(x)}{x}(\bar{y} - y) + h(x)(\bar{y} - y) \ln \frac{1}{\bar{y} - y}. \tag{1.14.4}$$

Then, the initial value problem (1.1.1) with $(x_0, y_0) = (0, 0)$ has at most one solution in $[0, a]$.

Proof. Let $y(x)$ and $\bar{y}(x)$ be two solutions of (1.1.1) in the interval $[0, a]$. Then, the function $\phi(x) = \bar{y}(x) - y(x)$ satisfies $\phi(0) = \phi'(0) = 0$. We assume that $\phi(x) > 0$ for $x_1 < x \le x_2 \le a$, and $\phi(x) = 0$ for $0 \le x \le x_1$. Since for all $x \in (x_1, x_2]$, $\phi(x) > 0$ the inequality (1.14.4) implies that

$$\phi'(x) = f(x, \bar{y}(x)) - f(x, y(x)) \le \frac{1 + \varepsilon(x)}{x}\phi(x) + h(x)\phi(x) \ln \frac{1}{\phi(x)},$$

which is the same as

$$-\frac{d}{dx} \ln \frac{1}{\phi(x)} - h(x) \ln \frac{1}{\phi(x)} \le \frac{1 + \varepsilon(x)}{x}.$$

Further, since by assumption, $\lim_{x\to 0+} \int_x^\delta h(t)dt$ is finite, $\int_0^x h(t)dt$ is continuous and $\frac{d}{dx} \int_0^x h(t)dt = h(x)$. Therefore, it follows that

$$-\frac{d}{dx}\left[\exp\left(\int_0^x h(t)dt\right) \ln \frac{1}{\phi(x)}\right] \le \frac{1 + \varepsilon(x)}{x} \exp\left(\int_0^x h(t)dt\right)$$

and hence an integration from x to δ gives

$$\exp\left(\int_0^x h(t)dt\right)\ln\frac{1}{\phi(x)} - \exp\left(\int_0^\delta h(t)dt\right)\ln\frac{1}{\phi(\delta)} \tag{1.14.5}$$
$$\leq \int_x^\delta \frac{1}{t}\exp\left(\int_0^t h(s)ds\right)dt + \int_x^\delta \frac{\varepsilon(t)}{t}\exp\left(\int_0^t h(s)ds\right)dt.$$

If $\phi(x_1) = 0$ for $x_1 > 0$, then as $x \to x_1$, the left side of (1.14.5) tends to ∞, whereas the right side remains finite. This contradiction forces that x_1 must be 0. Thus, if $\phi(x) > 0$ for $x_1 < x \leq x_2$, then it is positive for $0 < x \leq x_2$ also. Now, since $\exp(\int_0^x h(t)dt) > 1$, for small x, conditions (1.14.2) and (1.14.3) give

$$\int_x^\delta \frac{1}{t}\exp\left(\int_0^t h(s)ds\right)dt$$
$$\leq \int_x^\delta \frac{1}{t}\left[1 + \int_0^t h(s)\exp\left(\int_0^s h(\tau)d\tau\right)ds\right]dt$$
$$\leq \ln\delta + \ln\frac{1}{x} + \exp\left(\int_0^\delta h(t)dt\right)\int_x^\delta \frac{\exp\left(\int_0^t h(s)ds\right)}{t}dt$$
$$< \exp\left(\int_0^x h(t)dt\right)\ln\frac{1}{x} + \ln\delta + e^N \times \bar{N}.$$

Also, in view of (1.14.1) and (1.14.2), we have

$$\int_x^\delta \frac{\varepsilon(t)}{t}\exp\left(\int_0^t h(s)ds\right)dt \leq \exp\left(\int_0^\delta h(t)dt\right)\int_x^\delta \frac{\varepsilon(t)}{t}dt \quad < e^N M.$$

Using the above estimates in (1.14.5), we obtain

$$\exp\left(\int_0^x h(t)dt\right)\ln\frac{x}{\phi(x)} < \ln\delta + e^N\left(M + \bar{N} + \ln\frac{1}{\phi(\delta)}\right) \quad < G,$$

where G is positive and depends only on δ. Therefore, it follows that

$$\ln\frac{x}{\phi(x)} < G\exp\left(-\int_0^x h(t)dt\right) \quad < G.$$

In this inequality as $x \to 0^+$ the right side remains finite, whereas for the left side we find that

$$\lim_{x\to 0^+}\ln\frac{x}{\phi(x)} = -\ln\lim_{x\to 0^+}\frac{\phi(x)}{x} = -\ln\lim_{x\to 0^+}\frac{\phi'(x)}{1} = \infty.$$

This contradiction completes the proof. ∎

Corollary 1.14.2. Let the function $f(x,y)$ be continuous in $\bar{S}$, where $(x_0, y_0) = (0,0)$ and for all (x,y), $(x,\bar{y})$ in $\bar{S}$ the following inequality is satisfied

$$(1.14.6) \qquad |f(x,y) - f(x,\bar{y})| \le \frac{1+\varepsilon(|x|)}{|x|}|y-\bar{y}| + h(|x|)|y-\bar{y}| \ln \frac{1}{|y-\bar{y}|},$$

where the functions $\varepsilon(x)$ and $h(x)$ satisfy the conditions of Theorem 1.14.1. Then, the initial value problem (1.1.1) with $(x_0, y_0) = (0,0)$ has at most one solution in $|x| \le a$.

Proof. Let $\phi(x) = \bar{y}(x) - y(x)$ in a small interval $(x_1, x_2]$ so that

$$\left|\frac{d\phi(x)}{dx}\right| \le \frac{1+\varepsilon(|x|)}{|x|}\phi(x) + h(|x|)\phi(x) \ln \frac{1}{\phi(x)}.$$

If we set $\xi = |x|$, then since $\frac{d\xi}{dx} = \pm 1$ it follows that

$$\left|\frac{d\phi(x)}{dx}\right| = \left|\frac{d\bar{\phi}(\xi)}{d\xi} \times \frac{d\xi}{dx}\right| = \left|\frac{d\bar{\phi}(\xi)}{d\xi}\right| \ge \frac{d\bar{\phi}(\xi)}{d\xi}.$$

Hence, in the interval (ξ_1, ξ_2) where $\xi_1 = \min(|x_1|, |x_2|)$ and $\xi_2 = \max(|x_1|, |x_2|)$ it follows that

$$\frac{d\bar{\phi}(\xi)}{d\xi} \le \frac{1+\varepsilon(\xi)}{\xi}\bar{\phi}(\xi) + h(\xi)\bar{\phi}(\xi) \ln \frac{1}{\bar{\phi}(\xi)}.$$

But, this implies as in the proof of Theorem 1.14.1 that $\phi(x) \equiv 0$. ∎

1.15 KAMKE'S UNIQUENESS THEOREM

The main result of this section is more general than those of Perron and Iyanaga, also it includes as special cases many known criteria.

Theorem 1.15.1 (Kamke's Uniqueness Theorem). Assume that

(i) the function $g(x,z)$ is continuous and nonnegative in $x_0 < x \le x_0 + a$, $0 \le z \le 2b$ and for every x_1, $x_0 < x_1 < x_0 + a$, $z(x) \equiv 0$ is the only differentiable function in $x_0 < x < x_1$ and continuous in $x_0 \le x \le x_1$ for which $z'_+(x_0) = \lim_{x \to x_0^+} \frac{z(x) - z(x_0)}{x - x_0}$ exists; and

$$z'(x) = g(x, z(x)),\ x_0 < x < x_1 \tag{1.15.1}$$

$$z(x_0) = z'_+(x_0) = 0, \tag{1.15.2}$$

(ii) the function $f(x,y)$ is continuous in $\bar{S}_+$ and for all $(x,y),(x,\bar{y}) \in \bar{S}_+$ where $x \neq x_0$ the inequality (1.11.2) is satisfied.

Then, the initial value problem (1.1.1) has at most one solution in $[x_0, x_0 + a]$.

To prove this result we need the following :

Lemma 1.15.2. Let the function $g(x,z)$ satisfy the hypothesis (i) of Theorem 1.15.1. Further, let the function $g_1(x,z)$ be continuous and nonnegative in $x_0 \leq x \leq x_0 + a,\ 0 \leq z \leq 2b,\ g_1(x,0) \equiv 0$, and

$$g_1(x,z) \leq g(x,z),\ x \neq x_0. \tag{1.15.3}$$

Then, for every $x_1,\ x_0 < x_1 < x_0 + a,\ z(x) \equiv 0$ is the only differentiable function in $x_0 \leq x \leq x_1$, which satisfies

$$z' = g_1(x,z),\ z(x_0) = 0. \tag{1.15.4}$$

Proof. It suffices to show that the maximal solution $r(x)$ of (1.15.4) is identically zero. On the contrary, suppose that there exists a $\sigma,\ x_0 < \sigma < x_0 + a$, such that $r(\sigma) > 0$. Because of the inequality (1.15.3), we have

$$r'(x) \leq g(x, r(x)),\ x_0 < x \leq \sigma.$$

If $\rho(x)$ is the minimal solution of $z' = g(x,z),\ z(\sigma) = r(\sigma)$ then an application of Theorem 1.11.3 gives that

$$\rho(x) \leq r(x) \tag{1.15.5}$$

as far as $\rho(x)$ exists to the left of σ. The solution $\rho(x)$ by using standard arguments can be continued up to $x = x_0$. If $\rho(\tau) = 0$, for some $\tau, x_0 < \tau < \sigma$, we can effect the continuation by defining $\rho(x) = 0$ for $x_0 < x < \tau$. Otherwise, (1.15.5) ensures the possibility of continuation. Since $r(x_0) =$

0, $\lim_{x\to x_0^+}\rho(x) = 0$, and we define $\rho(x_0) = 0$. Furthermore, since $g_1(x,z)$ is continuous at $(x_0,0)$ and $g_1(x_0,0) = 0$, $r'_+(x_0)$ exists and is equal to zero. Thus, because of (1.15.5), it follows that $\rho'_+(x_0)$ exists and $\rho'_+(x_0) = 0$. But, we have assumed that $g(x,z)$ satisfies the hypothesis of Theorem 1.15.1. Hence, $\rho(x) \equiv 0$. This contradicts the fact that $\rho(\sigma) = r(\sigma) > 0$. Therefore, $r(x) \equiv 0$. ∎

Proof of Theorem 1.15.1. Define the function

$$g_f(x,z) = \sup_{|y-\bar{y}|=z} |f(x,y) - f(x,\bar{y})| \tag{1.15.6}$$

in $x_0 \le x \le x_0 + a$, $0 \le z \le 2b$. Since $f(x,y)$ is continuous in $\bar{S}_+$, $g_f(x,z)$ is continuous in $x_0 \le x \le x_0 + a$, $0 \le z \le 2b$. In view of (1.15.6) it is clear that (1.11.2) holds with the function $g_f(x,z)$. Moreover, $g_f(x,z) \le g(x,z)$ for $x_0 < x \le x_0 + a$, $0 \le z \le 2b$. Thus, Lemma 1.15.2 is applicable with $g_1(x,z) \le g_f(x,z)$, and therefore $g_f(x,z)$ satisfies the assumptions of Theorem 1.11.4. Hence, the result follows. ∎

Corollary 1.15.3. In Theorem 1.15.1 condition (1.15.2) cannot be replaced by

$$\lim_{x\to x_0^+} z(x) = \lim_{x\to x_0^+} z'(x) = 0. \tag{1.15.7}$$

Proof. We shall consider $(x_0, y_0) = (0,0)$ and $a = 1$. Let $\phi(x)$ be a function defined in $(0,1]$ and satisfy the following conditions :

(i) $x^2 < \phi(x) \le \frac{4}{3}x^2$,

(ii) $\phi'(x)$ is continuous,

(iii) $\phi'(x) \ge 2x$; and

(iv) $\lim_{x\to x_0^+} \phi'(x)$ does not exist.

The construction of such a function is not difficult. For $x > 0$ and $z \ge 0$, let

$$g(x,z) = \frac{\phi'(x)}{\phi(x)} z. \tag{1.15.8}$$

From (i), (ii) and (iii) it is clear that this function $g(x,z)$ is continuous and nonnegative for $0 < x \le 1,\ z \ge 0$. Further, from (i) and (iii), for $0 < x \le 1,\ z \ge 0$ we have

$$g(x,z) \ge \frac{3z}{2x}. \tag{1.15.9}$$

Since all solutions of the differential equation (1.15.1) with $g(x,z)$ defined in (1.15.8) are of the form $z(x) = c\phi(x)$, from (i) it follows that all solutions of (1.15.1) satisfy (1.15.2). Therefore, the hypotheses of Theorem 1.15.1 are not satisfied. However, from (iv) we see that (1.15.7) is satisfied only for the solution $z(x) \equiv 0$.

Now in the strip $0 \le x \le 1, |y| < \infty$ we define a continuous function

$$f(x,y) = \begin{cases} \dfrac{4y}{3x} & 0 < x \le 1, 0 \le y \le x^{4/3} \\ \dfrac{4}{3}x^{1/3} & 0 < x \le 1, x^{4/3} < y < \infty \\ 0 & 0 \le x \le 1, -\infty < y \le 0. \end{cases} \tag{1.15.10}$$

For this function, in view of (1.15.9), it follows that for $x \ne 0$

$$|f(x,y) - f(x,\bar{y})| \le \frac{4}{3x}|y - \bar{y}| \le \frac{3}{2x}|y - \bar{y}| \le g(x, |y - \bar{y}|).$$

The initial value problem (1.9.4) with this function $f(x,y)$ has two distinct solutions $y(x) \equiv 0$ and $\bar{y}(x) = x^{4/3}$ in $[0,1]$. ∎

Corollary 1.15.4. The function $g(x,z) = \lambda(x)z$, where $\lambda(x) \ge 0$ and continuous for $x_0 < x \le x_0 + a$ satisfies the conditions of Theorem 1.15.1 provided

$$\limsup_{x \to x_0^+} [1 + \lambda(x)]e^{-\mu(x)} > 0, \tag{1.15.11}$$

where

$$\mu(x) = \int_x^{x_0+a} \lambda(t)dt,\ x \ne x_0. \tag{1.15.12}$$

Proof. Consider the differential equation

$$z' = \lambda(x)z. \tag{1.15.13}$$

The nontrivial solutions of (1.15.13) are nonvanishing constant multiples of the function $e^{-\mu(x)}$, where $\mu(x)$ is given in (1.15.12). The derivative of this function is $\lambda(x)e^{-\mu(x)}$. Since $\lambda(x) \geq 0$, it follows from the assumption (1.15.11) that every solution $z(x) \not\equiv 0$ of (1.15.13) violates at least one of the two limiting conditions (1.15.2). ∎

If $\lambda(x) = L \geq 0$, then $\mu(x) = L(x_0 + a - x)$ so that the condition (1.15.11) becomes $\limsup_{x\to x_0^+}(1 + L)e^{-L(x_0+a-x)} > 0$, which certainly holds. Thus, in view of Corollary 1.15.4 the Lipschitz Uniqueness Theorem 1.2.4 is a particular case of Theorem 1.15.1. Similarly, if $\lambda(x) = k(x - x_0)^{-1}, (k \geq 0)$ then $\mu(x) = k\ln(a(x - x_0)^{-1})$ and the condition (1.15.11) reduces to $\limsup_{x\to x_0^+}(1 + k(x - x_0)^{-1})\frac{(x-x_0)^k}{a^k} > 0$, which is true if $k \leq 1$. Therefore, once again in view of Corollary 1.15.4 Nagumo's Uniqueness Theorem 1.6.2 is included in Theorem 1.15.1. We further note that the condition (1.15.11) is definitely satisfied provided

$$(1.15.14) \qquad \limsup_{x\to x_0^+} e^{-\mu(x)} > 0, \quad \text{i.e.,} \ \liminf_{x\to x_0^+} \mu(x) < \infty$$

which is the same as

$$(1.15.15) \qquad \lim_{x\to x_0^+} \int_x^{x_0+a} \lambda(t)dt < \infty.$$

Of course, the function $\lambda(x) = k(x - x_0)^{-1}$, $0 < k \leq 1$ fails to satisfy the condition (1.15.15).

Corollary 1.15.5. For the condition (1.15.11) to be satisfied it is sufficient that

$$(1.15.16) \qquad \ln(x - x_0) + \int_x^{x_0+a} \lambda(t)dt \not\to \infty, \quad \text{as} \quad x \to x_0^+.$$

Proof. If (1.15.15) holds then (1.15.16) is obvious. Thus, it suffices to exclude the case (1.15.15). This in view of (1.15.14) means that $e^{-\mu(x)} \to 0$ as $x \to x_0^+$. Hence the condition for the applicability of the one - sided version of l'Hospital's rule

$$\limsup_{x\to x_0^+} \frac{F'(x)}{G'(x)} \geq \limsup_{x\to x_0^+} \frac{F(x)}{G(x)}$$

is satisfied by $F(x) = e^{-\mu(x)}$ and $G(x) = (x - x_0)$. Therefore, it follows that

$$\limsup_{x \to x_0^+} \lambda(x) e^{-\mu(x)} \geq \limsup_{x \to x_0^+} e^{-\mu(x)} (x - x_0)^{-1}.$$

Thus, the condition (1.15.11) is certainly satisfied if

$$\limsup_{x \to x_0^+} e^{-\mu(x)} (x - x_0)^{-1} > 0. \tag{1.15.17}$$

But, (1.15.17) is equivalent to $\liminf_{x \to x_0^+} e^{\mu(x)} (x - x_0) < \infty$, which in turn implies that

$$-\infty \leq \liminf_{x \to x_0^+} \left[\ln(x - x_0) + \int_x^{x_0 + a} \lambda(t) dt \right] < \infty. \tag{1.15.18}$$ ∎

Corollary 1.15.6. If $x_0 = 0$, then the function $g(x, z) = \lambda(x)\phi(z)$ is admissible in Theorem 1.15.1 provided that $\lambda(x) \geq 0$ is continuous for $0 < x \leq a$; $\phi(z)$ is continuous for $0 \leq z \leq 2b, \phi(0) = 0, \phi(z) > 0$ for $0 < z \leq 2b$; and either of the following holds

$$\limsup_{x \to 0^+} \int_x \left[\frac{1}{\phi(t)} - \lambda(t) \right] dt = \infty; \quad \text{or} \tag{1.15.19}$$

$$\limsup_{x \to 0^+} \int_x \left[\frac{1}{\phi(t)} - \lambda(t) \right] dt > -\infty \quad \text{and } \phi(z) \leq z. \tag{1.15.20}$$

Proof. From Theorem 1.15.1 it suffices to show that the only solution of the differential equation $z' = \lambda(x)\phi(z)$ satisfying (1.15.2) is $z(x) \equiv 0$. Assume the contrary, i.e., there is a solution $z(x)$ such that $z(\xi) \neq 0$, where $0 < \xi < a$. Let c be the first zero of $z(x)$ to the left of ξ. Then, $z(x) > 0$ for $0 \leq c < x \leq \xi$, and $z'(x) = \lambda(x)\phi(z(x))$. If $c = 0$, then by the hypothesis $z(c) = z'_+(c) = 0$. If $c > 0$, then $z'(c) = \lambda(c)\phi(0) = 0$. Hence, for all sufficiently small positive ε, $z(c + \varepsilon) < \varepsilon$ and

$$\int_\varepsilon^{z(\xi)} \frac{dt}{\phi(t)} \leq \int_{z(c+\varepsilon)}^{z(\xi)} \frac{dt}{\phi(t)} = \int_{c+\varepsilon}^{\xi} \lambda(t) dt.$$

Therefore, we find that

$$\int_\xi^{z(\xi)} \frac{dt}{\phi(t)} \leq \int_{c+\varepsilon}^{\xi} \lambda(t) dt - \int_\varepsilon^{\xi} \frac{dt}{\phi(t)} \leq \int_\varepsilon^{\xi} \left[\lambda(t) - \frac{1}{\phi(t)} \right] dt. \tag{1.15.21}$$

But this contradicts $z(\xi) \neq 0$, since (1.15.19) implies that the right side of (1.15.21) tends to $-\infty$ as $\varepsilon \to 0^+$. If $c > 0$, then from (1.15.20) also the same contradiction holds. If $c = 0$, then since

$$\int_{z(\varepsilon)}^{\varepsilon} \frac{dt}{\phi(t)} = \int_{\varepsilon}^{\xi} \left[\lambda(t) - \frac{1}{\phi(t)}\right] dt - \int_{\xi}^{z(\xi)} \frac{dt}{\phi(t)}$$

in view of $z'_+(0) = 0$ and (1.15.20) it follows that the left side tends to ∞ and the right side remains bounded as $\varepsilon \to 0^+$. ∎

The following particular cases of Corollary 1.15.6 give several known uniqueness criteria.

(i) **Generalized Lipschitz Uniqueness**

$$\lambda(x) \quad \text{satisfies (1.15.15);} \quad \phi(z) = z.$$

(ii) **Osgood's Uniqueness**

$$\lambda(x) = \lambda, \quad \text{a constant;} \quad \limsup_{x\to 0^+} \int_x \frac{dt}{\phi(t)} = \infty.$$

(Thus, as in Theorem 1.4.4 we note that in Theorem 1.4.2 the function g need not be nondecreasing.)

(iii) **Montel - Tonelli's Uniqueness**

$$\limsup_{x\to 0^+} \int_x \lambda(t)dt < \infty; \quad \limsup_{x\to 0^+} \int_x \frac{dt}{\phi(t)} = \infty.$$

(iv) **Nagumo's Uniqueness**

$$\lambda(x) = \frac{1}{x}; \quad \phi(z) = z.$$

(v) **van Kampen's Uniqueness**

$$\lambda(x) = \frac{1}{x} + g(x); \quad \phi(z) = z,$$

where $g(x)$ is continuous and integrable for $0 < x \leq a$.

(vi) **Iyanaga's Uniqueness**

$$\lambda(x) = a_1x + a_2x^2 + \cdots + a_nx^n; \quad \phi(z) = b_1z + b_2z^2 + \cdots + b_mz^m, \quad 0 \le \frac{b_1}{a_1} \le 1.$$

Corollary 1.15.7. The Krasnosel'skii - Krein Uniqueness Theorem 1.7.1 is a particular case of Theorem 1.15.1.

Proof. It suffices to note that in Theorem 1.15.1 the function $g(x,z) = \min\{cz^\alpha, kz/(x-x_0)\}$ for $x > x_0$ with $c > 0, k > 0, 0 < \alpha < 1$ and $k(1-\alpha) < 1$ satisfies all the required hypotheses. ∎

Corollary 1.15.8. Kooi's Uniqueness Theorem 1.8.1 is a particular case of Theorem 1.15.1.

Proof. It suffices to note that in Theorem 1.15.1 the function $g(x,z) = \min\{c(x-x_0)^{-\beta}z^\alpha, kz/(x-x_0)\}$ for $x > x_0$ with $c > 0, k > 0, 0 < \alpha < 1, \beta < \alpha$ and $k(1-\alpha) < 1-\beta$ satisfies all the required hypotheses. ∎

1.16 SHEN'S UNIQUENESS THEOREM

A nontrivial generalization of Kamke's Uniqueness Theorem 1.15.1 is embodied in the following :

Theorem 1.16.1 (Shen's Uniqueness Theorem). Assume that

(i) the function $g(x,z)$ is continuous and nonnegative in $0 < x \le a$, $0 < z \le 2b$ and for every ξ, $0 < \xi < a$ the differential equation (1.13.1) has a sequence of positive solutions $z_{n,\xi}(x)$ satisfying :

(a) either $\overline{\lim}_{x\to 0}\ z_{n,\xi}(x) > 0$, or $\lim_{x\to 0}\ z_{n,\xi}(x) = 0$ and $\overline{\lim}_{x\to 0}\ \frac{z_{n,\xi}(x)}{x} > 0$; and

(b) $\lim_{n\to 0}\ z_{n,\xi}(\xi) = 0$,

(ii) the function $f(x,y)$ is continuous in $\bar{S}_+$ with $(x_0,y_0) = (0,0)$ and for all $0 < x \le a$, $-b \le y < \bar{y} \le b$ the inequality (1.13.2) is satisfied.

Then, the initial value problem (1.9.4) has at most one solution in $[0, a]$.

Proof. As usual suppose that $y(x)$ and $\bar{y}(x)$ are two solutions of the initial value problem (1.9.4) in the interval $[0, a]$. Let $\xi \in (0, a)$ be such that $y(\xi) \neq \bar{y}(\xi)$. Without loss of generality we may assume that $\bar{y}(x) \geq y(x)$ in $(0, a)$. In fact, if these $y(x)$ and $\bar{y}(x)$ do not satisfy this inequality then we can choose the maximal and the minimal solutions of (1.9.4). Let $\phi(x) = \bar{y}(x) - y(x)$, then $\phi(x) \geq 0,\ x \in (0, a),\ \phi(0) = 0,\ \phi(\xi) > 0$, and

$$\begin{aligned} \phi'(x) &= \bar{y}'(x) - y'(x) = f(x, \bar{y}(x)) - f(x, y(x)) \\ &= f(x, y(x) + \phi(x)) - f(x, y(x)). \end{aligned}$$

Let $h(x, \phi) = f(x, y(x) + \phi) - f(x, y(x))$, then $\phi(x)$ satisfies the differential equation $\phi'(x) = h(x, \phi(x))$.

Since $\xi \in (0, a)$ and $\phi(\xi) > 0$ from the given hypothesis on the function $g(x, z)$ the differential equation (1.13.1) must have a positive solution $z_{n_0,\xi}(x)$ such that

(a) either $\overline{\lim}_{x \to 0} z_{n_0,\xi}(x) > 0$, or $\lim_{x \to 0} z_{n_0,\xi}(x) = 0$ and $\overline{\lim}_{x \to 0} \frac{z_{n_0,\xi}(x)}{x} > 0$; and

(b) $z_{n_0,\xi}(\xi) < \phi(\xi)$.

Now we shall show that (a) ensures the existence of an $\eta \in (0, \xi)$ such that $z_{n_0,\xi}(\eta) > \phi(\eta)$. For this, if $\overline{\lim}_{x \to 0} z_{n_0,\xi}(x) > 0$, then for sufficiently small positive η it is impossible to have $z_{n_0,\xi}(\eta) \leq \phi(\eta)$. Since otherwise, in the limit we will have $z_{n_0,\xi}(0) \leq \phi(0) = 0$, which is certainly not true. If $\lim_{x \to 0} z_{n_0,\xi}(x) = 0$ and $\overline{\lim}_{x \to 0} \frac{z_{n_0,\xi}(x)}{x} > 0$ then we need to consider $\frac{z_{n_0,\xi}(\eta)}{\eta} \leq \frac{\phi(\eta)}{\eta}$. However, since $\phi'(0) = 0$ it follows that $\overline{\lim}_{\eta \to 0} \frac{z_{n_0,\xi}(\eta)}{\eta} \leq \phi'(0) = 0$. But, this contradicts $\overline{\lim}_{x \to 0} \frac{z_{n_0,\xi}(x)}{x} > 0$.

On the other hand due to (b) we have $z_{n_0,\xi}(\xi) < \phi(\xi)$ and therefore there must be a $\mu \in (\eta, \xi)$ such that $z_{n_0,\xi}(\mu) = \phi(\mu)$.

Now let $\nu = \text{glb}\{x : z_{n_0,\xi}(x) = \phi(x),\ x \in (\eta, \xi)\}$, then for $x \in (\eta, \nu)$ it follows that $z_{n_0,\xi}(x) > \phi(x)$, and $z_{n_0,\xi}(\nu) = \phi(\nu)$. Thus, we have

$$\frac{z_{n_0,\xi}(x) - z_{n_0,\xi}(\nu)}{x - \nu} < \frac{\phi(x) - \phi(\nu)}{x - \nu}, \quad x \in (\eta, \nu).$$

In the above inequality as $x \to \nu$, we find that $z'_{n_0,\xi}(\nu) \le \phi'(\nu)$. But, since

$$z'_{n_0,\xi}(\nu) = g(\nu, z_{n_0,\xi}(\nu)) = g(\nu, \phi(\nu))$$

and on the other hand in view of (1.13.2)

$$\phi'(\nu) = h(\nu, \phi(\nu)) < g(\nu, \phi(\nu))$$

we get the contradiction $g(\nu, \phi(\nu)) < g(\nu, \phi(\nu))$. Therefore, $\phi(x) \equiv 0$, or $y(x) = \bar{y}(x)$, $x \in [0, a]$. ∎

If the function $g(x, z)$ satisfies the condition (i) of Theorem 1.15.1 then the condition (i) of Theorem 1.16.1 is also satisfied. In order to show this, we note that the differential equation (1.13.1) has the solutions $y_{n,\xi}(x)$ satisfying $y_{n,\xi}(\xi) = \frac{1}{n}$. Obviously this sequence of functions satisfies the condition (b). To show that it satisfies the condition (a) also, we note that if $\overline{\lim}_{x \to 0} z_{n,\xi}(x) > 0$ is not true then $z_{n,\xi}(0) = 0$. Hence, we must have $\overline{\lim}_{x \to 0} \frac{z_{n,\xi}(x)}{x} > 0$. Because, if this is not true then necessarily $z'_{n,\xi}(0) = 0$. But, then from the hypothesis (i) of Theorem 1.15.1, $z_{n,\xi}(x) \equiv 0$, and this contradicts the initial condition $y_{n,\xi}(\xi) = \frac{1}{n}$.

Example 1.16.1. The differential equation

$$(1.16.1) \qquad z' = g(x, z) = \begin{cases} 0, & z \ge 2x^3 \\ 3(x^2 - (z - x^3)^{2/3}), & z \le 2x^3 \end{cases}$$

has the following sequence of positive solutions

$$z_{n,\xi}(x) = \begin{cases} x^3 + \left(\frac{1}{n} - x\right)^3, & x \ge \frac{1}{2n} \\ \frac{1}{4n^3}, & x \le \frac{1}{2n} \end{cases}$$

and this sequence satisfies the condition (i) of Theorem 1.16.1 (in condition (a) $\overline{\lim}_{x \to 0} z_{n,\xi}(x) > 0$ holds). However, for the differential equation (1.16.1), $z(x) = x^3$ is a nontrivial solution which satisfies the conditions $z(0) = z'(0) = 0$, and hence for (1.16.1) the condition (i) of Theorem 1.15.1 is not satisfied. ∎

Example 1.16.2. The differential equation

(1.16.2) $z' = g(x,z)$

$$= \begin{cases} \left(\dfrac{x^2+(x^4+4xz)^{1/2}}{x}\right)\left(\dfrac{x^2+(x^4+4xz)^{1/2}}{4x}-x\right), & z \geq 2x^3 \\ 3(x^2-(z-x^3)^{2/3}), & z \leq 2x^3 \end{cases}$$

has the following sequence of positive solutions

$$z_{n,\xi}(x) = \begin{cases} x^3+\left(\dfrac{1}{n}-x\right)^3, & x \geq \dfrac{1}{2n} \\ -\dfrac{1}{n}\left(x-\dfrac{1}{2n}\right)^2+\dfrac{1}{4n^3}, & x \leq \dfrac{1}{2n} \end{cases}$$

and this sequence satisfies the condition (i) of Theorem 1.16.1 (in condition (a) $\overline{\lim}_{x\to 0} \frac{z_{n,\xi}(x)}{x} > 0$ holds). However, for the differential equation (1.16.2), $z(x) = x^3$ is a nontrivial solution which satisfies the conditions $z(0) = z'(0) = 0$, and hence for (1.16.2) the condition (i) of Theorem 1.15.1 is not satisfied. ∎

1.17 MIKOLAJSKA'S UNIQUENESS THEOREM

Here we shall present a far reaching extension of Witte's Uniqueness Theorem 1.10.1. For this, we begin with the following result which provides sufficient conditions on the function $g(x,z)$ so that the condition (i) of Theorem 1.15.1 is guaranteed.

Theorem 1.17.1. Assume that

(i) the function $h(x,z)$ is continuous for $0 < x \leq a$, $0 \leq z < \infty$, increasing with respect to z and $h(x,0) \equiv 0$,

(ii) $\rho(x,x_1,c)$, $x_1 \in (0,a]$ is the left minimal solution of the problem

(1.17.1) $$z' = h(x,z), \quad z(x_1) = c > 0,$$

(iii) the function $g(x, z)$ is continuous for $0 < x \le a,\ 0 \le z \le 2b$ and that

$$g(x, z) \le h(x, z), \tag{1.17.2}$$

(iv) for each solution $z(x)$ of the equation (1.13.1) defined in $(0, x_1]$, $0 < x_1 \le a$ such that $\lim_{x\to 0} z(x) = 0$ there exists a function $\mu(x, z(\cdot)) \ge 0$ which is continuous in $[0, a]$, $\mu(0, z(\cdot)) = 0$, and a number δ, $0 < \delta \le a$ such that

$$g(x, z(x)) < h(x, \rho(x, x_1, \mu(x, z(\cdot)))), \quad 0 < x \le \delta. \tag{1.17.3}$$

Then, each solution of (1.13.1) defined in the interval $(0, \alpha), 0 < \alpha < a$ such that $\lim_{x\to 0} z(x) = 0$ is equal to 0 for $x \in (0, \alpha)$.

Proof. Suppose there exists a solution $z_1(x)$ of (1.13.1) which is defined in $(0, \alpha]$, $0 < \alpha < a$ such that $z_1(x) \not\equiv 0$, $\lim_{x\to 0} z_1(x) = 0$. Then, there exists $\xi_0 \in (0, \alpha]$ such that $c_0 = z_1(\xi_0) > 0$. Because of (1.17.2) we have $z_1'(x) \le h(x, z_1(x))$, $x \in (0, \xi_0]$ such that $z_1(x) \ge 0$ and consequently

$$z_1(x) \ge \rho(x, \xi_0, c_0) \ge 0, \quad 0 < x \le \xi_0. \tag{1.17.4}$$

From the hypothesis (iv) it follows that there exists $\delta_0 > 0$ such that

$$z_1'(x) < h(x, \rho(x, \xi_0, \mu(x, z_1(\cdot)))), \quad 0 < x \le \delta_0. \tag{1.17.5}$$

Consider $\varepsilon \in (0, c_0)$. Then, there exists $x_2 \in (0, \delta_0)$ such that $0 \le \mu(x, z_1(\cdot)) \le \varepsilon$, $0 < x \le x_2$.

Since $\rho(x, x_1, \zeta)$ is increasing with respect to ζ it follows that

$$\rho(x, \xi_0, \mu(x, z_1(\cdot))) \le \rho(x, \xi_0, \varepsilon) \le \rho(x, \xi_0, c_0), \quad 0 < x \le x_2. \tag{1.17.6}$$

Now we will show that there exists $x_1 \in (0, x_2]$ such that

$$\rho_0 = z_1(x_1) - \rho(x_1, \xi_0, \varepsilon) > 0. \tag{1.17.7}$$

From (1.17.4) and (1.17.6) it follows that for each $x \in (0, x_2]$, $z_1(x) - \rho(x, \xi_0, \varepsilon) \ge 0$. Suppose that $z_1(x) - \rho(x, \xi_0, \varepsilon) = 0$, $0 < x \le x_2$. Then, $z_1'(x) = \rho'(x, \xi_0, \varepsilon)$

for $0 < x \leq x_2$, and hence $h(x, \rho(x, \xi_0, \varepsilon)) \equiv g(x, z_1(x))$. Thus, in view of (1.17.5) we obtain

$$h(x, \rho(x, \xi_0, \varepsilon)) < h(x, \rho(x, \xi_0, \mu(x, z_1(\cdot)))), \quad 0 < x \leq x_2.$$

Hence, from (1.17.6) and the increasing nature of h we find that

$$h(x, \rho(x, \xi_0, \varepsilon)) < h(x, \rho(x, \xi_0, \varepsilon)), \quad 0 < x \leq x_2$$

which is impossible, and thus there exists $x_1 \in (0, x_2]$ such that ρ_0 defined in (1.17.7) is positive.

Next, in $(0, x_2]$ we have

$$z_1'(x) < h(x, \rho(x, \xi_0, \mu(x, z_1(\cdot)))) \leq h(x, \rho(x, \xi_0, \varepsilon))$$

and thus

$$\begin{aligned} z_1(x) &\geq \rho(x, \xi_0, \varepsilon) + z_1(x_1) - \rho(x_1, \xi_0, \varepsilon) \\ &= \rho(x, \xi_0, \varepsilon) + \rho_0, \ 0 < x \leq x_1 \end{aligned}$$

i.e., $z_1(x) \geq \rho_0 > 0, \ 0 < x \leq x_1$. But, then $0 = \lim_{x \to 0} z_1(x) = \rho_0 > 0$, and hence $z_1(x) = 0, \ 0 < x \leq a$. ∎

If $g(x, z)$ is continuous for $0 < x \leq a, \ 0 \leq z \leq 2b$ then the conditions (i) - (iv) in Theorem 1.17.1 are necessary for the uniqueness of the solution $z(x)$ of (1.13.1) satisfying $\lim_{x \to 0} z(x) = 0$. This means that if $z(x) = 0$ is the unique solution of the equation (1.13.1) satisfying $\lim_{x \to 0} z(x) = 0$ then there exists a function $h(x, z)$ continuous for $0 < x \leq a, \ 0 \leq z < \infty$, continuously differentiable with respect to z, increasing with respect to z, and there exists $\mu(x)$ continuous in $[0, a]$, $\mu(x) > 0$ for $(0, a]$, $\mu(0) = 0$ such that

$$0 = g(x, 0) < h(x, \rho(x, x_1, \mu(x, 0)))$$

for each $x_1 \in (0, a]$. For this, it is sufficient to take for $h(x, z)$ any function which is continuous for $0 < x \leq a, \ 0 \leq z < \infty$, continuously differentiable with respect to z, increasing with respect to z and such that

$$h(x, z) \geq z + \max_{0 \leq t \leq z} |g(x, t)|.$$

It can easily be seen that such a function exists. Consider any $\mu(x)$ continuous in $[0,a]$ such that $\mu(0) = 0 < \mu(x)$, $0 < x \le a$. From the uniqueness of the solutions of the equation $z' = h(x,z)$ it follows that $\rho'(x, x_1, \mu(x)) > 0$, $0 < x \le a$, and hence

$$h(x, \rho(x, x_1, \mu(x))) > 0 = g(x, z(x)) = g(x, 0), \quad 0 < x \le a.$$

We further note that in Theorem 1.17.1 the inequality (1.17.3) cannot be replaced by the weak inequality. To show this we have the following :

Example 1.17.1. Let in (1.13.1) the function $g(x,z)$ be defined as follows

$$g(x,z) = \begin{cases} 2x, & z \ge x^2,\ 0 < x \le 1 \\ 2[x - \sqrt{x^2 - z}], & 0 \le z \le x^2. \end{cases}$$

We set $h(x,z) = g(x,z)$, $z \ge 0$. The function $h(x,z)$ thus defined is increasing in z and continuous for $0 < x \le 1$, $z \ge 0$. For (x_1, z_1) such that $x_1^2 \le z_1$ we have $\rho(x, x_1, z_1) = x^2 - x_1^2 + z_1$, and hence $\lim_{x\to 0} \rho(x, x_1, z_1) = z_1 - x_1^2$ for $z_1 \ge x_1^2$. For $0 \le z_1 < x_1^2$ we have $z' = 2x - 2\sqrt{x^2 - z}$ for x such that $z \le x^2$ from which it is easy to obtain

$$z(x) = 2\left(x_1 - \sqrt{x_1^2 - z_1}\right)(x - x_1) + z_1, \ x_1 \ge x \ge x_1 - \sqrt{x_1^2 - z_1}$$

and hence

$$\rho(x, x_1, c) = \begin{cases} 2\left(x_1 - \sqrt{x_1^2 - c}\right)(x - x_1) + c, & x_1 - \sqrt{x_1^2 - z_1} \le x \le 1 \\ x^2, & 0 < x \le x_1 - \sqrt{x_1^2 - z_1} \end{cases}$$

i.e., $\rho(x, x_1, c) \to 0$ as $x \to 0$ provided $c \le x_1^2$. Each solution of (1.13.1) such that $\lim_{x\to 0} z(x) = 0$, $z(x) \ne 0$ is of the form $z(x) = \rho(x, x_1, c_0)$, $0 < c_0 \le x_1^2$ and $z' = 2x$, $0 < x \le x_1 - \sqrt{x_1^2 - c_0} = \delta_0$.

We set

$$\mu(x, z(\cdot)) = \begin{cases} x_1^2 - (x - x_1)^2, & 0 \le x \le \delta_0 \\ c_0, & \delta_0 \le x \le x_1 \end{cases}$$

so that

$$\rho(x, x_1, \mu(x, z(\cdot))) = \rho(x, x_1, c_0), \ \delta_0 \le x \le x_1$$

and for $0 < x \leq \delta_0$,

$$x_1 - \sqrt{x_1^2 - \mu(x, z(\cdot))} = x_1 - \sqrt{x_1^2 - x_1^2 + (x - x_1)^2} = x.$$

Consequently, we find that $\rho(x, x_1, \mu(x, z(\cdot))) = x^2$ from which we obtain

$$h(x, \rho(x, x_1, \mu(x, z(\cdot)))) = 2x = z'(x), \ 0 < x \leq \delta_0$$

and $\lim_{x \to 0} \mu(x, z(\cdot)) = \lim_{x \to 0}(x_1^2 - (x - x_1)^2) = 0.$ ∎

If the left solutions of the equation $z' = h(x, z)$ are unique for $0 < x \leq a, \ z \geq 0$ then it is possible to replace the inequality (1.17.3) be the weaker inequality

$$g(x, z(x)) \leq h(x, \rho(x, x_1, \mu(x, z(\cdot)))), \ 0 < x < \delta_0. \tag{1.17.8}$$

Indeed in this case

$$0 < \rho(x_1, \xi_0, \varepsilon) < \rho(x_1, \xi_0, c_0) \leq z_1(x_1)$$

hence $\rho_0 > 0$ for each $x_1 \in (0, \xi_0]$ and consequently the proof in this case is simpler than that of Theorem 1.17.1.

The following result is an application of Theorem 1.17.1.

Theorem 1.17.2. Let $g(x, z)$ be continuous for $0 < x \leq a, \ 0 \leq z \leq 2b$, increasing in z and that

$$0 \leq g(x, z) \leq g(x) z^2, \ 0 < x \leq a, \ z^2 \leq (2b)^2. \tag{1.17.9}$$

Further, let for each $x_1 \in (0, a]$ there exists $x_2 \in (0, x_1)$ such that

$$g(x, 2b) < \frac{g(x)}{[2 \int_x^{x_1} g(t) dt]^2}, \ 0 < x \leq x_2. \tag{1.17.10}$$

Then, for (1.13.1), $z(x) \equiv 0$ is the only solution satisfying the condition $\lim_{x \to 0} z(x) = 0$.

Proof. In this case we have $h(x, z) = g(x) z^2, \ 0 < x \leq a$ and hence

$$\rho(x, x_1, c) = \frac{c}{1 - c \int_{x_1}^{x} g(t) dt}, \ 0 < x \leq x_1. \tag{1.17.11}$$

Suppose that $z(x)$ is a solution of (1.13.1) such that $0 < z(x_1) = z_1 \leq 2b$ and that $\lim_{x\to 0} z(x) = 0$. Set $\mu(x) = \max(x, 2z(x))$. From (1.17.9) it follows that

$$\mu(x) \geq 2z(x) \geq 2\sqrt{\frac{g(x,z(x))}{g(x)}}, \quad 0 < x \leq x_1.$$

But, because of (1.17.10) and the fact that $g(x,z)$ is increasing, we have

$$\sqrt{\frac{g(x,z(x))}{g(x)}} \leq \sqrt{\frac{g(x,2b)}{g(x)}} < \frac{1}{2\int_x^{x_1} g(t)dt}, \quad 0 < x \leq x_2$$

and hence

$$1 - \int_x^{x_1} g(t)dt \sqrt{\frac{g(x,z(x))}{g(x)}} > \frac{1}{2}$$

from which we obtain, finally

$$2\sqrt{\frac{g(x,z(x))}{g(x)}} \geq \frac{\sqrt{\frac{g(x,z(x))}{g(x)}}}{1 - \int_x^{x_1} g(t)dt \sqrt{\frac{g(x,z(x))}{g(x)}}}$$

so that

$$\mu(x) > \frac{\sqrt{\frac{g(x,z(x))}{g(x)}}}{1 - \int_x^{x_1} g(t)dt \sqrt{\frac{g(x,z(x))}{g(x)}}}, \quad 0 < x \leq x_2.$$

Consequently,

$$g(x,z(x)) < \left[\frac{\mu(x)}{1 + \mu(x)\int_x^{x_1} g(t)dt}\right]^2 g(x), \quad 0 < x \leq x_2.$$

Because of (1.17.9) and (1.17.11) we therefore have

$$g(x,z(x)) < h(x, \rho(x, x_1, \mu(x))).$$

From the definition of $\mu(x)$ and the fact that $\lim_{x\to 0} z(x) = 0$ it follows that $\lim_{x\to 0}\mu(x) = 0$. Thus, the conditions of Theorem 1.17.1 are satisfied, and hence it follows that $z(x) = 0, \ 0 < x \leq x_1$. ∎

It is clear that in the above theorem the increasing nature of g is not necessary. In fact, it suffices to replace (1.17.10) by

$$|g(x,z)| < \frac{g(x)}{[\int_x^{x_1} g(t)dt]^2}, \quad 0 < x \leq x_2 < x_1, \ |z| \leq 2b$$

and (1.17.9) by $|g(x,z)| \le g(x)z^2$.

Theorem 1.17.3 (Mikolajska's Uniqueness Theorem). Assume that

(i) $f(x,y)$ is continuous for $0 < x \le a,\ |y - y_0| \le b_1$,

(ii) the function $h_1(x,z)$ is continuous for $0 < x \le a,\ 0 \le z < \infty$, increasing with respect to z and $h_1(x,0) \equiv 0$,

(iii) $\rho_1(x,x_1,c),\ x_1 \in (0,a],\ c \ge 0$ is the minimal solution of the initial value problem

$$z' = h_1(x,z), \quad z(x_1) = c \tag{1.17.12}$$

(iv) for all $(x,y),\ (x,\bar{y})$ such that $0 < x \le a,\ y_0 - b_1 \le y,\ \bar{y} \le y_0 + b_1$ the following inequality holds

$$|f(x,y) - f(x,\bar{y})| \le h_1(x,|y-\bar{y}|) \tag{1.17.13}$$

(v) for each $\alpha \in (0,a]$ and each pair of solutions $y(x),\ \bar{y}(x)$ of $y' = f(x,y)$ defined in the interval $(0,\alpha]$ such that $\lim_{x\to 0} y(x) = \lim_{x\to 0}\bar{y}(x) = y_0$ there exists a $\delta_1 \in (0,\alpha]$ and a function $\mu_1(x,y,\bar{y}) \ge 0$, continuous in $(0,\delta_1],\ \mu_1(0,y,\bar{y}) = 0$, such that

$$|f(x,y(x)) - f(x,\bar{y}(x))| < h_1(x,\rho_1(x,\alpha,\mu_1(x,y,\bar{y}))), \quad 0 < x \le \delta_1. \tag{1.17.14}$$

Then, the initial value problem

$$y' = f(x,y), \quad \lim_{x\to 0} y(x) = y_0 \tag{1.17.15}$$

has at most one solution $y(x)$ in $(0,\alpha_0],\ 0 < \alpha_0 \le a$.

Proof. Suppose $y(x)$ is a solution of (1.17.15) in the interval $(0,\lambda]$. Then, there exists $0 < \lambda_0 \le \lambda$ such that $|y(x) - y_0| \le b_1/2,\ 0 < x \le \lambda_0$. Suppose $w(x)$ is another solution of (1.17.15) in the interval $(0,\bar{\lambda}]$, where $\bar{\lambda} \le \lambda_0$. In the interval $(0,\bar{\lambda}]$ we set $y_1(x) = \min(y(x),w(x))$ and $y_2(x) = \max(y(x),w(x))$. Then, $y_1(x)$ and $y_2(x)$ are solutions of (1.17.15) and $y_1(x) \le y_2(x),\ 0 < x \le \bar{\lambda}$.

Let τ be the largest s such that $|y_i(x) - y_0| \leq b_1/2,\ 0 < x \leq s;\ i = 1,2$. We shall show that $y_1(x) = y_2(x)$ in $(0,\tau]$ from which it will follow that $|y_i(\tau) - y_0| = |y(\tau) - y_0| \leq b_1/2,\ i = 1,2$ and hence $\tau = \bar{\lambda}$.

Consider the function $z(x) = y_2(x) - y_1(x)$. It satisfies the equation

$$z' = f(x, z + y_1(x)) - f(x, y_1(x)),\ 0 < x \leq \tau.$$

We define

$$g(x,z) = \begin{cases} f(x, z + y_1(x)) - f(x, y_1(x)), & 0 < x \leq \tau,\ 0 < z \leq b_1/2 \\ f(\tau, z + y_1(\tau)) - f(\tau, y_1(\tau)), & \tau < x \leq a,\ 0 < z \leq b_1/2. \end{cases}$$

Clearly, $z' = g(x,z)$ satisfies the hypotheses of Theorem 1.17.1 with $b = b_1/4$, $\delta = \min(\tau, \delta_1)$, and

$$h(x,z) = \begin{cases} h_1(x,z), & 0 < x \leq \tau \\ h_1(\tau, z), & \tau < x \leq a. \end{cases}$$

Thus, each solution $z(x) \geq 0$ of $z' = g(x,z)$ such that $\lim_{x\to 0} z(x) = 0$ is equal to 0 in $(0,a]$ from which it follows that $y_1(x) = y_2(x),\ 0 < x \leq \tau$. This proves the result with $\alpha_0 = \tau$. ∎

Corollary 1.17.4. If the conditions of Theorem 1.17.3 are satisfied except that $\rho(x, x_1, c),\ x_1 \in (0,a],\ c \geq 0$ is the only solution of (1.17.12) then the inequality (1.17.14) can be replaced by the weaker inequality, i.e.,

$$(1.17.16)\quad |f(x, y(x)) - f(x, \bar{y}(x))| \leq h_1(x, \rho_1(x, \alpha, \mu_1(x, y, \bar{y}))),\ 0 < x \leq \delta_1.$$ ∎

If $y_0 = 0$ then the condition (1.17.16) can be simplified. Indeed, in this case it is sufficient to assume

$$|f(x, y(x))| \leq h_1(x, \rho_2(x, \mu_2(x, y(\cdot)))),$$

where $\mu_2(x, y(\cdot)) \to 0$ as $x \to 0$ and $h_1(x, \rho_2(x, c_1)) + h_1(x, \rho_2(x, c_2)) \leq h_1(x, \rho_2(x, c_1 + c_2))$, $\rho_2(x,c)$ is the solution of the initial value problem $z' = h_1(x,z),\ z(a) = c$.

In particular, if $h_1(x,z) = h(x)z$, $h(x) \geq 0$ then we have $\rho_2(x,c) = c\exp\left(\int_a^x h(t)dt\right)$, $h_1(x,\rho_2(x,c_1)) + h_1(x,\rho_2(x,c_2)) = h(x)(c_1+c_2)\exp\left(\int_a^x h(t)dt\right) = h_1(x,\rho_2(x,c_1+c_2))$.

From these observations it is clear that Witte's Uniqueness Theorem 1.10.1 is a particular case of the Corollary 1.17.4.

1.18 BRAUER'S UNIQUENESS THEOREM

The motivation of the following result is due to the Krasnosel'skii - Krein uniqueness theorem.

Theorem 1.18.1 (Brauer's Uniqueness Theorem). Suppose that

(i) the functions $A(x)$ and $B(x)$ are continuous and nonnegative in $[x_0, x_0+a]$ such that $A(x_0) = B(x_0) = 0, B(x) > 0,\ x > x_0$ and

$$\lim_{x \to x_0^+} \frac{A(x)}{B(x)} = 0, \tag{1.18.1}$$

(ii) the functions $g_1(x,z)$, $g_2(x,z)$ are continuous and nonnegative for $x_0 < x \leq x_0 + a,\ 0 \leq z \leq 2b$,

(iii) all the solutions $z(x)$ of

$$z' = g_1(x,z) \tag{1.18.2}$$

with $z(x_0) = 0$ obey $z(x) \leq A(x)$ in $[x_0, x_0 + a]$,

(iv) the only solution $w(x)$ of

$$w' = g_2(x,w) \tag{1.18.3}$$

in $[x_0, x_0 + a]$ such that

$$\lim_{x \to x_0^+} \frac{w(x)}{B(x)} = 0 \tag{1.18.4}$$

is the trivial solution; and

(v) the function $f(x, y)$ is continuous in $\bar{S}_+$ and for $(x, y), (x, \bar{y}) \in \bar{S}_+$ where $x \neq x_0$

$$|f(x,y) - f(x,\bar{y})| \leq \begin{cases} g_1(x, |y - \bar{y}|) \\ g_2(x, |y - \bar{y}|). \end{cases} \tag{1.18.5}$$

Then, the initial value problem (1.1.1) has at most one solution in $[x_0, x_0 + a]$.

To prove this result it is convenient to prove the following :

Lemma 1.18.2. Let the functions $A(x), B(x), g_1(x, z)$ and $g_2(x, z)$ satisfy the hypotheses of Theorem 1.18.1. Further, let the function $g(x, z)$ be continuous and nonnegative in $x_0 \leq x \leq x_0 + a$, $0 \leq z \leq 2b$, $g(x, 0) \equiv 0$, and for $x \neq x_0$

$$g(x, z) \leq \begin{cases} g_1(x, z) \\ g_2(x, z). \end{cases} \tag{1.18.6}$$

Then, for every x_1, $x_0 < x_1 \leq x_0 + a$, $z(x) \equiv 0$ is the only differentiable function in $x_0 \leq x < x_1$, which satisfies

$$z' = g(x, z), \quad z(x_0) = 0. \tag{1.18.7}$$

Proof. We shall show that the maximal solution $r(x)$ of (1.18.7) is identically zero. On the contrary, suppose that there exists a σ, $x_0 < \sigma < x_0 + a$ such that $r(\sigma) > 0$. Now proceeding as in Lemma 1.15.2 and using (1.18.6) and (1.18.7), we obtain

$$\rho_2(x) \leq r(x) \tag{1.18.8}$$

as far as $\rho_2(x)$ exists to the left of σ, where $\rho_2(x)$ is the minimal solution of (1.18.3) such that $\rho_2(\sigma) = r(\sigma)$. As before, we can continue $\rho_2(x)$ up to x_0 by defining $\rho_2(x_0) = 0$. Since $\rho_2(x) \not\equiv 0$, we have $\lim_{x \to x_0^+} \rho_2(x)/B(x) \neq 0$, which in view of (1.18.8) implies that $\lim_{x \to x_0^+} r(x)/B(x) \neq 0$. This together with (1.18.1) shows that there exists a x_2 such that

$$r(x_2) > A(x_2). \tag{1.18.9}$$

Let $\rho_1(x)$ be the minimal solution of (1.18.2) such that $\rho_1(x_2) = r(x_2)$. Then it can be shown, arguing similarly, that $\rho_1(x)$ can be continued up to x_0, $\rho_1(x_0) = 0$, and

$$0 \le \rho_1(x) \le r(x), \quad x_0 \le x \le x_2. \tag{1.18.10}$$

Since, by hypothesis (iii) of Theorem 1.18.1 all solutions $z(x)$ of (1.18.2) must obey $z(x) \le A(x), x_0 \le x \le x_0 + a$ we must have

$$\rho_1(x) \le A(x), \quad x_0 \le x \le x_0 + a.$$

But this is not possible because of (1.18.9) and the fact that $\rho_1(x_2) = r(x_2)$. Hence, $\rho_1(x_0) > 0$, which implies in view of (1.18.10) that $0 < \rho_1(x_0) \le r(x_0)$, contradicting the assumption $r(x_0) = 0$. Therefore, $r(x) \equiv 0$. ∎

Proof of Theorem 1.18.1. Consider the function $g(x,z) = g_f(x,z)$, where $g_f(x,z)$ is defined in (1.15.6). Now combining the arguments in the proofs of Theorem 1.15.1 and Lemma 1.18.2, it is clear that $g(x,z)$ verifies the conditions of Theorem 1.11.4, which is sufficient to establish the uniqueness of the solutions. ∎

Corollary 1.18.3. The Krasnosel'skii - Krein Uniqueness Theorem 1.7.1 is a particular case of Theorem 1.18.1.

Proof. It suffices to note that the functions $A(x) = [c(1-\alpha)(x-x_0)]^{(1-\alpha)^{-1}}$, $B(x) = (x-x_0)^k$, $g_1(x,z) = cz^\alpha$ and $g_2(x,z) = kz/(x-x_0)$ where $c > 0$, $k > 0$, $0 < \alpha < 1$ and $k(1-\alpha) < 1$ are admissible in Theorem 1.18.1. ∎

Corollary 1.18.4. Kooi's Uniqueness Theorem 1.8.1 is a particular case of Theorem 1.18.1.

Proof. It suffices to note that the functions $A(x) = [c(1-\alpha)(1-\beta)^{-1}(x-x_0)^{1-\beta}]^{(1-\alpha)^{-1}}$, $B(x) = (x-x_0)^k, g_1(x,z) = cz^\alpha(x-x_0)^{-\beta}$ and $g_2(x,z) = kz/(x-x_0)$ where $c > 0$, $k > 0$, $0 < \alpha < 1$, $\beta < \alpha$ and $k(1-\alpha) < 1-\beta$ are admissible in Theorem 1.18.1. ∎

Corollary 1.18.5. Kamke's Uniqueness Theorem 1.15.1 is a particular case of Theorem 1.18.1.

Proof. Define the function

$$g(x,z) = \min\{g_1(x,z), g_2(x,z)\}. \tag{1.18.11}$$

This function $g(x,z)$ satisfies (1.18.6). Thus, it suffices to show that no nontrivial solution of (1.15.1) fulfills the limiting conditions (1.15.2). The assumption that there exists a differentiable function $z(x)$ satisfying the differential equation (1.15.1) and the conditions (1.15.2) for which $z(\sigma) > 0,\ x_0 < \sigma < x_0 + a$ leads to the contradiction that $z(x_0) > 0$ on following the proof of Lemma 1.18.2. ∎

1.19 WEND'S UNIQUENESS THEOREM

From Example 1.3.1 it is clear that in Peano's Uniqueness Theorem 1.3.1 the nonincreasing nature of the function $f(x,y)$ with respect to y cannot be replaced by nondecreasing. Here, we shall show that under suitable restrictions one - sided uniqueness of (1.1.1) still holds when $f(x,y)$ is nondecreasing in both x and y.

Theorem 1.19.1 (Wend's Uniqueness Theorem). Let $f(x,y)$ be continuous and nondecreasing in both x and y in $\bar{S}_+$. Further, let $f(x,y)$ be nonnegative in $\bar{S}_+$ and $f(x,y_0) \neq 0$ for all $x \in (x_0, x_0 + a]$. Then, the initial value problem (1.1.1) has at most one solution in $[x_0, x_0 + a]$.

Proof. We may assume without loss of generality that $x_0 = 0$, since otherwise a translation would give us this case. From the given hypotheses if $y(x)$ is a solution of (1.1.1) then $y'(x) = f(x, y(x)) > 0$ in $(0, a]$. This implies that $y(x)$ is strictly increasing in x for $0 < x \le a$. Since $y(x)$ is continuous and increasing, for $\delta > 0$ sufficiently small, there exists a η such that $0 < \eta < a$ and $y(\eta) = y_0 + \delta$. We define $y_\delta(x) = y(x + \eta),\ 0 \le x \le a - \eta$. Then, $y_\delta(0) = y(\eta) = y_0 + \delta$. Using the hypothesis that $f(x,y)$ is nondecreasing in x

and the fact that $y_\delta'(x) = y'(x+\eta) = f(x+\eta, y_\delta(x))$, it follows that

$$\begin{aligned} y_\delta(x) &= y_\delta(0) + \int_0^x f(t+\eta, y_\delta(t))dt \\ &= y_0 + \delta + \int_0^x f(t+\eta, y_\delta(t))dt, \end{aligned}$$

which is the same as

$$\begin{aligned} (1.19.1) \quad y_\delta(x) - y(x) &= \delta + \int_0^x (f(t+\eta, y_\delta(t)) - f(t, y(t)))dt \\ &\geq \delta + \int_0^x (f(t, y_\delta(t)) - f(t, y(t)))dt. \end{aligned}$$

Suppose $y_\delta(t) > y(t)$ for $0 \leq t < x$ and $y_\delta(x) = y(x)$, then (1.19.1) gives $f(t, y_\delta(t)) - f(t, y(t)) < 0$ for some t, $0 < t < x$, which is a contradiction since f is nondecreasing in y. Thus, it follows that $y_\delta(x) > y(x)$ for $0 \leq x \leq a - \eta$. Now let $r(x)$ be the maximal solution of $y' = f(x,y), y(0) = y_0$. Then, by the same argument for $0 < \delta' < \delta$, we have

$$y(x) \leq r(x) < y_{\delta'}(x) < y_\delta(x), \; 0 \leq x \leq a - \eta.$$

But $y_\delta(x) \to y(x)$ as $\eta \to 0$ and $\delta \to 0$. So, $y(x) \equiv r(x)$, and hence (1.1.1) has at most one solution in $[x_0, x_0 + a]$. ∎

From Theorem 1.2.6 it is clear that if $f(x,y) = f(x)g(y)$ then in Theorem 1.19.1 the nondecreasing condition with respect to x as well as y can be dropped.

1.20 van KAMPEN'S UNIQUENESS THEOREM

So far all the results we have proved impose conditions on the function $f(x,y)$. In the following result the conditions are imposed on the family of the solutions of (1.1.1).

Theorem 1.20.1 (van Kampen's Uniqueness Theorem). Let $f(x,y)$ be continuous in $\bar{S}_+$. Further, let there exist a function $\eta(x, x_1, y_1)$ in $x_0 \leq x, x_1 \leq x_0 + a$, $|y_1 - y_0| \leq \beta (< b)$ with the following properties :

(i) for a fixed (x_1, y_1), $y(x) = \eta(x, x_1, y_1)$ is a solution of

$$y' = f(x,y), \quad y(x_1) = y_1 \tag{1.20.1}$$

(ii) $\eta(x, x_1, y_1)$ is uniformly Lipschitz continuous with respect to y_1; and

(iii) no two solution arcs $y(x) = \eta(x, x_1, y_1)$, $y(x) = \eta(x, x_2, y_2)$ pass through the same point (x, y) unless $\eta(x, x_1, y_1) = \eta(x, x_2, y_2)$ for $x_0 \leq x \leq x_0 + a$.

Then, $y(x) = \eta(x, x_0, y_0)$ is the only solution of (1.1.1) in $[x_0, x_0 + a]$.

Proof. Let $y(x)$ be any solution of (1.1.1). We will show that $y(x) = \eta(x, x_0, y_0)$ for small $x - x_0 \geq 0$. Condition (ii) implies that there exists a constant K such that

$$|\eta(x, x_1, y_1) - \eta(x, x_1, y_2)| \leq K|y_1 - y_2| \tag{1.20.2}$$

for $x_0 \leq x$, $x_1 \leq x_0 + a$ and $|y_1 - y_0| \leq \beta, |y_2 - y_0| \leq \beta$. Let $|f(x,y)| \leq M$ in $\bar{S}_+$. Then, any solution $y(x)$ of (1.1.1) satisfies $|y(x) - y_0| \leq M(x - x_0) \leq \frac{1}{2}\beta$ if $x_0 \leq x \leq x_0 + \frac{\beta}{2M}$. Thus, $\eta(x, s, y(s))$ is defined and $|\eta(x, s, y(s)) - y(s)| \leq M|x - s| \leq \frac{1}{2}\beta$ if $x_0 \leq x, s \leq x_0 + \frac{\beta}{2M}$. Hence, $|\eta(x, s, y(s)) - y_0| \leq \beta$ if $x_0 \leq x, s \leq x_0 + \gamma$, where $\gamma = \min(a, \frac{\beta}{2M})$. Condition (iii) means that any point on any of the arcs $y(x) = \eta(x, x_1, y_1)$ can be used to determine this arc. Thus, (1.20.2) with $y_1 = y(x_1)$ and $y_2 = \eta(x_1, s, y(s))$ gives

$$|\eta(x, x_1, y(x_1)) - \eta(x, s, y(s))| \leq K|y(x_1) - \eta(x_1, s, y(s))| \tag{1.20.3}$$

if $x_0 \leq x, x_1, s \leq x_0 + \gamma$. Now, let x be fixed in $[x_0, x_0 + \gamma]$. We will show that

$$\tau(x) = \eta(x, x_0, y_0) - y(x) = 0. \tag{1.20.4}$$

For this, we put

$$\sigma(s) = \eta(x, x_0, y_0) - \eta(x, s, y(s)) \tag{1.20.5}$$

for $x_0 \leq s \leq x$ $(\leq x_0 + \gamma)$, so that $\sigma(x_0) = 0$ and $\sigma(x) = \tau(x)$. Then, (1.20.3) and (1.20.5) imply that

$$|\sigma(x_1) - \sigma(s)| \leq K|y(x_1) - \eta(x_1, s, y(s))|. \tag{1.20.6}$$

Since $y(x) = \eta(x,s,y(s))$ is a solution of $y' = f(x,y)$ through the point $(s,y(s))$, it is seen that $\eta(x_1,s,y(s)) = y(s)+(x_1-s)[f(s,y(s))+o(1)]$ as $x_1 \to s$. Also, $y(x_1) = y(s)+(x_1-s)[f(s,y(s))+o(1)]$ as $x_1 \to s$. Hence, (1.20.6) gives that $\sigma(x_1)-\sigma(s) = Ko(1)|x_1-s|$ as $x_1 \to s$; i.e., $\frac{d\sigma}{ds}$ exists and is 0. Thus, $\sigma(s)$ is the constant $\sigma(x_0) = 0$ for $x_0 \le s \le x$. In particular, $\tau(x) = \sigma(x)$ satisfies (1.20.4), as was to be proved. ∎

The existence of a continuous function $\eta(x,x_1,y_1)$ satisfying the conditions (i) and (iii) of Theorem 1.20.1 is not sufficient for the uniqueness of the solutions of (1.1.1). For this, we note that the continuous function $y(x) = \eta(x,x_1,y_1) = [y_1^{1/3}+\frac{1}{3}(x-x_1)]^3$ is a solution of $y' = y^{2/3}$, $y(x_1) = y_1$. Obviously, this function $\eta(x,x_1,y_1)$ is not uniformly Lipschitz continuous with respect to y_1. Further, $\eta(x^*,x_1,y_1) = \eta(x^*,x_2,y_2)$ holds provided that $y_1^{1/3}-y_2^{1/3} = \frac{1}{3}(x_1-x_2)$. But, this condition implies that $\eta(x,x_1,y_1) = \eta(x,x_2,y_2)$ for all $x \ge 0$. Of course, the initial value problem $y' = y^{2/3}$, $y(0) = 0$ has an infinite number of solutions.

If $f(x,y)$ satisfies the uniform Lipschitz condition (1.2.1), then for small $\beta > 0$ a function $\eta(x,x_1,y_1)$ satisfying the conditions of Theorem 1.20.1 exists. However, the converse is not true, i.e., the existence of $\eta(x,x_1,y_1)$ satisfying the conditions (i) - (iii) of Theorem 1.20.1 does not imply that $f(x,y)$ is uniformly Lipschitz continuous with respect to y. For this it is sufficient to consider the differential equation $y' = g(y)$, where $g(y) > 0$ is continuous for all y.

A one - sided analog of van Kampen's uniqueness theorem is the following :

Theorem 1.20.2. Let $f(x,y)$ be continuous in $\bar{S}_+$. Further, let there exist a function $\eta(x,x_1,y_1)$ in $x_0 \le x_1 \le x \le x_0+a$, $|y_1-y_0| \le \beta(< b)$ with the following properties :

(i) for a fixed (x_1,y_1), $y(x) = \eta(x,x_1,y_1)$ is a solution of (1.20.1); and

(ii) there exists a constant K such that for $\max(x_1,x_2) \le x^* \le x \le x_0+a$ and $|y_1-y_0| \le \beta$, $|y_2-y_0| \le \beta$

$$|\eta(x,x_1,y_1)-\eta(x,x_2,y_2)| \le K|\eta(x^*,x_1,y_1)-\eta(x^*,x_2,y_2)|.$$

Then, $y(x) = \eta(x, x_1, y_1)$ is the only solution of (1.20.1) for sufficiently small intervals $[x_1, x_1+\varepsilon]$, $\varepsilon > 0$ to the right of x_1 (but not necessarily to the left of x_1). ∎

1.21 YOSIE'S UNIQUENESS THEOREM

So far, we have discussed several sufficient conditions for the uniqueness of the solutions of the initial value problem (1.1.1). Here we shall give the first known result which provides the necessary as well as sufficient conditions. To begin with, we note that if $f(x,y)$ is continuous in $\bar{S}_+$ then for the initial value problem (1.1.1) Theorem 1.11.1 ensures the existence of the maximal solution $r(x)$ and the minimal solution $\rho(x)$ in the interval $[x_0, x_0+\alpha]$. For simplicity, hereafter we shall assume that $\alpha = a$. Thus, if the initial value problem has any other solution $y(x)$ in the interval $[x_0, x_0+a]$, then it is necessary that

$$\rho(x) \le y(x) \le r(x), \quad x \in [x_0, x_0+a]. \tag{1.21.1}$$

Therefore, a necessary and sufficient condition for the uniqueness of the solutions of the initial value problem (1.1.1) is

$$\rho(x) = r(x), \quad x \in [x_0, x_0+a]. \tag{1.21.2}$$

This condition can be written in a different form, for this we need the following :

Definition 1.21.1. Suppose that the function $y(x)$ is continuous in the interval $[x_0, x_0+a]$, and $y'(x)$ exists. If $y(x)$ is such that $y(x_0) = y_0$, and satisfies the inequality

$$y'_+(x) < f(x, y(x)), \quad x \in [x_0, x_0+a]$$

then it is said to be a lower - function with respect to the initial value problem (1.1.1). On the other hand, if

$$y'_+(x) > f(x, y(x)), \quad x \in [x_0, x_0+a]$$

then $y(x)$ is said to be an upper - function.

Theorem 1.21.1 [92]. Let $\phi(x)$ and $\psi(x)$ be lower - and upper - functions with respect to the initial value problem (1.1.1) in the interval $[x_0, x_0 + a]$. If $y(x)$ is any solution of (1.1.1) in the interval $[x_0, x_0 + a]$, then

$$\phi(x) < y(x) < \psi(x), \quad x \in (x_0, x_0 + a]. \qquad (1.21.3) \quad ■$$

Thus, in view of (1.21.1) and (1.21.3) and the fact that $\rho(x)$ and $r(x)$ are solutions of (1.1.1) it follows that

$$\phi(x) < \rho(x) \le y(x) \le r(x) < \psi(x), \quad x \in (x_0, x_0 + a]. \qquad (1.21.4)$$

Therefore, if for every $\varepsilon > 0$ there exists a pair of lower - and upper - functions $\phi(x)$ and $\psi(x)$ with respect to the initial value problem (1.1.1) such that $0 < \psi(x) - \phi(x) < \varepsilon$ in the interval $(x_0, x_0 + a]$ then, the condition (1.21.2) must be satisfied. Hence, the following result holds.

Theorem 1.21.2 (Yosie's Uniqueness Theorem). The initial value problem (1.1.1) has at most one solution in the interval $[x_0, x_0 + a]$ if and only if for every $\varepsilon > 0$ there exists a pair of lower - and upper - functions $\phi(x)$ and $\psi(x)$ with respect to the initial value problem (1.1.1) such that $0 < \psi(x) - \phi(x) < \varepsilon$ in the interval $(x_0, x_0 + a]$. ■

Note that by definition, although, for an arbitrary but fixed function $\psi(x)$

$$\psi'_+(x) - f(x, \psi(x)) > 0, \quad x \in [x_0, x_0 + a] \qquad (1.21.5)$$

the lower limit of $\psi'_+(x) - f(x, \psi(x))$ may be zero. Such an upper - function will now be replaced by $\bar{\psi}(x)$ which is close to $\psi(x)$, but $\bar{\psi}'_+(x) - f(x, \bar{\psi}(x))$ is greater than a fixed positive number for all $x \in [x_0, x_0 + a]$.

If the lower limit of $\psi'_+(x) - f(x, \psi(x))$ is zero, then on the curve $y = \psi(x)$, $x_0 \le x \le x_0 + a$ we can find a sequence of points $\{P_1, P_2, \cdots\}$ for which the lower limit of this difference is zero. Let P be the limit of this sequence. Obviously, at P this difference is non - zero since (1.21.5) holds everywhere on the curve $y = \psi(x)$ in the interval $[x_0, x_0 + a]$.

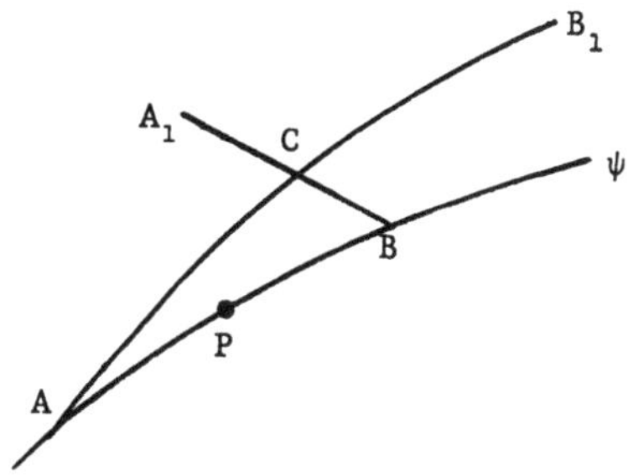

Assume that P is different from the end points of the curve $y = \psi(x)$, $x_0 \le x \le x_0 + a$. Then, we take a sufficiently small piece AB of this curve which contains P. In the following we shall denote by x_A, x_B, f_A, f_B etc. the values of x, y, f etc. at the points A, B and so on. We also assume that $x_A < x_P < x_B$. Since the difference $(\psi'_+ - f)_B$ is non - zero we can set $(\psi'_+ - f)_B \ge \eta_2$, where η_2 is a sufficiently small positive number.

The line BA_1

$$y = y_B + (f_B + \eta_2)(x - x_B) \tag{1.21.6}$$

through the point B is above the curve $y = \psi(x)$ if $x_B - x$ is sufficiently small and positive. This is because ψ'_+ is always greater than $f(x, \psi)$, and hence also greater than $f(x, \psi) + \eta_2$ for sufficiently small η_2. From the continuity of the function f it follows that

$$\psi'_+(x) > f_B + \eta_2$$

for sufficiently small $x_B - x$. We also assume that $x_B - x_A$ is small enough so that the line (1.21.6) is above the curve $y = \psi(x)$.

If η_1 is small and positive then the curve AB_1

$$y = \psi(x) + \eta_1(x - x_A) \tag{1.21.7}$$

passes through the point A and is above the curve $y = \psi(x)$, as long as $x - x_A > 0$. One can bring (1.21.7) close enough to the curve $y = \psi(x)$ by choosing η_1 small enough. The difference of the ordinates AB_1 and A_1B

$$[\psi(x) + \eta_1(x - x_A)] - [y_B + (f_B + \eta_2)(x - x_B)]$$

is obviously a continuous function of x, and is negative at A and is positive at B because AB_1 and A_1B in the interval (x_A, x_B) are above the curve $y = \psi(x)$. Hence, it vanishes at least once in the interval (x_A, x_B), which means that AB_1 and A_1B have at least one common point C where $x_A < x_C < x_B$. If η_1 is small enough than C is arbitrarily close to the point B, along the line A_1B.

Since $f(x,y)$ is continuous one can choose small convex area G around B in which for every point Q the condition

$$|f_Q - f_B| < \eta_3 < \eta_2 \tag{1.21.8}$$

is satisfied. Here η_3 is arbitrarily small.

Now we take η_1 so small that the point C is in G. Then, the equation of the dotted line ACB can be written as

$$y = \begin{cases} \psi(x) + \eta_1(x - x_A), & x_A \le x \le x_C \\ y_B + (f_B + \eta_2)(x - x_B), & x_C \le x \le x_B. \end{cases}$$

Along AC one has

$$\begin{aligned} y'_+ - f(x,y) &= \psi'_+(x) + \eta_1 - f(x, \psi(x) + \eta_1(x - x_A)) \\ &= [\psi'_+(x) - f(x,\psi)] + \eta_1 + [f(x,\psi) - f(x, \psi + \eta_1(x - x_A))] \\ &> \eta_1 + [f(x,\psi) - f(x,\psi + \eta_1(x - x_A))]. \end{aligned}$$

From the continuity of $f(x,y)$, for an arbitrarily small ρ $(0 < \rho < \eta_1)$ one can set

$$|f(x,\psi) - f(x, \psi + \eta_1(x - x_A))| < \eta_1 - \rho, \; x_A \le x \le x_C$$

provided $(x_B - x_A)$ is small enough.

Then, it follows that

$$y'_+ - f(x,y) > \rho, \; x_A \le x \le x_C$$

and therefore along CB in view of (1.21.8) and the fact that C is in G, one has

$$\begin{aligned} y'_+ - f(x,y) &= f_B + \eta_2 - f(x, y_B + (f_B + \eta_2)(x - x_B)) \\ &> \eta_2 - \eta_3. \end{aligned}$$

Thus, along the dotted line ACB the difference $y'_+ - f(x,y)$ is always positive. Moreover, it is greater than a fixed positive value, because ρ and η_3 can always be chosen so that

$$0 < \rho < \eta_1, \quad 0 < \eta_3 < \eta_2.$$

Finally, we remark that the dotted line ACB can be pushed arbitrarily close to the curve $y = \psi(x)$. Thus, if the piece AB of the curve $y = \psi(x)$ is replaced by the dotted line ACB then, one can avoid the point at which the limit of $y'_+ - f(x,y)$ equals zero.

The above method with slight modifications is equally valid if P equals to A or B.

If such points P are everywhere dense on a part of the curve, then we divide it into several parts such that every part is small enough. Since the value of $y'_+ - f(x,y)$ is positive at the end points of these parts, one can replace each part (as above) by the dotted line on which the lower limit of $y'_+ - f(x,y)$ is not equal to zero, and which is close enough to the curve $y = \psi(x)$.

From these considerations in the uniqueness condition we may assume that the lower - and upper - functions are such that in the interval $x_0 \le x \le x_0 + a$ the differences $\psi'_+(x) - f(x,\psi(x))$ and $f(x,\phi(x)) - \phi'_+(x)$ have positive lower limits, i.e., there exist μ_1 and μ_2 positive numbers such that

$$\begin{aligned} &\psi'_+(x) - f(x,\psi(x)) \ge \mu_1 \\ &\phi'_+(x) - f(x,\phi(x)) \le -\mu_2, \qquad x_0 \le x \le x_0 + a. \end{aligned} \tag{1.21.9}$$

Now let A and B be two points on the curve $y = \psi(x)$ and let ρ be a positive number less than μ_1.

Let G be the convex area containing the line AB. If G is small enough then since $f(x,y)$ is continuous, for each pair of points M and N in G we have

$$|f_M - f_N| < \rho < \mu_1. \tag{1.21.10}$$

Let

$$t = \frac{y_B - y_A}{x_B - x_A},$$

then on the piece AB of the curve $y = \psi(x)$ we can find two points P and Q such that

$$(\psi'_+(x))_P \geq t \geq (\psi'_+(x))_Q \tag{1.21.11}$$

otherwise the two lines starting from A would not meet again for $x > x_A$.

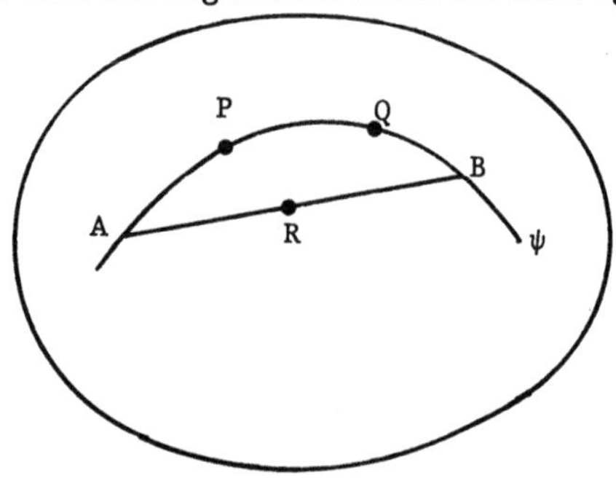

Let R be any point of AB then by (1.21.10), we have

$$f_Q > f_R - \rho.$$

Moreover, from (1.21.9) and (1.21.11) it follows that

$$(\psi'_+)_Q - f_Q \geq \mu_1 \quad \text{and} \quad t - (\psi'_+)_Q \geq 0.$$

From these three inequalities, we find that

$$t - f_Q \geq \mu_1 \quad \text{and} \quad t - f_R > \mu_1 - \rho.$$

Along AB then one has the inequality

$$y'_+ - f(x,y) > \mu_1 - \rho > 0. \tag{1.21.12}$$

If we replace the curve piece AB of $y = \psi(x)$ by the line AB then for the curve $y = \bar{\psi}(x)$ obtained in this way we have the inequality $\bar{\psi}'_+ - f(x,\bar{\psi}) > 0$. This inequality characterizes the upper - function. Of course, the line AB can be chosen arbitrarily close to $\psi(x)$ if $|x_A - x_B|$ is small enough.

Because of this reason one can replace the function $\psi(x)$ from the uniqueness condition by such a function $\bar{\psi}(x)$. A similar remark holds for the function $\phi(x)$.

Now we divide the curves $y = \psi(x)$ and $y = \phi(x)$ in to a finite number of small pieces, and replace each piece by its chord so that one gets two polygons passing through the point (x_0, y_0). Thus, we can reformulate Theorem 1.21.2 as follows :

Theorem 1.21.3. The initial value problem (1.1.1) has at most one solution in the interval $[x_0, x_0 + a]$ if and only if for every $\varepsilon > 0$ there exists a pair of polygons $y = \bar{\phi}(x)$ and $y = \bar{\psi}(x)$ such that

$$\begin{aligned} & 0 < \bar{\psi}(x) - \bar{\phi}(x) < \varepsilon, && x \in (x_0, x_0 + a] \\ & \bar{\psi}'_+(x) - f(x, \bar{\psi}(x)) > 0, && x \in [x_0, x_0 + a] \\ & \bar{\phi}'_+(x) - f(x, \bar{\phi}(x)) < 0, && x \in [x_0, x_0 + a] \\ & \bar{\psi}(x_0) = \bar{\phi}(x_0) = y_0 \end{aligned} \tag{1.21.13}$$

and these polygons consist of a finite number of chords. The number of these chords depends on ε. ∎

The functions $\bar{\phi}(x)$ and $\bar{\psi}(x)$ are continuous in the interval $x_0 \leq x \leq x_0 + a$ whereas their derivatives make finite jumps at the vertex of these polygons $y = \bar{\phi}(x)$ and $y = \bar{\psi}(x)$. If the vertex of $y = \bar{\psi}(x)$ are $E_1, E_2, \cdots$ then from (1.21.12) it follows that

$$\bar{\psi}'_+ - f(x, \bar{\psi}) > \mu, \quad x_0 \leq x \leq x_0 + a \tag{1.21.14}$$

where μ is a small positive number.

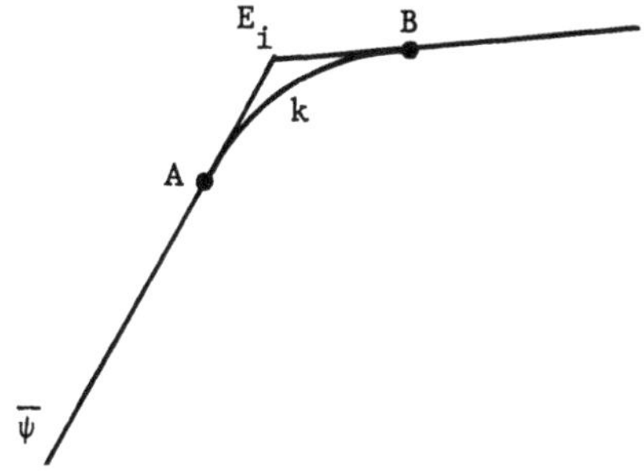

For the vertex E_i, we choose two points A and B on the polygon $y = \bar{\psi}(x)$ having the same distance from E_i. We connect these points by a circular arc

$y = k(x)$ which meets the line AE_i at A and E_iB at B. In the interval $x_A \leq x \leq x_B$ the difference $|f(x,k) - f(x,\bar{\psi})|$ will be small provided A and B are close enough. Now we set

$$|f(x,k) - f(x,\bar{\psi})| < \varepsilon_i, \; x_A \leq x \leq x_B$$

where $\varepsilon_i > 0$ and less than μ. Then, from (1.21.14) we get

$$\bar{\psi}'_+ - f(x,k) > \mu + \theta\varepsilon_i > 0, \; -1 < \theta < 1. \tag{1.21.15}$$

Let t_1 and t_2 be the direction coefficients of the lines AE_i and E_iB and let t be the minimum of these two, then obviously $k'_+ = \frac{dk}{dx}$ has the value between t_1 and t_2 at each point of the interval $x_A < x < x_B$. Hence, $k'_+ - t > 0$ for $x_A < x < x_B$ and 0 at x_A and x_B. Thus, from (1.21.15), we have

$$t - f(x,k) > \mu + \theta\varepsilon_i$$

and hence

$$k'_+ - f(x,k) > \mu + \theta\varepsilon_i > 0.$$

Hence, we can replace the dotted line AB by the circular arc $y = k(x)$ so that the obtained curve $y = \xi_i(x)$ has all the properties of an upper - function. Following the same at every E_i leads to an upper - function $\xi(x)$ which in the whole interval $x_0 \leq x \leq x_0 + a$ has continuous derivative and is close to the curve $y = \bar{\psi}(x)$. Similarly, one can construct a lower - function which is close to $y = \bar{\phi}(x)$ and is continuously differentiable.

Since the derivative $\xi'(x)$ is continuous in $[x_0, x_0 + a]$, for a given small $\bar{\varepsilon} > 0$ we can construct a polynomial $Q(x)$ such that

$$|\xi'(x) - Q(x)| < \frac{\bar{\varepsilon}}{a}.$$

Thus, by an integration it follows that

$$|\xi(x) - P(x)| < \bar{\varepsilon}, \; x_0 \leq x \leq x_0 + a$$

where

$$P(x) = \int_{x_0}^{x} Q(t)dt + \xi(x_0).$$

Since $P(x)$ is also a polynomial and $P'(x) = Q(x)$ it is clear that $P(x_0) = \xi(x_0) = y_0$. Therefore, the uniqueness result can be restated as follows :

Theorem 1.21.4. The initial value problem (1.1.1) has at most one solution in the interval $[x_0, x_0 + a]$ if and only if for every $\varepsilon > 0$ there exists a pair of polynomials $P_1(x)$ and $P_2(x)$ such that

$$(1.21.16)\quad \begin{aligned} &0 < P_1(x) - P_2(x) < \varepsilon, && x \in (x_0, x_0 + a] \\ &P'_{1+}(x) - f(x, P_1(x)) > 0, && x \in [x_0, x_0 + a] \\ &P'_{2+}(x) - f(x, P_2(x)) < 0, && x \in [x_0, x_0 + a] \\ &P_1(x_0) = P_2(x_0) = y_0. && \blacksquare \end{aligned}$$

Remark 1.21.1. We note that the fourth condition in (1.21.13) as well as (1.21.16) is not necessary. In fact these conditions can be replaced by

$$(1.21.17)\quad \begin{aligned} &\bar{\psi}(x_0) \geq y_0 \geq \bar{\phi}(x_0) \quad \text{and} \\ &P_1(x_0) \geq y_0 \geq P_2(x_0), \end{aligned}$$

respectively.

To show this, let P_1Q_1 and P_2Q_2 be the curves $y = P_1(x)$ and $y = P_2(x)$, and let P be the point (x_0, y_0). Let M be a positive number such that $|f(x,y)| < M$, $(x,y) \in \bar{S}_+$. Take two lines PR_1 and PR_2 through P which correspond to upper - and lower - functions and whose direction coefficients are M and $-M$. Then,

$$y' - f(x,y) = M - f(x,y) > 0 \quad \text{on} \quad PR_1,$$

and

$$y' - f(x,y) = -M - f(x,y) < 0 \quad \text{on} \quad PR_2$$

holds. Let R_1 and R_2 be the points where PR_1 and PR_2 intersect P_1Q_1 and P_2Q_2, respectively. Then, the curves PR_1Q_1 and PR_2Q_2 give the required upper - and lower - functions. ∎

Example 1.21.1. If the function $f(x,y)$ is monotonically decreasing in y, then the initial value problem (1.1.1) has at most one solution in $[x_0, x_0 + a]$. For this, let $y(x)$ be a solution of (1.1.1), then consider the curves $y = y(x) \pm \nu$ $(\nu > 0)$, $x_0 \le x \le x_0 + a$. Since by assumption we have

$$f(x, y + \nu) < f(x,y)$$

$$f(x, y - \nu) > f(x,y), \quad x_0 \le x \le x_0 + a$$

it follows that

$$(y+\nu)'_+ = y'_+ = f(x,y) > f(x, y+\nu)$$

$$(y-\nu)'_+ = y'_+ = f(x,y) < f(x, y-\nu), \quad x_0 \le x \le x_0 + a.$$

If $\nu > 0$ is sufficiently small then the curves $y = y(x) \pm \nu$ are very close, moreover

$$y(x_0) + \nu > y(x_0) = y_0 > y(x_0) - \nu.$$

Therefore, in view of Theorem 1.21.2 and the Remark 1.21.1, the conclusion follows.

Example 1.21.2. If the function $f(x,y)$ satisfies the Lipschitz condition (1.2.1) in $\bar{S}_+$ then the initial value problem (1.1.1) has at most one solution in $[x_0, x_0 + a]$. For this, once again let $y(x)$ be a solution of (1.1.1). Since $y(x)$ is continuously differentiable in $[x_0, x_0 + a]$, for every $\nu_1 > 0$ there exists a polynomial $P(x)$ such that

$$|y(x) - P(x)| < \nu_1, \quad \text{and} \quad |y'(x) - P'(x)| < \nu_1. \tag{1.21.18}$$

By the continuity of $f(x,y)$ we can set

$$|f(x,y) - f(x,P)| < \nu_2, \tag{1.21.19}$$

where ν_2 tends to zero if ν_1 does. From (1.21.18) and (1.21.19) and $y' = f(x, y)$ it follows that

$$(1.21.20) \qquad |P'(x) - f(x, P(x))| < \nu_1 + \nu_2, \quad x_0 \le x \le x_0 + a.$$

Now consider the polynomial $P_1(x) = P(x) + \alpha(x - x_0)$, $\alpha > 0$. For sufficiently small α the curve $y = P_1(x)$ is in $\bar{S}_+$, and hence by (1.2.1), we find that

$$(1.21.21) \quad |f(x, P) - f(x, P_1)| < L|P - P_1| < L\alpha|x - x_0|, \; x_0 \le x \le x_0 + a.$$

From (1.21.20) and (1.21.21) it follows that

$$(1.21.22) \qquad \begin{aligned} P'(x) &= f(x, P) \pm \theta_1(\nu_1 + \nu_2) \\ f(x, P_1) &= f(x, P) \pm \theta_2 L\alpha|x - x_0|, \; |\theta_1|, \; \theta_2| < 1. \end{aligned}$$

Now we choose an interval (x_0, x_1) so small that $L|x_1 - x_0| < 1$ and ν_1 and ν_2 small enough so that

$$\nu_1 + \nu_2 + L\alpha|x_1 - x_0| < \alpha.$$

Then, from (1.21.22) we get

$$(1.21.23) \qquad |P'(x) - f(x, P_1(x))| < \alpha, \quad x_0 \le x \le x_0 + a.$$

However, since $P_1'(x) = P'(x) + \alpha$, we have $P_1'(x) > f(x, P_1(x))$. Similarly, one can find a polynomial $P_2(x) = P(x) - \beta(x - x_0)$, $\beta > 0$ such that $P_2'(x) < f(x, P_2(x))$. Finally, we note that for a given $\varepsilon > 0$ the difference $P_1(x) - P_2(x) = (\alpha + \beta)(x - x_0)$ in the interval (x_0, x_1) can be made less than ε, only if α and β are small. The conclusion now follows from Theorem 1.21.4 and the Remark 1.21.1.

Finally, we remark that on following as in Example 1.21.2 the conclusion of Theorem 1.4.4 can also be deduced from Theorem 1.21.4 and the Remark 1.21.1.

1.22 KITAGAWA'S UNIQUENESS THEOREM

Let the function $f(x,y)$ be continuous for $x_0 \leq x \leq x_0 + a$, $\mu(x) \leq y \leq \nu(x)$, where $\mu(x)$ and $\nu(x)$ are two continuous functions in the interval $[x_0, x_0 + a]$ and satisfy the following conditions :

1°. $\mu(x_0) = y_0 = \nu(x_0)$,

2°. $\mu'_+(x) \leq f(x, \mu(x))$; and

3°. $\nu'_+(x) \geq f(x, \nu(x))$.

Then, it is well known that for the initial value problem (1.1.1) there exist maximal and minimal solutions $\rho(x)$ and $r(x)$ and sequences $\{\phi_n(x)\}$ and $\{\psi_n(x)\}$ of lower - and upper - functions which converge uniformly to $\rho(x)$ and $r(x)$, respectively. Here we shall combine this result with Yosie's Uniqueness Theorem 1.22.2 to deduce a property of the maximal and minimal solutions when the solutions of (1.1.1) are not unique in the interval $[x_0, x_0 + a]$. Further, as an immediate consequence of this property, we shall obtain two sets of sufficient conditions for the uniqueness of the solutions of (1.1.1). For this, throughout it is necessary to assume that

$$D_+\mu(x_0) < f(x_0, y_0) < D_+\nu(x_0). \tag{1.22.1}$$

Let $n(u, v, x_1)$ be an auxiliary neighborhood of the minimal solution $\rho(x)$, defined by

$$\begin{cases} x_0 \leq x \leq x_1 \leq x_0 + a \\ u(x) < y < v(x), \end{cases}$$

where $u(x)$, $v(x)$ are two continuous functions in $[x_0, x_1]$ such that

1. $D_+\mu(x_0) \leq D_+u(x_0) < f(x_0, y_0) < D_+v(x_0) \leq D_+\nu(x_0)$,
2. $\mu(x) \leq u(x) < \rho(x) < v(x) \leq \nu(x)$, $x \in (x_0, x_1]$; and
3. $u(x_0) = v(x_0) = y_0$.

Similarly, let $N(U, V, X_1)$, be an auxiliary neighborhood of the maximal solution $r(x)$, defined by

$$\begin{cases} x_0 \leq x \leq X_1 \leq x_0 + a \\ U(x) < y < V(x), \end{cases}$$

where $U(x)$, $V(x)$ are two continuous functions in $[x_0, X_1]$ such that

1. $D_+\mu(x_0) \leq D_+U(x_0) < f(x_0, y_0) < D_+V(x_0) \leq D_+\nu(x_0)$,
2. $\mu(x) \leq U(x) < r(x) < V(x) \leq \nu(x)$, $x \in (x_0, X_1]$; and
3. $U(x_0) = V(x_0) = y_0$.

Theorem 1.22.1. If the initial value problem (1.1.1) has more than one solutions in the interval $[x_0, x_0 + a]$, then the following hold

(i) for any $n(u, v, x_1)$ there exists at least one point ξ such that

$$\begin{vmatrix} f(\xi, y) & y & 1 \\ f(\xi, y_\rho) & y_\rho & 1 \\ f(\xi, \bar{y}) & \bar{y} & 1 \end{vmatrix} < 0$$

where $u(\xi) < y < y_\rho = \rho(\xi) < \bar{y} < v(\xi)$; and

(ii) for any $N(U, V, X_1)$ there exists at least one point Ξ such that

$$\begin{vmatrix} f(\Xi, Y) & Y & 1 \\ f(\Xi, Y_r) & Y_r & 1 \\ f(\Xi, \bar{Y}) & \bar{Y} & 1 \end{vmatrix} > 0$$

where $U(\Xi) < Y < Y_r = r(\Xi) < \bar{Y} < V(\Xi)$.

To prove this result we need the following :

Lemma 1.22.2. For any small $\varepsilon > 0$ and any given neighborhood $n(u, v, x_1)$ there exists a lower - function $\bar{\alpha}(x)$ such that, for

$$\begin{cases} \rho(x) - \bar{\alpha}(x) = d(x) \\ \min[\varepsilon,\ \rho(x) - u(x),\ v(x) - \rho(x)] = b(x) \end{cases}$$

it follows that

(i) $0 < d(x) < b(x)$, $x_0 < x \leq x_1$,

(ii) $d(x_0) = b(x_0) = 0$,

(iii) $D_+d(x_0) < D_+b(x_0)$; and

(iv) $u(x) < \bar{\alpha}(x) = \rho(x) - d(x) < \rho(x) < \rho(x) + d(x) < v(x)$, $x_0 < x \leq x_1$.

Proof. First we shall show that (iv) is a consequence of (i) - (iii). For this, from (i) we have $0 < d(x) = \rho(x) - \bar{\alpha}(x) < b(x) \leq \rho(x) - u(x)$, and hence $u(x) < \bar{\alpha}(x) = \rho(x) - d(x) < \rho(x) < \rho(x) + d(x) < \rho(x) + b(x) \leq \rho(x) + v(x) - \rho(x) = v(x)$.

Next, since the function $f(x,y)$ is continuous at the point (x_0, y_0), we can find positive numbers δ_1 and η such that

$$f(x,y) > f(x_0,y_0) - \eta > f(x_0,y_0) - D_+b(x_0)$$

in the domain $x_0 \leq x < x_0 + \delta_1,\ \rho(x) \leq y \leq r(x)$.

Now consider the function $L(x) = y_0 + (f(x_0,y_0) - \eta)(x - x_0)$, which has the following properties :

1. $L(x_0) = y_0$,
2. $L'(x) < f(x, L(x))$; and
3. $\rho(x) - b(x) < L(x)$ in $x_0 < x \leq x_0 + \delta$ $(\leq x_0 + \delta_1)$.

On the other hand from the existence of the sequence of lower - functions uniformly converging to the minimal solution $\rho(x)$ there must exist a lower - function $\bar{\phi}(x)$ such that

$$0 < \rho(x) - \bar{\phi}(x) < \min_{x_0+\delta \leq x \leq x_1} [b(x), \{f(x_0,y_0) - \eta\}\delta].$$

Now we define $\bar{\alpha}(x)$ as follows : If $\bar{\phi}(x) > L(x)$ in $(x_0, x_0 + \delta)$, then $\bar{\alpha}(x) = \bar{\phi}(x)$; and if there exists x_2 such that

$$\begin{cases} \bar{\phi}(x) > L(x),\ x_0 < x_2 \leq x_0 + \delta \\ \bar{\phi}(x_2) = L(x_2) \end{cases}$$

then

$$\bar{\alpha}(x) = \begin{cases} L(x),\ x_0 \leq x \leq x_2 \\ \bar{\phi}(x),\ x_2 \leq x \leq x_1. \end{cases}$$

This $\bar{\alpha}(x)$ is a required lower - function. ∎

Proof of Theorem 1.22.1. The proof of only (i) will be given, whereas the proof of (ii) is similar. For this, we shall show that if (i) is not true, i.e., if for every ξ in $(x_0, x_1]$ and every y and $\bar{y}$ in $(u(\xi), v(\xi))$

$$\begin{vmatrix} f(\xi,y) & y & 1 \\ f(\xi,y_\rho) & y_\rho & 1 \\ f(\xi,\bar{y}) & \bar{y} & 1 \end{vmatrix} \geq 0$$

then we can find, for any given positive number ε, a pair of lower - and upper - functions $\phi(x)$, $\psi(x)$ such that $0 < \psi(x) - \phi(x) < 2\varepsilon$ in $(x_0, x_1]$. But, then in view of Theorem 1.22.2 this will lead to the uniqueness of the solutions of (1.1.1).

To prove the existence of such a pair, we define the functions $p(x)$, $q(x)$, $\psi(x)$ and y and $\bar{y}$ in terms of the lower - function $\bar{\alpha}(x)(= \rho(x) - d(x))$ as follows :

1. $\psi(x) = \rho(x) + d(x)$ in $[x_0, x_1]$,

2. $f(x, \bar{\alpha}(x)) = \bar{\alpha}'_+(x) + p(x)$, where $p(x) > 0$ in $[x_0, x_1]$ since $\bar{\alpha}(x)$ is a lower - function,

3. $f(x, \psi(x)) = \psi'_+(x) - q(x)$; and

4. $y = \bar{\alpha}(\xi) = \rho(\xi) - d(\xi)$, $\bar{y} = \psi(\xi) = \rho(\xi) + d(\xi)$.

Now we will show that $\psi(x)$ is a required upper - function, i.e., it satisfies the following conditions :

(a) $0 < \psi(x) - \bar{\alpha}(x) < 2\varepsilon$ for $x_0 < x \leq x_1$, where ε is a small positive number; and

(b) $\psi'_+(x) - f(x, \psi(x)) = q(x) > 0$, for $x_0 \leq x \leq x_1$.

Since, in $(x_0, x_1]$

$$0 < \psi(x) - \bar{\alpha}(x) = (\rho(x) + d(x)) - (\rho(x) - d(x)) = 2d(x) < 2\varepsilon$$

(a) follows. To show (b), we note that

$$\begin{vmatrix} f(\xi, y) & y & 1 \\ f(\xi, y_\rho) & y_\rho & 1 \\ f(\xi, \bar{y}) & \bar{y} & 1 \end{vmatrix} = \begin{vmatrix} f(\xi, \bar{\alpha}(\xi)) & \bar{\alpha}(\xi) & 1 \\ f(\xi, \rho(\xi)) & \rho(\xi) & 1 \\ f(\xi, \psi(\xi)) & \psi(\xi) & 1 \end{vmatrix}$$

$$= \begin{vmatrix} \bar{\alpha}'_+(\xi) + p(\xi) & \rho(\xi) - d(\xi) & 1 \\ \rho'(\xi) & \rho(\xi) & 1 \\ \psi'_+(\xi) - q(\xi) & \rho(\xi) + d(\xi) & 1 \end{vmatrix} = - \begin{vmatrix} -d'_+(\xi) + p(\xi) & -d(\xi) \\ d'_+(\xi) - q(\xi) & d(\xi) \end{vmatrix}$$

$$= d(\xi)(q(\xi) - p(\xi)) \geq 0$$

by the hypothesis. As $d(\xi) > 0$ in $(x_0, x_1]$, it follows that $q(\xi) \geq p(\xi) > 0$ for $x_0 \leq \xi \leq x_1$. ∎

Theorem 1.22.3 (Kitagawa's Uniqueness Theorem). Let in addition to the above hypotheses one of the following conditions hold

(1) $f(x, y)$ is concave function of y; or

(2) $f(x, y)$ is convex function of y.

Then, the initial value problem (1.1.1) has at most one solution in $[x_0, x_0 + a]$.

Proof. The conclusion (i) of Theorem 1.22.1 for $y < y_\rho < \bar{y}$ can be written as

$$f(\xi, y)(y_\rho - \bar{y}) + f(\xi, y_\rho)(\bar{y} - y) + f(\xi, \bar{y})(y - y_\rho) < 0.$$

If $f(x, y)$ is concave function of y, then we have

$$f(\xi, y_\rho) = f\left(\xi, y\frac{\bar{y} - y_\rho}{\bar{y} - y} + \bar{y}\frac{y_\rho - y}{\bar{y} - y}\right)$$

$$\geq f(\xi, y)\frac{\bar{y} - y_\rho}{\bar{y} - y} + f(\xi, \bar{y})\frac{y_\rho - y}{\bar{y} - y},$$

which is the same as

$$f(\xi, y)(y_\rho - \bar{y}) + f(\xi, y_\rho)(\bar{y} - y) + f(\xi, \bar{y})(y - y_\rho) \geq 0.$$

If $f(x, y)$ is convex function of y, then a similar contradiction follows from the conclusion (ii) of Theorem 1.22.1. ∎

Example 1.22.1. For the initial value problem

$$(1.22.2) \qquad \begin{aligned} y' &= f(x, y) = \frac{4x^3 y}{x^4 + y^2} \\ y(0) &= 0 \end{aligned}$$

it is easily seen that $\rho(x) = -x^2$ and $r(x) = x^2$ are the minimal and the maximal solutions for $x > 0$. As a function of y it is clear that $\dfrac{4x^3 y}{x^4 + y^2}$ is concave for $y < -\sqrt{3}x^2$ or $0 < y < \sqrt{3}x^2$, and convex for $-\sqrt{3}x^2 < y < 0$ or $y > \sqrt{3}x^2$.

If we take the functions $u(x), v(x)$ such that

$$\begin{cases} u(x) < \rho(x) = -x^2 < v(x), \ x \in (0, x_1] \\ u(0) = \rho(0) = v(0) = 0 \\ D_+u(0) < \rho'(0) = 0 < D_+v(0) \end{cases}$$

then there exists ξ such that $u(\xi) < 0$, and $v(\xi) > 0$. Since $f(\xi, y)$ is not always concave with respect to y in $(u(\xi), v(\xi))$, the conclusion (i) of Theorem 1.22.1 is justified. The conclusion (ii) of Theorem 1.22.1 can be similarly verified. ∎

Example 1.22.2. Consider the initial value problem

$$\begin{aligned} & y' = f(x,y) = b(1 + |y - ax|^{1/3}), \ b, a > 0 \\ (1.22.3) \qquad & \\ & y(0) = 0. \end{aligned}$$

If $b > a$, then we can take $\mu(x) = ax$ and $\nu(x) = (a + b)x$. For sufficiently small x_1, we have

$$\mu'(x) = a < b = f(x, \mu(x))$$

$$\nu'(x) = a + b > b(1 + |bx|^{1/3}) = f(x, \nu(x)), \quad 0 \le x \le x_1$$

and this $f(x, y)$ is a concave function of y, when b and $y - ax$ are positive. In this case the initial value problem (1.22.3) has a unique solution follows from Theorem 1.22.3.

If $b < a$, then we choose η such that $a - \eta < b$ and $\mu(x) = (a - \eta)x$ and $\nu(x) = ax$. For such a choice, we have

$$\mu'(x) = a - \eta < b(1 + |-\eta x|^{1/3}) = f(x, \mu(x))$$

$$\nu'(x) = a > b = f(x, \nu(x))$$

and this $f(x, y)$ is a convex function of y, when b and $ax - y$ are positive. In this case also the initial value problem (1.22.3) has a unique solution follows from Theorem 1.22.3.

If $b = a$, then Theorem 1.22.3 is not applicable. However, in this case the initial value problem (1.22.3) has at least two solutions $y(x) = ax$ and $\bar{y}(x) = ax + (\frac{2}{3}ax)^{3/2}$. ∎

Example 1.22.3. The condition (1.22.1) is essential for the Theorem 1.22.3. To show this we consider the initial value problem

$$y' = f(x,y) = (1-(y-1)^2)^{1/2} \tag{1.22.4}$$
$$y(0) = 0$$

Clearly, this $f(x,y)$ is continuous and is a convex function of y in the domain D defined by $x \ge 0$, $|y-1| \le 1$. For this problem, let $\mu(x) = 0$ and $\nu(x) = \begin{cases} x, 0 \le x \le 2 \\ 2, x \ge 2 \end{cases}$ so that

$$D_+\mu(x) = f(x,\mu(x)) = 0.$$

For the initial value problem (1.22.4), $\rho(x) = 0$ and $r(x) = \begin{cases} 1-\cos x, 0 \le x \le \pi \\ 2, x \ge \pi \end{cases}$ are the minimal and the maximal solutions. ∎

1.23 TAMARKINE'S NONUNIQUENESS THEOREM

In Section 1.2 we have seen that if the function $f(x,y)$ does not satisfy the Lipschitz condition (1.2.1), then the initial value problem (1.1.1) may have several solutions. In the year 1922, Tamarkine [156] pointed out that the same situation prevails if in Osgood's condition (1.4.3) the integral $\int_\varepsilon \frac{dz}{g(z)}$ converges as $\varepsilon \to 0$. Specifically, his result is

Theorem 1.23.1 (Tamarkine's Nonuniqueness Theorem). Let $f(x,y)$ be continuous in $\bar{S}$ with $(x_0,y_0) = (0,0)$, and for all $(x,y) \in \bar{S}$

$$|f(x,y) - f(x,y_0(x))| \ge g(|y - y_0(x)|),$$

where $y_0(x)$ is a solution of (1.9.4); $g(z)$ is an increasing continuous function for $z \ge 0$, $g(0) = 0$ and the integral $\int_\varepsilon \frac{dz}{g(z)}$ converges as $\varepsilon \to 0$. Then, the initial value problem (1.9.4) has at least two solutions in $|x| \le a$.

Proof. The proof is parallel to a more general result which we shall prove in the next section. ∎

Example 1.23.1. For the initial value problem

$$y' = y^{1/2}, \; y(0) = 0 \tag{1.23.1}$$

$y_0(x) \equiv 0$ is a solution. Since the function $f(x,y) = y^{1/2}$ satisfies all the conditions of Theorem 1.23.1 with $g(z) = z^{1/2}$, and the integral $\int_\epsilon z^{-1/2}dz$ converges, (1.23.1) has at least two solutions. ∎

1.24 LAKSHMIKANTHAM'S NONUNIQUENESS THEOREM

If certain conditions of Kamke's uniqueness theorem are violated, then the initial value problem (1.1.1) has more than one solutions. Indeed, this is the content of the following result in which we shall assume that $(x_0, y_0) = (0,0)$.

Theorem 1.24.1 (Lakshmikantham's Nonuniqueness Theorem). Assume that

(i) the function $g(x,z)$ is continuous in $0 < x \le a$, $0 \le z \le 2b$, $g(x,0) \equiv 0$, and $g(x,z) > 0$ for $z > 0$,

(ii) for each x_1, $0 < x_1 < a$, $z(x) \not\equiv 0$ is a differentiable function in $0 < x < x_1$, and continuous in $0 \le x \le x_1$ for which $z'_+(0)$ exists; and

$$\begin{aligned} z'(x) &= g(x, z(x)), \; 0 < x < x_1 \\ z(0) &= z'_+(0) = 0, \end{aligned}$$

(iii) the function $f(x,y)$ is continuous in $\bar{S}_+$ with $(x_0, y_0) = (0,0)$ and for all $(x,y), (x,\bar{y}) \in \bar{S}_+$, $x \ne 0$

$$|f(x,y) - f(x,\bar{y})| \ge g(x, |y - \bar{y}|). \tag{1.24.1}$$

Then, the initial value problem (1.9.4) has at least two solutions in $[0,a]$.

Proof. Let us first assume that $f(x,0) \equiv 0$, so that, putting $\bar{y} = 0$ in (1.24.1) leads to the inequality $|f(x,y)| \ge g(x,|y|)$. Since $f(x,y)$ is continuous and

$g(x, z) > 0$ for $z > 0$, it follows that either $f(x, y) < 0$ or $f(x, y) > 0$ for $y \neq 0$. This implies that

$$f(x,y) \geq g(x, |y|), \quad \text{or} \tag{1.24.2}$$

$$f(x,y) \leq -g(x, |y|). \tag{1.24.3}$$

By hypothesis, there exists a σ, $0 < \sigma < a$ such that $z(\sigma) > 0$. Let $\rho(x)$ be the minimal solution of $y' = f(x, y), y(\sigma) = z(\sigma)$. Then, using an argument similar to that in the proof of Lemma 1.15.2, and the inequality (1.24.2), it can be shown that $\rho(x) \leq z(x)$ to left of σ as far as $\rho(x)$ exists. Moreover, $\rho(x)$ can be continued up to $x = 0$ and $0 \leq \rho(x) \leq z(x)$, $0 \leq x \leq \sigma$. Since $z(0) = z'_+(0) = 0$, we obtain from this relation that $\rho(0) = \rho'_+(0) = 0$. This proves that the initial value problem (1.9.4) has a solution $\rho(x)$, not identically zero. On the other hand, since $f(x, 0) \equiv 0$, (1.9.4) admits the identically zero solution. Hence, we have two different solutions for the problem (1.9.4). Corresponding to the case (1.24.3), we can employ a similar reasoning to arrive at the same conclusion.

We shall now remove the restriction that $f(x, 0) \equiv 0$. Let $y_0(x)$ be a solution of (1.9.4), existing in $0 \leq x \leq a$. Using the transformation $w = y - y_0(x)$, we get

$$w' = y' - y_0'(x) = f(x, y) - f(x, y_0(x)) = f(x, w + y_0(x)) - f(x, y_0(x)) \tag{1.24.4}$$

$$= F(x, w), \quad \text{say}.$$

Evidently, $F(x, 0) \equiv 0$. It follows, therefore that $w(x) \equiv 0$ is one solution of (1.24.4) passing through $(0, 0)$. But the previous considerations show that (1.24.4) has a solution $w(x)$ which is different from the identically zero solution. This implies that $y(x) = w(x) + y_0(x)$ is not identically equal to $y_0(x)$. ∎

Corollary 1.24.2. Tamarkine's Nonuniqueness Theorem 1.23.1 is a particular case of Theorem 1.24.1.

Proof. It suffices to note that the initial value problem $z' = g(z)$, $z(0) = 0$ has a unique nontrivial solution $z(x) = G^{-1}(x)$, where $G(z) = \int_0^z \frac{dt}{g(t)}$. ∎

1.25 SAMIMI'S NONUNIQUENESS THEOREM

The following result is a generalization of Lakshmikantham's nonuniqueness theorem.

Theorem 1.25.1 (Samimi's Nonuniqueness Theorem). Let in Theorem 1.24.1 the condition (ii) be replaced by

(ii)′ for each x_1, $0 < x_1 < a$, $z(x) \not\equiv 0$ is a differentiable function in $0 < x < x_1$ and continuous in $0 \le x \le x_1$ for which $\lim_{x\to 0^+} \frac{z(x)}{B(x)} = 0$, where the function $B(x)$ is continuous in $[0, a]$ and $B(x) > 0$ for $x > 0$, and

$$\begin{aligned} z'(x) &= g(x, z(x)),\ 0 < x < x_1 \\ z(0) &= 0. \end{aligned}$$

Then, the initial value problem (1.9.4) has at least two solutions $y(x)$ and $\bar{y}(x)$ in $[0, a]$ such that $\lim_{x\to 0^+} \frac{|y(x)-\bar{y}(x)|}{B(x)} = 0$.

Proof. Imitating the proof of Theorem 1.24.1 we arrive at $0 \le \rho(x) \le z(x)$, $0 \le x \le \sigma$ which in view of $\rho(0) = 0$ implies that $z(0) = 0$. Thus, from the condition (ii)′ it follows that

$$0 \le \lim_{x\to 0^+} \frac{\rho(x)}{B(x)} \le \lim_{x\to 0^+} \frac{z(x)}{B(x)} = 0$$

and hence, we have $\lim_{x\to 0^+} \frac{\rho(x)}{B(x)} = 0$. This proves that the initial value problem (1.9.4) has a solution $\rho(x)$, not identically zero such that $\lim_{x\to 0^+} \frac{\rho(x)}{B(x)} = 0$. Since $f(x, 0) \equiv 0$, (1.9.4) admits the identically zero solution $y_0(x)$ which obviously satisfies $\lim_{x\to 0^+} \frac{y_0(x)}{B(x)} = 0$. Hence, $\lim_{x\to 0^+} \frac{|y_0(x)-\rho(x)|}{B(x)} = 0$. The restriction $f(x, 0) \equiv 0$ can easily be removed as in Theorem 1.24.1. ∎

CHAPTER 2

FIRST ORDER DIFFERENTIAL EQUATIONS (CONTD.)

2.1 INTRODUCTION

Once again we shall consider the initial value problem (1.1.1), but instead of assuming that the function $f(x,y)$ is continuous we shall assume that it is only defined in the domain $D = J \times T \subseteq R^2$, where J is one of the intervals $[x_0, x_0+a], [x_0, x_0+a), (x_0, x_0+a], (x_0, x_0+a)$, and $y_0 \in T \subseteq R$. In such a case by a **classical solution** of (1.1.1) in the interval $\{x_0\} \bigcup J$, we mean a function $y(x)$ satisfying :

(i) $y(x_0) = y_0$,

(ii) for all $x \in J$, the points $(x, y(x)) \in D$,

(iii) $y(x)$ is continuous in $\{x_0\} \bigcup J$,

(iv) $y'(x)$ exists in $J^o = (x_0, x_0 + a)$; and

(v) $y'(x) = f(x, y(x))$.

It is clear that if $f(x,y)$ is continuous in D then a classical solution of (1.1.1) is the same as an ordinary solution.

We shall also consider the case when the function $f(x,y)$ satisfies the Caratheodory conditions, i.e., it is defined in D; measurable in x for each fixed y; continuous in y for each fixed x, and there exists an (Lebesgue) integrable function $M(x)$ in the interval $\bar{J} = [x_0, x_0 + a]$ such that

$$|f(x,y)| \leq M(x); \ (x,y) \in D.$$

In such a case (or otherwise) by a **extended solution** of (1.1.1) in the interval $\bar{J}$, we mean a function $y(x)$ satisfying the above conditions (i) and (ii), and

(vi) $y(x)$ is absolutely continuous in $\bar{J}$; and

(vii) $y'(x) = f(x, y(x))$ almost everywhere in $\bar{J}$, i.e., for all $x \in \bar{J}$ except on a set of measure zero.

Obviously, all the uniqueness theorems we shall prove here will have wider applications than those results discussed in Chapter 1.

2.2 LIPSCHITZ UNIQUENESS THEOREM

We begin with the observation that for classical solutions of (1.1.1) the one - sided Lipschitz Uniqueness Theorem 1.2.5 can be stated as follows :

Theorem 2.2.1. Let $f(x,y)$ be defined and satisfy the one - sided Lipschitz condition (1.2.7) in

$$S_+ : x_0 < x < x_0 + a, \ |y - y_0| < b.$$

Then, the initial value problem (1.1.1) has at most one classical solution in $[x_0, x_0 + a)$.

Proof. From the remark following the proof of Theorem 1.2.5, it suffices to note that in the second proof of Theorem 1.2.4 the main tool, namely, the mean value theorem is equally valid if the continuous differentiability of the functions is replaced by the existence of their derivatives. ∎

Let

$$c(x) = \begin{cases} 0, & x \neq a/2, \ 0 < x < a \\ 1, & x = a/2. \end{cases}$$

From Theorem 2.2.1 it is clear that $y(x) = 0, \ x \in [0, a)$ is the only classical solution of the initial value problem $y' = c(x)y, \ y(0) = 0$. Evidently, for this problem Theorem 1.2.5 is not applicable. Similarly, for the initial value problem $y' = c(x), \ y(0) = 0$ Theorem 2.2.1 is applicable, while Theorem 1.2.5 is powerless. It is easy to see that this problem has no classical solution.

Our next result is for the initial value problem (1.2.10) and it accommodates both the Theorems 1.2.8 and 1.4.3.

Theorem 2.2.2. Let the function $g(y)$ be continuous in the interval $|y - y_0| \leq b$, and $g(y_0) = 0$. Then, the initial value problem (1.2.10) has a unique extended solution $y(x) \equiv y_0$ unless there exists an ε, $y_0 < \varepsilon \leq y_0 + b$ such that either $g(y) \geq 0$ for $y_0 \leq y \leq \varepsilon$ and $1/g(y)$ is (Lebesgue) integrable in $[y_0, \varepsilon]$ or $g(y) \leq 0$ for $-\varepsilon \leq y \leq y_0$ and $1/g(y)$ is (Lebesgue) integrable in $[-\varepsilon, y_0]$.

Proof. First we shall prove the following elementary facts :

(i) If $g(y)$ is continuous for $y_0 \leq y \leq y_0 + b$ and $y(x)$ is a extended solution of (1.2.10) in $[x_0, x_0 + a]$, then $y(x)$ is monotone in $[x_0, x_0 + a]$.

For this, if the extended solution $y(x)$ is not monotone, then there will be two numbers $x_1 \neq x_2$ in $[x_0, x_0 + a]$ such that $y(x_1) = y(x_2)$ and $y'(x_1) > 0, \ y'(x_2) < 0$. But, this contradicts (1.2.10), since (1.2.10) implies that y' is a single - valued function of y.

(ii) If $g(y)$ is continuous for $y_0 \leq y \leq y_0 + b$ and if $g(y)$ changes sign in every interval $(y_0, y_0 + \varepsilon)$, $\varepsilon > 0$ then $y(x) \equiv y_0$ is the only extended solution of (1.2.10).

The continuity of $g(y)$ implies that $g(y_0) = 0$ and, therefore $y(x) \equiv y_0$ is a extended solution of (1.2.10). Suppose $y(x) \not\equiv y_0$ is a extended solution of (1.2.10) in $[x_0, x_0 + a]$. Then, for some x_1 where $x_0 < x_1 \leq x_0 + a$, the function value $y(x_1) > y_0$. But, $g(y)$ changes sign in the interval $(y_0, y(x_1))$, which means by (1.2.10) that $y(x)$ is not monotone on $x_0 \leq x \leq x_1$. This contradicts (i) and so (ii) follows.

Now suppose that $y(x) \neq y_0$ is a extended solution of (1.2.10) in $[x_0, x_0+a]$. The statement (ii) implies that there exists a number ε, $y_0 < \varepsilon \leq y_0 + b$ such that $g(y)$ does not change sign in $(y_0, y_0 + \varepsilon)$. We claim that $g(y) \geq 0$ for $y_0 \leq y \leq \varepsilon$. Since, otherwise, $g(y) \leq 0$ for $y_0 \leq y \leq \varepsilon$, which by (1.2.10) and (i) means that $y(x)$ is nonincreasing in $[x_0, x_0 + a]$. Hence, $y(x) \not\equiv y_0$ implies that $y(x) < y_0$ for some value of x. But this contradicts the assumption that $g(y)$ is defined only for $y_0 \leq y \leq y_0 + b$.

Now we shall show that $\int_{y_0}^{\varepsilon} \frac{dz}{g(z)} < \infty$, where $\varepsilon = y(x_0 + a)$. Since $y(x) \not\equiv y_0$, this ε satisfies $y_0 < \varepsilon \leq y_0 + b$. Since a monotone function $y(x)$ maps the set of x - values on which the derivative $y'(x)$ vanishes on a y - set of measure zero, from (1.2.10), the zeros of $g(y)$, where $0 \leq y \leq \varepsilon$, form a y - set, say, F of measure zero. The continuity of $g(y)$ implies that F is a closed set. Let E denote the open set which is the complement of F in (y_0, ε). Let $E = \sum(y_n, \varepsilon_n)$ be the decomposition of E into disjoint open intervals and let the numbers a_n, b_n be defined by $y(a_n) = y_n$, $y(b_n) = \varepsilon_n$. It follows from the definition of the open set E, that $g(y) > 0$ for $y_n < y < \varepsilon_n$, and from Theorem 1.4.3 it follows that the integral

$$\int_{y_n}^{\varepsilon_n} \frac{dz}{g(z)}$$

converges and has the value $b_n - a_n$. Hence,

$$\int_E \frac{dz}{g(z)} = \sum(b_n - a_n).$$

Since the disjoint open intervals (a_n, b_n) are contained in $(y_0, y_0 + b)$ and since complement F of E in (y_0, ε) is a zero set, it follows that $\int_{y_0}^{\varepsilon} \frac{dz}{g(z)} \leq a < \infty$. This completes the necessity part.

To show the sufficiency part, we note that the conditions in the theorem imply that the y - set on which $g(y)$ vanishes is a zero set in (y_0, ε). Hence,

$$x(y) = \int_{y_0}^{y} \frac{dz}{g(z)} \tag{2.2.1}$$

is a strictly increasing function of y and so possesses an inverse function $y(x)$, where $x_0 \le x \le c$ and $c = x(\varepsilon)$. Then, (2.2.1) and $g(y) \ge 0,\ y_0 \le y \le \varepsilon$ and the continuity of $g(y)$ imply that

$$y'(x) = 1/x'(y) = g(y(x)),\ x_0 \le x \le c. \quad \blacksquare$$

2.3 NAGUMO'S UNIQUENESS THEOREM

In view of Example 1.6.2 the one - sided Nagumo's Uniqueness Theorem 1.6.3 for the classical solutions takes the following form :

Theorem 2.3.1. Let $f(x,y)$ be defined in $\bar{S}_+$ and continuous in (x,y) at (x_0,y_0), i.e., $\lim_{(x,y)\to(x_0,y_0)} f(x,y) = f(x_0,y_0)$. Further, let for all (x,y), $(x,\bar{y})$ in $\bar{S}_+$ with $\bar{y} \ge y$ it satisfies one - sided Nagumo condition (1.6.8). Then, the initial value problem (1.1.1) has at most one classical solution in $[x_0, x_0 + a]$. $\blacksquare$

The initial value problem (1.6.4) has an infinite number of classical solutions $y(x) = cx$, where $0 \le c \le 1$. However, if the term "classical solution" is relaxed to mean that, in addition, $y(x)$ has a finite right hand derivative at $x = x_0$ which satisfies the differential equation, i.e., $y'_+(x_0) = f(x_0,y_0)$; then, in this modified sense of the word, there is only one classical solution of (1.6.4), namely $y(x) \equiv 0$. The Theorem 2.3.1 holds in this modified sense also, however, then the hypothesis concerning the continuity of $f(x,y)$ is now superfluous. Because, if $y(x)$ and $\bar{y}(x)$ are two solutions of (1.1.1) then from $y'_+(x_0) = \bar{y}'_+(x_0)$ it directly follows that $\lim_{x\to x_0} \frac{|y(x)-\bar{y}(x)|}{x-x_0} = 0$.

2.4 KRASNOSEL'SKII - KREIN'S UNIQUENESS THEOREM

One - sided analog of Theorem 1.7.1 is the following :

Theorem 2.4.1 (Krasnosel'skii - Krein's Uniqueness Theorem). Let $f(x,y)$ be defined in $\bar{S}_+$ and for all $(x,y),(x,\bar{y})$ in $\bar{S}_+$ with $\bar{y} \geq y$ it satisfies

$$(2.4.1) \qquad f(x,\bar{y}) - f(x,y) \leq k(x-x_0)^{-1}(\bar{y}-y), \ x \neq x_0, \ k > 0$$

and the Hölder condition

$$(2.4.2) \qquad f(x,\bar{y}) - f(x,y) \leq c(\bar{y}-y)^{\alpha}, \ c > 0, \ 0 < \alpha < 1, \ k(1-\alpha) < 1.$$

Then, the initial value problem (1.1.1) has at most one classical solution in $[x_0, x_0 + a]$.

Proof. We shall prove this result by employing the mean value theorem. If the conclusion is not true then the initial value problem (1.1.1) has at least two classical solutions $y(x)$ and $\bar{y}(x)$ in the interval $[x_0, x_0 + a]$. Since the functions $\min(y(x), \bar{y}(x))$ and $\max(y(x), \bar{y}(x))$ are also solutions of (1.1.1) without loss, we can assume that $y(x) \leq \bar{y}(x)$ for all $x \in [x_0, x_0 + a]$. For each number β, with $0 < \beta < 1/(1-\alpha)$, we define the auxiliary function g_β, on the interval $[x_0, x_0 + a]$ by

$$g_\beta(x) = \begin{cases} \dfrac{\bar{y}(x) - y(x)}{(x-x_0)^\beta}, & x > x_0 \\ 0 & x = x_0. \end{cases}$$

We shall first show by the mathematical induction that the function $g_\beta(x)$ is continuous at $x = x_0$, for the particular sequence of values $\beta_m = \sum_{i=0}^{m} \alpha^i$, where $m = 0, 1, 2, \cdots$; this will imply the desired continuity for $0 < \beta < 1/(1-\alpha)$. For $x > x_0$, by the mean value theorem, we have

$$\frac{\bar{y}(x) - y(x)}{(x-x_0)} = \bar{y}'(\bar{x}) - y'(\bar{x}) = f(\bar{x}, \bar{y}(\bar{x})) - f(\bar{x}, y(\bar{x})),$$

for some $\bar{x}$, with $x_0 < \bar{x} < x$. Now, by (2.4.2) we have that

$$f(\bar{x}, \bar{y}(\bar{x})) - f(\bar{x}, y(\bar{x})) \leq c(\bar{y}(\bar{x}) - y(\bar{x}))^{\alpha}.$$

Since $\bar{y}(x_0) = y(x_0)$, this completes the proof of the continuity of the function $g_\beta(x)$, for $\beta = \beta_0 = 1$. Suppose now that the desired continuity has been proved for $g_\beta(x)$, where $\beta = \beta_m = \sum_{i=0}^{m} \alpha^i$, and m is a nonnegative integer. Consider the function $g_\beta(x)$, where $\beta = \sum_{i=0}^{m+1} \alpha^i$. For $x > x_0$, we have, by the mean value theorem, that there is a number $\bar{x}$, with $x_0 < \bar{x} < x$, such that

$$\begin{aligned} g_\beta(x) &= \frac{1}{(x-x_0)} \times \frac{\bar{y}(x)-y(x)}{(x-x_0)^{\beta-1}} \\ &= \frac{\bar{y}'(\bar{x})-y'(\bar{x})}{(x-x_0)^{\beta-1}} \\ &= \frac{f(\bar{x},\bar{y}(\bar{x}))-f(\bar{x},y(\bar{x}))}{(x-x_0)^{\beta-1}} \\ &\le \frac{c(\bar{y}(\bar{x})-y(\bar{x}))^\alpha}{(x-x_0)^{\beta-1}} \\ &= c\left(\frac{\bar{x}-x_0}{x-x_0}\right)^{\beta-1} \times \frac{(\bar{y}(\bar{x})-y(\bar{x}))^\alpha}{(\bar{x}-x_0)^{\beta-1}} \\ &= c\left(\frac{\bar{x}-x_0}{x-x_0}\right)^{\beta-1} \times g^\alpha_{1+\alpha+\cdots+\alpha^m}(\bar{x}). \end{aligned}$$

The desired continuity of $g_\beta(x)$, at $x = x_0$, for $\beta = \sum_{i=0}^{m+1} \alpha^i$, now follows from the induction hypothesis on m. This implies that $g_\beta(x)$ is continuous at $x = x_0$, for all $0 < \beta < 1/(1-\alpha)$.

Since $y(x) \neq \bar{y}(x)$, and $\bar{y}(x) \geq y(x)$, on $[x_0, x_0+a]$, there exist two numbers x_1 and x_2 with the property that $x_0 \leq x_1 < x_2 \leq x_0 + a$, and $y(x_1) = \bar{y}(x_1)$, while $\bar{y}(x) > y(x)$ for $x_1 < x \leq x_2$. For any number β such that $0 < \beta < 1/(1-\alpha)$, we also have $g_\beta(x_1) = \frac{\bar{y}(x_1)-y(x_1)}{(x_1-x_0)^\beta} = 0$, while $g_\beta(x) = \frac{\bar{y}(x)-y(x)}{(x-x_0)^\beta} > 0$ for $x_1 < x \leq x_2$ (if $x_1 = x_0$, then $g_\beta(x_0)$ means that the simple zero).

Since $0 < k < 1/(1-\alpha)$, be the mean value theorem, applied to the function $g_k(x)$ in the interval $[x_1, x_2]$, there exists a number $\bar{x}$, with $x_1 < \bar{x} < x_2$, such that

$$\begin{aligned} 0 &< \frac{\bar{y}(x_2)-y(x_2)}{(x_2-x_0)^k} = \frac{\bar{y}(x_2)-y(x_2)}{(x_2-x_0)^k} - \frac{\bar{y}(x_1)-y(x_1)}{(x_1-x_0)^k} \\ &= (x_2-x_1)\frac{(\bar{x}-x_0)^k[\bar{y}'(\bar{x})-y'(\bar{x})]-k(\bar{x}-x_0)^{k-1}[\bar{y}(\bar{x})-y(\bar{x})]}{(\bar{x}-x_0)^{2k}} \\ &= (x_2-x_1)(\bar{x}-x_0)^{-k}\{[f(\bar{x},\bar{y}(\bar{x}))-f(\bar{x},y(\bar{x}))] \\ &\quad - k(\bar{x}-x_0)^{-1}[\bar{y}(\bar{x})-y(\bar{x})]\} \\ &\le 0, \end{aligned}$$

where the very last inequality follows from (2.4.1). ∎

2.5 BORŮVKA'S UNIQUENESS THEOREM

The following result is a generalization of Krasnosel'skii - Krein's Uniqueness Theorem 2.4.1.

Theorem 2.5.1 (Borůvka's Uniqueness Theorem). Let the function $f(x,y)$ be defined in $\bar{S}_+$, and $p(x,y,\bar{y})$, $q(x,y_1,z)$, $x_0 < x \le x_0+a$, $0 \le y \le \bar{y} \le 2b$, $0 \le y_1$, $z \le 2b$ be continuous functions such that

(i) $p(x,y,\bar{y}) > 0$ for $\bar{y} > y$, and $p(x,y,y) = 0$,

(ii) p_x, p_y and $p_{\bar{y}}$ exist and there hold $p_x(x,y,\bar{y}) + p_y(x,y,\bar{y})f(x,y) + p_{\bar{y}}(x,y,\bar{y})f(x,\bar{y}) \le q(x,y,p(x,y,\bar{y}))$,

(iii) for every two classical solutions $y(x)$ and $\bar{y}(x)$, $y(x) \le \bar{y}(x)$ of the initial value problem (1.1.1) in $[x_0, x_0+a]$, $\lim_{x\to x_0} p(x,y(x),\bar{y}(x)) = 0$; and

(iv) for every classical solution $y(x)$ of the initial value problem (1.1.1) in $[x_0, x_0+a]$ and for every $x_0 < x_1 < x_0+a$, the function $z(x) \equiv 0$ is the only solution of the differential equation $z' = q(x,y(x),z)$ in the interval $(x_0, x_1]$, for which $\lim_{x\to x_0} z(x) = 0$.

Then, the initial value problem (1.1.1) has at most one classical solution in $[x_0, x_0+a]$.

Proof. Suppose that $y(x)$ and $\bar{y}(x)$ are two classical solutions of the initial value problem (1.1.1) in the interval $[x_0, x_0+a]$. Let x_1, $x_0 < x_1 \le x_0+a$ be such that $y(x_1) < \bar{y}(x_1)$. As in the proof of Theorem 2.4.1 we can assume that $y(x) \le \bar{y}(x)$ for all x in $[x_0, x_0+a]$. Then, it follows that $p(x,y(x),\bar{y}(x)) \ge 0$, $p(x_1,y(x_1),\bar{y}(x_1)) > 0$. Since, $(p(x,y(x),\bar{y}(x)))' \le q(x,y(x),p(x,y(x),\bar{y}(x)))$ in view of Theorem 1.11.3 the initial value problem $z' = q(x,y(x),z)$, $z(x_1) = p(x_1,y(x_1),\bar{y}(x_1))$ has a solution $z(x)$ in $(x_0, x_1]$ such that $z(x) \le p(x,y(x),$

$\bar{y}(x))$. However, since $\lim_{x\to x_0} z(x) \leq \lim_{x\to x_0} p(x, y(x), \bar{y}(x)) = 0$, we have a contradiction to (iv). ∎

Corollary 2.5.2. Krasnosel'skii - Krein's Uniqueness Theorem 2.4.1 is a particular case of Theorem 2.5.1.

Proof. Let $p(x, y, \bar{y}) = \frac{\bar{y}-y}{(x-x_0)^k}$. The condition (i) then evidently holds. Furthermore, choose $q(x, y_1, z) = 0$. Then, the condition (ii) fulfills the inequality

$$-k\frac{\bar{y}-y}{(x-x_0)^{k+1}} + \frac{f(x,\bar{y})}{(x-x_0)^k} - \frac{f(x,y)}{(x-x_0)^k} \leq 0,$$

which is the same as (2.4.1). If the inequality (2.4.2) holds then for any two classical solutions $y(x)$ and $\bar{y}(x)$, $y(x) \leq \bar{y}(x)$ of the initial value problem (1.1.1) in $[x_0, x_0+a]$, it follows that $(\bar{y}(x)-y(x)) \leq ((1-\alpha)c(x-x_0))^{1/(1-\alpha)}$ and therefore $\lim_{x\to x_0} \frac{\bar{y}(x)-y(x)}{(x-x_0)^k} = 0$, and hence the condition (iii) is also satisfied. Finally, we note that the condition (iv) is obvious. ∎

More generally, we can choose $q(x, y_1, z) = Cz$. Hence, it follows that the Theorem 2.4.1 remains valid if the condition (2.4.1) is replaced by

$$f(x,\bar{y}) - f(x,y) \leq k(x-x_0)^{-1}(\bar{y}-y) + C(\bar{y}-y), \ x \neq x_0, \ k > 0.$$

2.6 WEND'S UNIQUENESS THEOREM

A generalization of Theorem 1.19.1 is the following :

Theorem 2.6.1 (Wend's Uniqueness Theorem). Let $\hat{S}_+$ be the subregion of $\bar{S}_+$ where $y \geq y_0$. Suppose the function $f(x, y)$ is defined in $\bar{S}_+$ and nondecreasing in x for (x, y) in $\hat{S}_+$. Further, suppose that for any extended solution $y(x)$ of (1.1.1), $f(x, y(x)) > 0$ for all $x \in (x_0, x_0 + a]$. Then, the initial value problem (1.1.1) has at most one extended solution in $[x_0, x_0 + a]$.

Proof. Assume that there are two extended solutions $y(x)$ and $\bar{y}(x)$ of (1.1.1). Then, $y(x)$ and $\bar{y}(x)$ are strictly increasing and absolutely continuous functions of x in $[x_0, x_0+a]$. Suppose that for some $x_1 \in (x_0, x_0+a)$, $y(x_1) < \bar{y}(x_1)$. In $[x_0, x_0+a]$, let the function $v(x)$ be defined by $\bar{y}(v(x)) = y(x)$. Since $y(x)$ and $\bar{y}(x)$ are strictly increasing and absolutely continuous, it follows that $v(x)$ is strictly increasing and absolutely continuous, $v(x_0) = x_0$ and $v'(x)$ exists almost everywhere in $[x_0, x_0+a]$. Thus, in this interval

$$y'(x) = \bar{y}'(v(x))v'(x) \tag{2.6.1}$$

almost everywhere. Since $v(x_1) < x_1$ if $y(x_1) < \bar{y}(x_1)$, from (2.6.1) and the hypotheses, we obtain

$$\begin{aligned} 0 < y'(x_1) &= \bar{y}'(v(x_1))v'(x_1) \\ &= f(v(x_1), \bar{y}(v(x_1)))v'(x_1) \\ &\le f(x_1, y(x_1))v'(x_1) \\ &\le y'(x_1)v'(x_1). \end{aligned}$$

Hence, it follows that $v'(x_1) \ge 1$.

Let $x^* = \inf\{x : x_0 \le x \le x_1 \text{ and for each } t,\ x < t \le x_1,\ y(t) < \bar{y}(t)\}$. By the above argument, $v'(x) \ge 1$ almost everywhere in $[x^*, x_1]$. So, we find that $v(x^*) \le v(x_1) + (x^* - x_1) = (v(x_1) - x_1) + x^* < x^*$, and therefore $y(x^*) = \bar{y}(v(x^*)) < \bar{y}(x^*)$. Hence, $x_0 < x^*$ and $y(x) < \bar{y}(x)$ immediately to the left of x^*. But this contradicts the definition of x^*. ∎

Corollary 2.6.2. Suppose the function $f(x, y)$ is defined in $\bar{S}_+$ and nondecreasing in both x and y in $\hat{S}_+$. Further, suppose that $f(x, y) \ge 0$ in $\bar{S}_+$ and $f(x, y_0) > 0$ for $x_0 < x \le x_0 + a$. Then, the initial value problem (1.1.1) has at most one extended solution in $[x_0, x_0+a]$. ∎

In the following examples we shall consider the rectangle $\bar{S}_+ : 0 \le x \le 1,\ |y| \le 1$.

Example 2.6.1. Consider the initial value problem

$$y' = f(x, y) = \begin{cases} 1, & y \le x \\ 2, & x < y \end{cases} \tag{2.6.2}$$
$$y(0) = 0.$$

This function $f(x,y)$ satisfies all the conditions of Theorem 2.6.1 except that it is decreasing in x. The initial value problem (2.6.2) has two extended solutions, the minimal solution $y(x) = x$ and the maximal solution $y(x) = 2x$. ∎

Example 2.6.2. The initial value problem

$$\begin{aligned} y' &= f(x,y) = \begin{cases} 0, & y \le 0 \\ 1, & y > 0 \end{cases} \\ y(0) &= 0 \end{aligned} \tag{2.6.3}$$

has many extended solutions, e.g., $y(x) \equiv 0$ and $y(x) = x$. Here, the function $f(x,y)$ is nonnegative and nondecreasing in both x and y but $f(x,0) = 0$. ∎

Example 2.6.3. For the initial value problem

$$\begin{aligned} y' &= f(x,y) = \begin{cases} 0, & y = \frac{1}{n};\ n = 1,2,\cdots \\ 1, & \text{otherwise} \end{cases} \\ y(0) &= 0 \end{aligned} \tag{2.6.4}$$

the maximal solution is $y(x) = x$. It also has the extended solution

$$\begin{aligned} y(x) &= x - \frac{1}{n(n+1)}, && \frac{1}{n} \le x \le \frac{1}{n(n+1)} + \frac{1}{n} \\ &= \frac{1}{n}, && \frac{1}{n(n+1)} + \frac{1}{n} \le x \le \frac{1}{(n-1)} \\ &= 0, && x = 0 \\ & && n = 2,3,\cdots. \end{aligned}$$

Here, the function $f(x,0) = 1$ for $x \ge 0$ and $f(x,y)$ is nondecreasing in x but not in y. ∎

Example 2.6.4. For the initial value problem

$$\begin{aligned} y' &= f(x,y) = \begin{cases} -1, & y < 0 \\ 1, & y \ge 0 \end{cases} \\ y(0) &= 0. \end{aligned} \tag{2.6.5}$$

it is clear that $y(x) = x$ and $y(x) = -x$ are both extended solutions. Of course, here also the function $f(x,y)$ does not satisfy all the conditions of Theorem 2.6.1. ∎

2.7 ZABREIKO'S UNIQUENESS THEOREM

Let $g(x,z)$ be defined for $0 < x \le a$, $|z| \le b$ and satisfy the Carathèodory conditions and the inequality $|g(x,z)| \le M(x)$, where $M(x)$ is (Lebesgue) integrable in every interval $[\varepsilon, a]$, $0 < \varepsilon < a$. Let $w(x)$, $0 < x \le a$ be some continuous function for which the differential equation (1.13.1) has at least one extended solution $z_0(x)$ in $(0,a]$ which satisfies the conditions $\lim_{x\to 0} z_0(x) = 0$, $z_0(x) \le w(x)$.

We introduce the function $w^*(x_0) = \sup z$, where sup is taken over the numbers $z \in [-b, w(x_0)]$ such that (1.13.1) has at least one extended solution $z(x)$ satisfying $z(0) = 0$, $z(x_0) = z$ and the inequality $z(x) \le w(x)$, $0 < x < x_0$. Clearly, $w^*(x)$ is defined for all $x \in (0,a]$ and satisfies the inequality $z_0(x) \le w^*(x) \le w(x)$, $0 < x \le a$. The function $w^*(x)$ will be called w - **separating function** for the equation (1.13.1).

Let $x_0 \in (0,a]$. Consider the set M_{x_0} of all extended solutions $z(x)$ of the equation (1.13.1) satisfying conditions

$$(2.7.1) \qquad z_0(x) \le z(x) \le w(x),\ 0 < x < x_0,\ z(0) = 0,\ z(x_0) = w^*(x_0).$$

This set is clearly nonempty and compact in the uniform metric, in each interval $[\varepsilon, a]$, $0 < \varepsilon < a$. Hence, the lub $z_{x_0}(x)$ of the set in $(0, x_0)$ is an extended solution of the equation (1.13.1); this extended solution satisfies (2.7.1). Continue this function $z_{x_0}(x)$ for $x > x_0$ taking its uniform upper extended solution of (1.13.1), by passing to the point $(x_0, w^*(x_0))$. The function $z_{x_0}(x)$ clearly has the following properties :

(a) $z_{x_0}(x) \le w^*(x)$, $0 < x \le x_0$, $z_{x_0}(x) \ge w^*(x)$, $x_0 \le x$; and

(b) if $w^*(x_0) < w(x_0)$ then $z_{x_0}(x) = w^*(x)$ in every interval $(\alpha, \beta) \subset [0,a]$ containing the point x_0 in which the inequality $w^*(x) < w(x)$ holds.

From these properties it follows that $w^*(x)$ is an extended solution of the equation (1.13.1) in each interval $(\alpha, \beta) \subset [0,a]$ where $w^*(x) < w(x)$. In particular, $w^*(x)$ is continuous at those points where $w^*(x_0) \le w(x_0)$.

Theorem 2.7.1. Suppose that $g(x,z) = h(x)g(z)$, where $h(x)$ is integrable in every interval $[\varepsilon, a]$, $0 < \varepsilon < a$, and $g(z) > 0$ is continuous for $z > 0$ and $g(0) = 0$. Further, suppose that one of the following conditions hold :

(i) $h(x)$ is nonnegative, or

(ii) $g(z)$ is Osgood type, i.e., $\int_0^{c_0} \frac{dt}{g(t)} = \infty$.

Suppose $w(x)$, $0 < x \leq a$ is nonnegative. Then, w - separating $w^*(x)$ for the equation (1.13.1) is given by

$$w^*(x) = \inf_{0<s<x} G^{-1}\left\{G[w(s)] - \int_s^x h(t)dt\right\}, \tag{2.7.2}$$

where $G(z) = \int_z^{c_0} \frac{dt}{g(t)}$.

Proof. Under the conditions of the theorem, equation (1.13.1) has the extended solution $z_0(x) \equiv 0$ satisfying the conditions $z_0(0) = 0$, $z_0(x) \leq w(x)$. Therefore, for the equation (1.13.1), w - separating function exists. Now let the right side of (2.7.2) be $\bar{w}(x)$, we shall show that $w^*(x) = \bar{w}(x)$. Let $z(x)$ be an extended solution of (1.13.1) satisfying the conditions $z(0) = 0$, $z(x_0) = c$, $z(x) \leq w(x)$, $0 < x < x_0$, where $c > 0$. Without loss of generality we can write $z(x) \equiv 0$ for $x \leq x_1$ and $z(x) > 0$ for $x > x_1$, where x_1 is some number in $[0, x_0)$; moreover, if (ii) holds, then x_1 is zero. In the interval $[x_1, x_0]$, it is clear that the following equality holds

$$G[z(x)] - G(c) = \int_x^{x_0} h(t)dt,$$

whereas in the interval $[0, x_1]$, the inequality

$$G[z(x)] - G(c) \leq \int_x^{x_0} h(t)dt.$$

From these relations it follows that

$$c \leq \inf_{0<s<x_0} G^{-1}\left\{G[w(s)] - \int_s^{x_0} h(t)dt\right\} = \bar{w}(x_0) \tag{2.7.3}$$

and hence $w^*(x_0) \leq \bar{w}(x_0)$. Conversely, let $c > 0$ satisfy (2.7.3). In this case, extended solution of the equation (1.13.1) defined by the equation

$$\int_{z(x)}^{c} \frac{dt}{g(t)} = \int_x^{x_0} h(t)dt,$$

clearly satisfies the inequality $z(x) \le w(x)$ for those values of x for which $z(x) > 0$. Extending this solution over the whole interval $[0, x_0]$ by zero, we obtain an extended solution of (1.13.1) satisfying $z(0) = 0$, $z(x_0) = c$, $0 \le z(x) \le w(x)$. Hence, $c \le w^*(x_0)$ and so $w^*(x_0) \ge \bar{w}(x_0)$. Thus, $w^*(x_0) = \bar{w}(x_0)$. Since, x_0 is any number in $(0, a]$, $w^*(x) = \bar{w}(x)$, $0 < x \le a$. ∎

In particular, if $g(x, z) = h(x)z^k$ then from Theorem 2.7.1 it follows that

(2.7.4)
$$w^*(x) = \begin{cases} \inf\limits_{0<s<x} w(s) \exp\left(\int_s^x h(t)dt\right), & k = 1 \\ \inf\limits_{0<s<x} \left\{w(s)^{1-k} + (1-k)\int_s^x h(t)dt\right\}^{\frac{1}{1-k}}, & k \neq 1. \end{cases}$$

A simple substitution of variables in Theorem 2.7.1 gives :

Theorem 2.7.2. Let $g(x, z) = h_1(x)z + h_2(x)g(\lambda(x)z)$, where $h_1(x)$ and $h_2(x)$ are integrable in every interval $[\varepsilon, a]$, $0 < \varepsilon < a$; $\lambda(x) = \exp\left(\int_x^a h_1(t)dt\right)$, $g(z)$ is continuous and positive for $z > 0$ and $g(0) = 0$. Further, let one of the following conditions be satisfied

(i) $h_2(x)$ is nonnegative, or

(ii) $g(z)$ is of Osgood type, i.e., $\int_0^{c_0} \frac{dt}{g(t)} = \infty$.

Suppose $w(x)$, $0 < x \le a$ is continuous and nonnegative. Then, w - separating $w^*(x)$ for the equation (1.13.1) is defined by

(2.7.5)
$$w^*(x) = \frac{1}{\lambda(x)} \inf_{0<s<x} G^{-1}\left\{G[\lambda(s)w(s)] - \int_s^x \lambda(t)h_2(t)dt\right\},$$

where $G(z) = \int_z^{c_0} \frac{dt}{g(t)}$. ∎

Theorem 2.7.3. Let $\mu(x)$, $\mu(0) = 0$ be absolutely continuous in $[0, a]$ and satisfy the inequality $\mu(x) \le w(x)$ for almost all $x \in (0, a]$, and

(2.7.6)
$$\mu'(x) \le g(x, \mu(x)).$$

Then, $\mu(x) \le w^*(x)$.

Proof. Assume that at some point $x_0 \in (0, a]$, $\mu(x_0) > w^*(x_0)$. Consider on the left of x_0 extended solution $z^*(x)$ of the equation (1.13.1) passing through $(x_0, \mu(x_0))$. Clearly, this extended solution for all x for which it is defined will satisfy the inequality $z^*(x) > w^*(x)$. On the other hand, in view of the conditions of the theorem and of the usual arguments on differential inequalities it does not go above $\mu(x)$, i.e., $z^*(x) \le \mu(x)$. Hence it may be extended for every $x > 0$. Since $\mu(0) = 0$, $z^*(0) = 0$. Thus, equation (1.13.1) has extended solution $z^*(x)$ satisfying the conditions

$$z^*(0) = 0,\ z^*(x_0) = \mu(x_0) > w^*(x_0),\ z^*(x) \le \mu(x) \le w(x)\ , 0 < x < x_0$$

But, this contradicts the definition of $w^*(x)$. ∎

The assumption of absolute continuity of $\mu(x)$ could be replaced by the assumption of its continuity if it is required to satisfy the inequality (2.7.6), or one of the more general inequalities

$$D^+\mu(x) \le g(x, \mu(x)),\ D_+\mu(x) \le g(x, \mu(x))$$

for $x \in (0, a]$, except for a countable set of values of x.

The following result follows from Theorem 2.7.3.

Theorem 2.7.4. Suppose $g(x, z)$ for all $x \in (0, a]$ is a nondecreasing function of z and $g(x, w(x))$ is integrable in $[0, a]$. Suppose the continuous function $\mu(x)$, defined for $0 \le x \le a$, satisfies the inequalities $\mu(x) \le w(x)$ and

$$\mu(x) \le \int_0^x g(t, \mu(t))dt. \tag{2.7.7}$$

Then, $\mu(x) \le v^*(x)$, where $v^*(x)$ is v - separating for the equation (1.13.1), and $v(x) = \int_0^x g(t, w(t))dt$. ∎

We shall now apply the above theorems to study the uniqueness of the extended solutions of (1.1.1) with $(x_0, y_0) = (0, 0)$. For this, the function $f(x, y)$ is assumed to be defined in $0 \le x \le a$, $|y| \le b$ and satisfies the Carathèodory conditions.

Theorem 2.7.5 (Zabreiko's Uniqueness Theorem). Let $g_i(x, z)$, $i = 1, 2, \cdots, n$

be defined for $0 < x \le a$, $|z| \le b$ and satisfy the Carathèodory conditions and the inequality $|g_i(x,z)| \le M(x)$, where $M(x)$ is integrable in every interval $[\varepsilon, a]$, $0 < \varepsilon < a$, also $g_i(x,0) = 0$. Suppose $w_0(x) = b$, $w_i(x)$ $(i = 1, 2, \cdots n)$ is w_{i-1} - separating for the equation

$$z' = g_i(x,z), \ i = 1, 2, \cdots, n \tag{2.7.8}$$

and $w_n(x) \equiv 0$. Further, let $f(x,y)$ satisfy

$$f(x,\bar{y}) - f(x,y) \le g_i(x, \bar{y} - y), \ 0 < x \le a, \ 0 \le \bar{y} - y \le w_{i-1}(x). \tag{2.7.9}$$

Then, the initial value problem (1.1.1) with $(x_0, y_0) = (0,0)$ has at most one extended solution in $[0, a]$.

Proof. In view of Theorem 2.7.3 the proof is obvious. ∎

Theorem 2.7.5 for $n = 2$ is similar to that of Brauer's Uniqueness Theorem 1.18.1. Next, we shall consider some particular cases of Theorem 2.7.5.

Let $h(x)$ be a nonnegative integrable function, k_0 some number in $[0, 1)$ and

$$w(x) = \left\{(1 - k_0) \int_0^x h(t)dt\right\}^{\frac{1}{1-k_0}}. \tag{2.7.10}$$

In our next result we shall assume that the function $f(x,y)$ satisfies either the condition (i) :

$$f(x,\bar{y}) - f(x,y) \le h(x)(\bar{y} - y)^{k_0}, \ 0 \le x \le a, \ 0 \le \bar{y} - y \le b, \tag{2.7.11}$$

or the condition (ii) :

$$f(x,\bar{y}) - f(x,y) \le \varepsilon(x_0, r_0) h(x)(\bar{y} - y)^{k_0}, \tag{2.7.12}$$

$$0 \le x \le x_0 \le a, \ 0 \le \bar{y} - y \le b$$

where $\lim_{x_0, r_0 \to 0} \varepsilon(x_0, r_0) = 0$. Note that neither of these conditions guarantees the uniqueness for (1.1.1).

Theorem 2.7.6. Let in addition to the above condition (i) or (ii) the function $f(x,y)$ satisfy

(2.7.13) $$f(x,\bar{y}) - f(x,y) \le k(x)(\bar{y} - y), \ 0 < x \le a, \ 0 \le \bar{y} - y \le b$$

where $k(x)$ is integrable in every interval $[\varepsilon, a]$, $0 < \varepsilon < a$. Further, let either condition (a) be satisfied

(2.7.14) $$\lim_{s\to 0}\left\{\int_s^a k(t)dt + \frac{1}{1-k_0}\ln\int_0^s h(t)dt\right\} = -\infty,$$

or the condition (b) :

(2.7.15) $$\limsup_{s\to 0}\left\{\int_s^a k(t)dt + \frac{1}{1-k_0}\ln\int_0^s h(t)dt\right\} < \infty.$$

Then, the initial value problem (1.1.1) has at most one extended solution in $[0, a]$. ∎

Theorem 2.7.6 contains Krasnosel'skii - Krein's Uniqueness Theorem 2.4.1 for extended solutions of (1.1.1).

Theorem 2.7.7. Let $f(x,y)$ satisfy the inequality

(2.7.16) $$f(x,\bar{y}) - f(x,y) \le h_1(x)(\bar{y} - y) + h_2(x)g(\lambda(x)(\bar{y} - y)),$$

$$0 < x \le a, \ 0 \le \bar{y} - y \le w(x),$$

where $h_1(x)$ and $h_2(x)$ are integrable in every interval $[\varepsilon, a]$, $0 < \varepsilon < a$ and $\lambda(x) = \exp\left(\int_x^a h_1(t)dt\right)$, and $g(z)$, $g(0) = 0$ is a continuous and positive function satisfying Osgood's condition $\lim_{z\to 0} G(z) = \infty$, where $G(z) = \int_z^{c_0} \frac{dt}{\psi(t)}$.

Further, let either condition (i) be satisfied

(2.7.17) $$\limsup_{s\to 0}\left\{G[\lambda(s)w(s)] - \int_s^a \lambda(t)h_2(t)dt\right\} = \infty,$$

or the condition (ii) :

(2.7.18) $$\limsup_{s\to 0}\left\{G[\varepsilon(s)\lambda(s)w(s)] - \int_s^a \lambda(t)h_2(t)dt\right\} = \infty,$$

for any positive function $\varepsilon(s)$ satisfying the condition $\lim_{s\to 0}\varepsilon(s) = 0$. Then, the initial value problem (1.1.1) has at most one extended solution in the interval $[0, a]$. ∎

2.8 LAKSHMIKANTHAM'S UNIQUENESS THEOREM

Let the function $f(x,y)$ be defined in $\bar{S}_+$ with $(x_0, y_0) = (0,0)$. Further, let $y(x)$ and $\bar{y}(x)$ be two classical solutions of (1.9.4) in $[0,a]$. The requirement $\lim_{x\to 0^+} \frac{|y(x)-\bar{y}(x)|}{x} = 0$ which is satisfied when $f(x,y)$ is continuous at $(0,0)$, is a necessary condition for the uniqueness of classical solutions. This condition can be generalized by the following :

$$\lim_{x\to 0^+} \frac{|y(x)-\bar{y}(x)|}{B(x)} = 0, \tag{2.8.1}$$

where the function $B(x)$ is continuous, positive in $(0,a]$, and $B(0^+) = 0$. That condition (2.8.1) is not sufficient for the uniqueness of the classical solutions of (1.9.4) is proved in the following :

Lemma 2.8.1. Suppose that the function $B(x)$ is continuous, positive in $(0,a]$, and $B(0^+) = 0$. Then, there exists an infinity of functions $f(x,y)$ such that the initial value problem (1.9.4) has more than one classical solution satisfying the condition (2.8.1).

Proof. We first construct a function $A(x)$ having a nonnegative derivative in $[0,a]$ and such that $\lim_{x\to 0^+} \frac{A(x)}{B(x)} = 0$. For this, we divide the interval $[0,a]$ into subintervals I_n such that

$$I_1 = (a/2, a), \ I_2 = (a/4, a/2), \ \cdots .$$

Suppose $b_n = \inf_{I_n} B(x)$. Find a positive linear function L_1 in I_1 such that

$$L_1(a) = b_1, \ L_1(a/2) \le \frac{1}{2} L_1(a).$$

Then, find L_2 in I_2 such that

$$\begin{aligned} L_2(a/2) &= b_2/2 \ \text{ if } \ b_2/2 \le L_1(a/2), \\ L_2(a/2) &\le \frac{1}{2} L_1(a/2) \ \text{ if } \ b_2/2 > L_1(a/2), \\ L_2(a/4) &\le \frac{1}{2} L_2(a/2). \end{aligned}$$

We continue this process and then connect the linear functions near the points

$a/2^n$ by suitable functions having nonnegative derivatives (for example, by arcs of parabolas). This modification gives us the function $A(x)$ with the required properties.

Having constructed the function $A(x)$, it is easy to define $f(x,y)$ by $f(x,y) = y^\alpha A'(x)$, $0 < \alpha < 1$. Then, $y(x) \equiv 0$ and $y(x) = ((1-\alpha)A(x))^{(1-\alpha)^{-1}}$ are classical solutions of (1.9.4). It is clear that any two classical solutions of this initial value problem satisfy (2.8.1). ∎

The following result is also immediate :

Lemma 2.8.2. Suppose that $f(x,y)$ is defined in $\bar{S}_+$ with $(x_0, y_0) = (0,0)$ and is continuous at $(0,0)$. Then, there exists a function $B(x)$ in $[0,a]$ such that (2.8.1) is satisfied.

Proof. Let $y(x)$ and $\bar{y}(x)$ be two classical solutions of (1.9.4) in $[0,a]$. We define $\phi(x) = |y(x) - \bar{y}(x)|$ and observe that $\phi(0) = 0$. Since $f(x,y)$ is continuous at $(0,0)$, we have $\lim_{x\to 0^+} \frac{\phi(x)}{x} = 0$. Setting $B(x) = \sup_{t \le x \le a} \frac{\phi(t)}{t}$, it can easily be verified that $\lim_{x\to 0^+} \frac{\phi(x)}{B(x)} = 0$. This is always possible if $\phi(x) \not\equiv 0$ in some neighborhood of the origin; otherwise, the existence of $B(x)$ is trivial. ∎

Before we prove the main result of this section, we note that the continuity requirement of $f(x,y)$ at $(0,0)$ is stronger than the condition (2.8.1). For this, we consider the initial value problem (1.6.4) for which the classical solutions are $y(x) = cx$, where $0 \le c \le 1$. Thus, we can take $B(x) = \sqrt{x}$ so that (2.8.1) is satisfied. The function $f(x,y)$ in this problem is not continuous at $(0,0)$.

Theorem 2.8.3 (Lakshmikantham's Uniqueness Theorem). Suppose that

(i) any two classical solutions $y(x)$ and $\bar{y}(x)$ of (1.9.4) satisfy (2.8.1), where $B(x)$ is continuous, positive in $(0,a]$ and $B(0^+) = 0$,

(ii) the function $g(x,z)$ is continuous and nonnegative for $0 < x \le a$, $0 \le z \le 2b$, and the only solution $z(x)$ of (1.13.1) in $[0,a]$ such that $\lim_{x\to 0^+} \frac{z(x)}{B(x)} = 0$ is the trivial solution; and

(iii) $f(x,y)$ is defined in $\bar{S}_+$ with $(x_0,y_0) = (0,0)$ and for all $(x,y), (x,\bar{y})$ in $\bar{S}_+$ where $x \ne 0$ the inequality (1.11.2) is satisfied.

Then, the initial value problem (1.9.4) has at most one classical solution in $[0, a]$.

Proof. Suppose there are two classical solutions $y(x)$ and $\bar{y}(x)$ of (1.9.4) in the interval $[0, a]$. Define $\phi(x) = |y(x) - \bar{y}(x)|$. Clearly, $\phi(0) = 0$ and from (1.11.2), we get

$$D^+\phi(x) \le |f(x, y(x)) - f(x, \bar{y}(x))| \le g(x, \phi(x)).$$

If we suppose that, for some σ, $0 < \sigma < a$, $\phi(\sigma) > 0$ then as in Lemma 1.15.2 we can show that $\rho(x) \le \phi(x)$, $x \le \sigma$ as far as $\rho(x)$ exists, where $\rho(x)$ is the minimal solution of $z' = g(x, z)$, $z(\sigma) = \phi(\sigma)$. Furthermore, as before, $\rho(x)$ can be continued up to $x = 0$ and $0 \le \rho(x) \le \phi(x)$, $0 \le x \le \sigma$. Then, because of the assumed condition (2.8.1), we have

$$0 \le \lim_{x \to 0+} \frac{\rho(x)}{B(x)} \le \lim_{x \to 0+} \frac{\phi(x)}{B(x)} = 0,$$

which, by hypothesis, implies that $\rho(x) \equiv 0$. This contradicts $\rho(\sigma) = \phi(\sigma) > 0$, and hence $\phi(x) = 0$, $x \in [0, a]$. ∎

Corollary 2.8.4. If $g(x, z) = zB'(x)/B(x)$ and $B(x) = x$ then Theorem 2.8.3 reduces to Nagumo's Uniqueness Theorem 1.6.2; and if $B(x) = e^{-1/x}$ then it gives Roger's Uniqueness Theorem 1.9.2. ∎

Corollary 2.8.5. If $B(x) = x$ and the function $f(x, y)$ is continuous in $\bar{S}_+$ with $(x_0, y_0) = (0, 0)$, then Theorem 2.8.3 reduces to Kamke's Uniqueness Theorem 1.15.1. ∎

Corollary 2.8.6. Theorem 2.8.3 contains Brauer's Uniqueness Theorem 1.18.1.

Proof. It is trivial to show that the extra condition in Theorem 1.18.1 implies the condition (2.8.1). ∎

2.9 BOWNDS' UNIQUENESS THEOREM

Here we shall use a certain factorization of the function $f(x, y)$ to obtain the uniqueness of the classical solutions of the initial value problem (1.1.1). The obtained result is particularly useful in situations where many of the standard Lipschitz type uniqueness theorems do not apply, i.e., those theorems which explicitly impose conditions on the difference $f(x,\bar{y}) - f(x,y)$.

For this, first we shall consider the following examples which will be used to demonstrate the applicability of the main result.

Example 2.9.1. The initial value problem

$$\begin{aligned} & y' = g(x,y) = (1-x)|y|^{1/2}, \; 0 < x < 1 \\ & y(0) = 0 \end{aligned} \tag{2.9.1}$$

has at least two classical solutions, $y(x) \equiv 0$ and $y(x) = \frac{1}{4}(x - \frac{x^2}{2})^2$ in $[0,1]$. ∎

Example 2.9.2. The initial value problem

$$\begin{aligned} & y' = f(x,y) = (1-x)(|y|^{1/2}+1), \; 0 < x < 1 \\ & y(0) = 0 \end{aligned} \tag{2.9.2}$$

will be shown to have at most one classical solution in $[0,1]$. ∎

For the function $f(x,y)$ in (2.9.2), it is clear that for any pair $y, \bar{y}$ and for any $x \in [0,1]$

$$f(x,\bar{y}) - f(x,y) = g(x,\bar{y}) - g(x,y).$$

So, if one attempts to check the uniqueness in Example 2.9.2 by using theorems we have provided earlier, then they are bound to fail because their assertion of uniqueness for Example 2.9.2 would be an assertion of uniqueness for Example 2.9.1.

Theorem 2.9.1 (Bownds' Uniqueness Theorem). Suppose that it is possible to find three functions f_1, f_2, f_3 in $\bar{S}_+$ such that $f = f_1 f_2 - f_3$ in $\bar{S}_+$, where f_1, f_3 are continuous in $\bar{S}_+$, and $\frac{\partial f_1(x,y)}{\partial x}$ exists, is nonnegative, and is dominated by an (Lebesgue) integrable function of y in S_+. Further, suppose that f_1, f_2 and f_3 have the following properties

(i) f_1 is strictly positive along the solution curves,

(ii) for each fixed $x \in (x_0, x_0 + a)$ and $y_0 - b \le y < \bar{y} \le y_0 + b$

$$f_3(x, \bar{y}) f_1(x, y) \ge f_3(x, y) f_1(x, \bar{y}); \quad \text{and}$$

(iii) f_2 satisfies one - sided generalized Lipschitz condition in S_+, i.e., there exists a nonnegative (Lebesgue) integrable function $\lambda(x)$, $x \in (x_0, x_0+a)$ such that for all $(x, y), (x, \bar{y})$ in S_+ with $y < \bar{y}$

$$f_2(x, \bar{y}) - f_2(x, y) \le \lambda(x)(\bar{y} - y). \tag{2.9.3}$$

Then, the initial value problem (1.1.1) has at most one classical solution in $[x_0, x_0 + a]$.

Proof. Suppose there are two classical solutions $y(x)$ and $\bar{y}(x)$ of (1.1.1) in the interval $[x_0, x_0 + a]$. Then, there exists a point $\bar{x} \in (x_0, x_0 + a)$ such that $\bar{y}(\bar{x}) - y(\bar{x}) > 0$. Let $\xi_0 = \sup\{x : x_0 \le x < \bar{x} \text{ and } y(x) = \bar{y}(x)\}$. Then, $y(\xi_0) = \bar{y}(\xi_0)$ and $\bar{y}(x) > y(x)$ in $(\xi_0, \bar{x})$. Now $f_1(\xi_0, y(\xi_0)) > 0$ and since f_1 is continuous there exists a ξ_1 such that for $\xi_0 \le x \le \xi_1$ and $y(x) \le y \le \bar{y}(x)$ it is true that $f_1(x, y)$ is bounded away from zero by a positive number, say, m.

Define a function $H : [\xi_0, \xi_1] \to R$ as follows

$$H(x) = \int_{y(x)}^{\bar{y}(x)} \frac{dt}{f_1(x, t)} + \int_{\xi_0}^{x} \left\{ \frac{f_3(t, \bar{y}(t))}{f_1(t, \bar{y}(t))} - \frac{f_3(t, y(t))}{f_1(t, y(t))} \right\} dt. \tag{2.9.4}$$

Clearly, $H(\xi_0) = 0$ and in view of the conditions (i) and (ii) it follows that

$$H(x) > 0 \quad \text{in } (\xi_0, \xi_1]. \tag{2.9.5}$$

Also, by the hypotheses on f_1, there exists an (Lebesgue) integrable function $g(y)$ in $|y - y_0| \le b$ so that

$$0 \le \frac{\partial f_1(x, y)}{\partial x} \le g(y) \quad \text{for all} \quad x \in (x_0, x_0 + a), \ y(x) \le y \le \bar{y}(x). \tag{2.9.6}$$

Because of (2.9.6), $H(x)$ may be differentiated in (ξ_0, ξ_1), to obtain

$$\begin{aligned} H'(x) = \int_{y(x)}^{\bar{y}(x)} \frac{\partial}{\partial x} \left[\frac{1}{f_1(x, t)} \right] dt + \frac{\bar{y}'(x)}{f_1(x, \bar{y}(x))} - \frac{y'(x)}{f_1(x, y(x))} \\ + \frac{f_3(x, \bar{y}(x))}{f_1(x, \bar{y}(x))} - \frac{f_3(x, y(x))}{f_1(x, y(x))} \end{aligned} \tag{2.9.7}$$

$$\leq \frac{\bar{y}'(x)+f_3(x,\bar{y}(x))}{f_1(x,\bar{y}(x))} - \frac{y'(x)+f_3(x,y(x))}{f_1(x,y(x))}$$
$$= \frac{f(x,\bar{y}(x))+f_3(x,\bar{y}(x))}{f_1(x,\bar{y}(x))} - \frac{f(x,y(x))+f_3(x,y(x))}{f_1(x,y(x))}.$$

In the above inequality, on using the factorization of f and the condition (iii), we obtain

$$H'(x) \leq f_2(x,\bar{y}(x)) - f_2(x,y(x)) \leq \lambda(x)(\bar{y}(x)-y(x)). \tag{2.9.8}$$

Now let M be the maximum of f_1 in $\xi_0 \leq x \leq \xi_1$, $y(x) \leq y \leq \bar{y}(x)$, then from (2.9.4) it follows that

$$H(x) \geq \int_{y(x)}^{\bar{y}(x)} \frac{dt}{M} = \frac{\bar{y}(x)-y(x)}{M}. \tag{2.9.9}$$

A combination of (2.9.8) and (2.9.9) leads to the differential inequality

$$H'(x) - M\lambda(x)H(x) \leq 0,\ x \in (\xi_0,\xi_1)$$

which may be written as

$$\frac{d}{dx}\left(H(x)\exp\left(-\int_{\xi_0}^{x} M\lambda(t)dt\right)\right) \leq 0, \quad \text{a.e. in} \quad (\xi_0,\xi_1).$$

Thus, we find that

$$D^+\left(-H(x)\exp\left(-\int_{\xi_0}^{x} M\lambda(t)dt\right)\right) \geq 0, \quad \text{a.e. in} \quad (\xi_0,\xi_1) \tag{2.9.10}$$

and

$$D^+\left(-H(x)\exp\left(-\int_{\xi_0}^{x} M\lambda(t)dt\right)\right) > -\infty \quad \text{for all} \quad (\xi_0,\xi_1). \tag{2.9.11}$$

Now inequalities (2.9.10) and (2.9.11) imply that

$$0 \ \leq\ \int_{\xi_0}^{x} \frac{d}{dt}\left(-H(t)\exp\left(-\int_{\xi_0}^{t} M\lambda(s)ds\right)\right)dt$$
$$= \ -H(x)\exp\left(-\int_{\xi_0}^{x} M\lambda(t)dt\right)$$

and hence $H(x) \leq 0$ in (ξ_0,ξ_1). But, this contradicts (2.9.5). ∎

Clearly Theorem 2.9.1 applies to (2.9.2) by taking

$$f_1 = 1+|y|^{1/2},\ f_2 = 1 \ \text{ and } \ f_3 = x(1+|y|^{1/2}).$$

It is important to note that in Theorem 2.9.1 none of the conditions (i) - (iii) can be dropped. For this, we consider the initial value problem

$$(2.9.12) \qquad y' = |y|^{1/2}, \ 0 < x < 1, \ y(0) = 0$$

which obviously has at least two classical solutions. For (2.9.12), we can choose

(i)′ $f_1 = |y|^{1/2}, \ f_2 = 1, \ f_3 = 0$, or

(ii)′ $f_1 = 1 + |y|^{1/2}, \ f_2 = 1, \ f_3 = 1$, or

(iii)′ $f_1 = 1, \ f_2 = |y|^{1/2}, \ f_3 = 0$.

For (i)′ all the conditions of theorem are satisfied except (i) since $f_1 = |y(x)|^{1/2} = 0$, when $x = 0$. For (ii)′ condition (ii) does not hold because for a fixed $x \in (0,1)$, and $-\infty < y < \bar{y} < \infty, 1 \times (1 + |y|^{1/2}) \not\geq 1 \times (1 + |\bar{y}|^{1/2})$. For (iii)′ condition (iii) is not satisfied because for $\bar{y} > 0$, $\frac{f_2(x,\bar{y}) - f_2(x,0)}{\bar{y} - 0} = \frac{1}{\bar{y}^{1/2}}$ is unbounded as $\bar{y} \to 0$.

If in Theorem 2.9.1 we take $f_1 = 1, \ f_2 = f$ and $f_3 = 0$ then all the conditions are satisfied provided f itself satisfies the one - sided generalized Lipschitz condition (2.9.3). Obviously, this conclusion is stronger than Theorems 1.2.5 and 2.2.1.

2.10 BOWNDS - METCALF'S UNIQUENESS THEOREM

The importance of the factorization of the function f has been demonstrated in the previous section. In particular, the Lipschitz condition need not be imposed on the entire function f, rather it suffices that only the factor f_2 satisfy the one - sided generalized Lipschitz condition. In this section we shall prove a similar result for the Nagumo condition.

Theorem 2.10.1 (Bownds - Metcalf's Uniqueness Theorem). In Theorem 2.9.1 condition (iii) can be replaced by (iii)′ f_2 satisfies a one - sided generalized Nagumo condition in S_+, i.e., there exists a constant $\alpha \geq 0$ such that for all $(x, y), (x, \bar{y})$ in S_+ with $y < \bar{y}$

$$(x - x_0)(f_2(x, \bar{y}) - f_2(x, y)) \leq \alpha(\bar{y} - y) \tag{2.10.1}$$

and $\lim_{(x,y)\to(x_0^+, y_0)}[f_2(x, y)/(x - x_0)^{M\alpha - 1}]$ exists, where $M = \max|f_1(x, y)|$ in $\bar{S}_+$.

Proof. Following as in Theorem 2.9.1, the inequality (2.9.8) takes the form

$$H'(x) \leq \alpha(\bar{y}(x) - y(x))/(x - x_0), \ x \in (\xi_0, \xi_1). \tag{2.10.2}$$

Thus, a combination of (2.10.2) and (2.9.9) leads to the differential inequality

$$H'(x) - M\alpha H(x)/(x - x_0) \leq 0, \ x \in (\xi_0, \xi_1).$$

Hence, for $0 < \varepsilon < \xi_1 - \xi_0$, we have

$$\frac{d}{dx}\left(H(x)\exp\left(-M\alpha\int_{\xi_0+\varepsilon}^{x}(t - x_0)^{-1}dt\right)\right) \leq 0, \ x \in [\xi_0 + \varepsilon, \xi_1).$$

Therefore, it follows that

$$0 \leq \frac{H(x)}{(x - x_0)^{M\alpha}} \leq \frac{H(\xi_0 + \varepsilon)}{(\xi_0 + \varepsilon - x_0)^{M\alpha}}, \ x \in [\xi_0 + \varepsilon, \xi_1). \tag{2.10.3}$$

If $\alpha = 0$, then (2.10.3), in view of $H(\xi_0) = 0$, implies that $H(x) = 0$ in (ξ_0, ξ_1). But, this contradicts (2.9.5). If $\alpha > 0$ and $\xi_0 > x_0$, then also the same conclusion holds. If $\alpha > 0$ and $\xi_0 = x_0$, then by the Cauchy mean value theorem, there exists a $0 < \eta < \varepsilon$ such that

$$0 \leq \frac{H(x_0 + \varepsilon)}{\varepsilon^{M\alpha}} = \frac{H'(x_0 + \eta)}{M\alpha\eta^{M\alpha - 1}}.$$

Now, on using the factorization of f in (2.9.7), we find

$$H'(x_0 + \eta) \leq f_2(x_0 + \eta, \bar{y}(x_0 + \eta)) - f_2(x_0 + \eta, y(x_0 + \eta)),$$

and since (iii)′ implies the existence of $\lim_{\eta\to 0^+}\frac{f_2(x_0+\eta, y(x_0+\eta))}{\eta^{M\alpha-1}}$ as well as $\lim_{\eta\to 0^+}\frac{f_2(x_0+\eta, \bar{y}(x_0+\eta))}{\eta^{M\alpha-1}}$ it follows that

$$0 \leq \frac{H(x)}{(x - x_0)^{M\alpha}} \leq \lim_{\varepsilon\to 0^+}\frac{H(x_0 + \varepsilon)}{\varepsilon^{M\alpha}} = 0, \ x \in [x_0 + \xi_1).$$

But, this also forces that $H(x) = 0$ in $[x_0, \xi_1)$. ∎

As a consequence of Theorem 2.10.1 we find that in Nagumo's Uniqueness Theorem 1.6.2 the constant k can be greater than one provided that f is sufficiently small at the origin. This sufficiency being determined by the actual value of the constant.

2.11 GARD'S UNIQUENESS THEOREM

In Sections 1.9 and 2.10 we have discussed the results of Roger and Bownds and Metcalf which generalize Nagumo's Uniqueness Theorem 1.6.2. The purpose of this section is to prove a further generalization of Nagumo's theorem which accommodates Roger as well as Bownds and Metcalf result. We shall also provide an example which dwells upon the importance of this new generalization.

Theorem 2.11.1 (Gard's Uniqueness Theorem). Let $\phi(x)$ be a continuous function defined in $[x_0, x_0 + a)$, and differentiable in $(x_0, x_0 + a)$ such that $\phi(x_0) = 0$, $\phi(x) > 0$ for $x > x_0$. Further, let the function $f(x, y)$ be defined in S_+ and

(i) for all $(x, y), (x, \bar{y}) \in S_+$ the following inequality holds

$$(2.11.1) \qquad (f(x,y) - f(x,\bar{y}))(y - \bar{y}) \le \frac{\phi'(x)}{\phi(x)}(y - \bar{y})^2, \quad \text{and}$$

(ii) there is an $L \in R$ such that

$$f(x,y) = L\phi'(x) + o(\phi'(x)) \quad \text{as } x \to x_0^+,\ y \to y_0.$$

Then, the initial value problem (1.1.1) has at most one classical solution in $[x_0, x_0 + a)$.

Proof. Let $y(x)$ and $\bar{y}(x)$ be two classical solutions of (1.1.1). We define a function

$$(2.11.2) \qquad v(x) = \begin{cases} \frac{1}{2}[(y(x) - \bar{y}(x))/\phi(x)]^2, & x \in (x_0, x_0 + a) \\ 0, & x = x_0. \end{cases}$$

Using (ii) and l'Hospital's rule, we obtain

$$\begin{aligned} &\lim_{x\to x_0^+} \frac{y(x) - \bar{y}(x)}{\phi(x)} \\ &= \lim_{x\to x_0^+} \frac{y'(x) - \bar{y}'(x)}{\phi'(x)} \\ &= \lim_{x\to x_0^+} \frac{f(x, y(x)) - f(x, \bar{y}(x))}{\phi'(x)} \\ &= \lim_{x\to x_0^+} \frac{[f(x, y(x)) - L\phi'(x)] - [f(x, \bar{y}(x)) - L\phi'(x)]}{\phi'(x)} \\ &= 0. \end{aligned}$$

Hence, the function $v(x)$ is continuous in $[x_0, x_0 + a)$. Now for $x > x_0$, using (i), we obtain

$$\begin{aligned} &v'(x) \\ &= \frac{(y(x) - \bar{y}(x))}{\phi(x)} \left[\frac{\phi(x)(f(x, y(x)) - f(x, \bar{y}(x))) - (y(x) - \bar{y}(x))\phi'(x)}{\phi^2(x)} \right] \\ &\leq \frac{\phi'(x)(y(x) - \bar{y}(x))^2}{\phi^3(x)} - \frac{(y(x) - \bar{y}(x))^2\phi'(x)}{\phi^3(x)} \\ &= 0. \end{aligned}$$

Thus, $v(x)$ is continuous and nonincreasing in $[x_0, x_0 + a)$. Since $v(x)$ is nonnegative, we find that $v(x) \equiv 0$, i.e., $y(x) = \bar{y}(x)$ in $[x_0, x_0 + a)$. ∎

Corollary 2.11.2. Roger's Uniqueness Theorem 1.9.2 is a particular case of Theorem 2.11.1.

Proof. For $\beta > 1$ it suffices to note that the function $\phi(x) = (\beta - 1)^{-1} \exp(-(x - x_0)^{1-\beta})$ satisfies all the hypotheses of Theorem 2.11.1 and the conditions (i) and (ii) become

(i) $(f(x,y) - f(x,\bar{y}))(y - \bar{y}) \le \dfrac{\beta - 1}{(x - x_0)^\beta}(y - \bar{y})^2$; $(x,y),(x,\bar{y}) \in S_+$, and

(ii) $\lim_{(x,y)\to(x_0^+,y_0)} f(x,y)(x - x_0)^\beta \exp((x - x_0)^{1-\beta})$ exists. ■

For $\beta = 2$ this reduces to the one - sided case of Roger's result.

Corollary 2.11.3. Bownds and Metcalf's Uniqueness Theorem 2.10.1 for $f_1 = 1$, $f_2 = f$ and $f_3 = 0$ is a particular case of Theorem 2.11.1.

Proof. For $\alpha > 1$ the function $\phi(x) = (x - x_0)^\alpha$ satisfies all the hypotheses of Theorem 2.11.1 and the conditions (i) and (ii) become

(i) $(f(x,y) - f(x,\bar{y}))(y - \bar{y}) \le \dfrac{\alpha}{(x - x_0)}(y - \bar{y})^2$; $(x,y),(x,\bar{y}) \in S_+$, and

(ii) $\lim_{(x,y)\to(x_0^+,y_0)} f(x,y)(x - x_0)^{1-\alpha}$ exists. ■

The following example illustrates a situation in which Theorem 2.11.1 is applicable but neither the Roger, nor the Bownds and Metcalf result can be used.

Example 2.11.1. Consider the initial value problem

$$y' = (-\ln x)y \ ,0 < x < a, \ y(0) = y_0 \tag{2.11.3}$$

where a is a positive number less than 1, to be more specifically determined below.

The unique solution of (2.11.3) is $y(x) = y_0 \exp(x(1 - \ln x))$. Since neither $[(-\ln x)y]x^2 e^{1/x}$, $[(-\ln x)y]x^{1-\alpha}$ $(\alpha > 1)$ have a finite limit as $x \to 0^+$, the results of Roger and Bownds and Metcalf can not be applied. To show that Theorem 2.11.1 does provide uniqueness, we observe that $x(\ln x)^2 \to 0$ as $x \to 0^+$. Now we choose $a < 1$, so that $x(\ln x)^2 \le 1$ for $x < a$. Let $\phi(x) = -(\ln x)^{-1}$, $\phi(0) = 0$. Then, $\phi'(x) = 1/x(\ln x)^2$ and $\phi'(x)/\phi(x) = -(x \ln x)^{-1}$. Now since $x(\ln x)^2 \le 1$ for $x < a$, we have

$$f(x,y) - f(x,\bar{y}) = (-\ln x)(y - \bar{y}) \le -\frac{1}{x \ln x}(y - \bar{y}); \ 0 < x < a, \ y > \bar{y}.$$

Further,

$$\frac{f(x,y)}{\phi'(x)} = \frac{(-\ln x)y}{1/x(\ln x)^2} = -x(\ln x)^3 y \to 0 \quad \text{as} \quad x \to 0^+.$$

Hence, the conditions of Theorem 2.11.1 are satisfied, and thus the problem (2.11.3) has a unique classical solution. ∎

2.12 MAJORANA'S NONUNIQUENESS THEOREM

Here we shall establish a necessary condition for the nonuniqueness of the initial value problem (1.1.1), and as a consequence deduce a new uniqueness criterion.

Theorem 2.12.1 (Majorana's Nonuniqueness Theorem). Let the function $f(x, y)$ be defined in $[x_0, x_0 + a] \times R$, continuous with respect to x, and such that $f(x, y_0) = 0$ for every $x \in [x_0, x_0 + a]$. Further, let the initial value problem (1.1.1) have two different classical solutions. Then, for every $\varepsilon > 0$ there exists $t \in [x_0, x_0 + a]$ such that the equation

$$u = y_0 + (t - x_0) f(t, u) \tag{2.12.1}$$

has at least two different roots u with $|u - y_0| < \varepsilon$.

Proof. From the hypotheses it is clear that $y(x) = y_0$ is a classical solution of (1.1.1), so we assume that there exists a classical solution $\bar{y}(x)$ of (1.1.1) which is different from y_0. Let $\varepsilon > 0$ be given. Since $u = y_0$ is a solution of (2.12.1) for every $t \in [x_0, x_0 + a]$, it is sufficient to show the existence of $t \in [x_0, x_0 + a]$ for which (2.12.1) is satisfied by some $u \neq y_0$, $|u - y_0| < \varepsilon$.

In the interval $[x_0, x_0 + a]$ we define the function

$$w(x) = \begin{cases} \dfrac{(\bar{y}(x) - y_0)}{(x - x_0)}, & x \neq x_0 \\ \\ 0, & x = x_0. \end{cases}$$

Of course this function $w(x)$ is continuous in $[x_0, x_0 + a]$, differentiable in $(x_0, x_0 + a)$, and for every $x \in (x_0, x_0 + a)$

$$w'(x) = \frac{1}{(x - x_0)^2}[(x - x_0)f(x, \bar{y}(x)) - \bar{y}(x) + y_0]. \tag{2.12.2}$$

Now, fix a point $x_2 \in (x_0, x_0 + a)$ with $\bar{y}(x_2) \neq y_0$ such that $|\bar{y}(x) - y_0| < \varepsilon$ for every $x \in [x_0, x_2]$, and denote $x_1 = \sup\{x \in [x_0, x_2] : w(x) = 0\}$. Obviously, $\bar{y}(x_1) = y_0$ and $\bar{y}(x) \neq y_0$ for every $x \in (x_1, x_2)$.

If there exists $t \in (x_1, x_2]$ such that $w'(t) = 0$, then from (2.12.2) for such a t, (2.12.1) is satisfied by $u = \bar{y}(t)$. Otherwise, if $w'(x) \neq 0$ for every $x \in (x_1, x_2]$ (hence $w'(x)$ has constant sign in $(x_1, x_2]$), then we take $u = \bar{y}(x_2)$ $(\neq y_0)$, and in the interval $[x_0, x_0 + a]$ define the function

$$g(x) = (x - x_0)f(x, u) - u + y_0.$$

If $w'(x) > 0$, $x \in (x_1, x_2]$, then $w(x_2) > w(x_1) = 0$, hence $u > y_0$ and consequently $g(x_0) = -u + y_0 < 0$. On the other hand, we have $g(x_2) = (x_2 - x_0)^2 w'(x_2) > 0$. Therefore, there exists $t \in (x_0, x_2)$ such that $g(t) = 0$. The case $w'(x) < 0$, $x \in (x_1, x_2]$ can be treated similarly. ∎

As a consequence of Theorem 2.12.1 we have the following uniqueness criterion.

Theorem 2.12.2. Let the function $f(x, y)$ be defined in $[x_0, x_0 + a] \times R$, continuous with respect to x, and such that $f(x, y_0) = 0$ for every $x \in [x_0, x_0 + a]$. Further, let there exist $\varepsilon > 0$ such that $u = y_0$ is the only root of (2.12.1) with $|u - y_0| < \varepsilon$ for every $t \in [x_0, x_0 + a]$. Then, for the initial value problem (1.1.1), $y(x) \equiv y_0$ is the only classical solution. ∎

If $\phi(x) \not\equiv y_0$ is a known classical solution of (1.1.1), then as in Theorem 1.24.1, we can use the transformation $w = y - \phi(x) + y_0$ to obtain the new problem

$$\begin{aligned} w' &= F(x, w) = f(x, w + \phi(x) - y_0) - f(x, \phi(x)) \\ w(x_0) &= y_0. \end{aligned} \tag{2.12.3}$$

Obviously, the initial value problem (2.12.3) admits the constant solution $w(x) = y_0$. Thus, replacing (2.12.1) with

$$u = y_0 + (t - x_0)F(t, u) \tag{2.12.4}$$

we can restate Theorems 2.12.1 and 2.12.2 involving the classical solution $\phi(x)$ instead of y_0.

Corollary 2.12.3. Let the function $f(x, y)$ be defined in $[x_0, x_0 + a] \times R$, continuous with respect to x, and such that $f(x, y_0) = 0$ for every $x \in [x_0, x_0 + a]$. Further, let the function $f(x, y)$ satisfy the Lipschitz condition (1.2.1) in $[x_0, x_0 + a] \times R$. Then, for the initial value problem (1.1.1), $y(x) \equiv y_0$ is the only solution.

Proof. For any two solutions u_1 and u_2 of (2.12.1) corresponding to the same t, we have

$$|u_1 - u_2| = (t - x_0)|f(t, u_1) - f(t, u_2)| \le L(t - x_0)|u_1 - u_2|.$$

Thus, $u_1 = u_2$ provided that $L(t - x_0) < 1$. Hence, by Theorem 2.12.2 it follows that (1.1.1) has a unique solution for $(t - x_0) < 1/L$. By iterating the application of Theorem 2.12.2, we get the uniqueness on the whole interval $[x_0, x_0 + a]$. ∎

Corollary 2.12.4. Let the function $f(x, y)$ be continuous in $[x_0, x_0 + a] \times R$, and $f(x, y_0) = 0$ for every $x \in [x_0, x_0 + a]$. Further, let the function $f(x, y)$ be nonincreasing in y for each fixed x in $[x_0, x_0 + a]$. Then, for the initial value problem (1.1.1), $y(x) \equiv y_0$ is the only classical solution.

Proof. Since for each $t \in [x_0, x_0 + a]$ the function

$$h(t, u) = y_0 + (t - x_0)f(t, u) - u$$

is decreasing in u, the result is a consequence of Theorem 2.12.2. ∎

Example 2.12.1. Consider the initial value problem

$$\begin{aligned} y' = f(x, y) &= \begin{cases} 1/(y_0 - y), & y \ne y_0 \\ 0, & y = y_0 \end{cases} \\ y(x_0) &= y_0. \end{aligned} \tag{2.12.5}$$

For this function $f(x,y)$ the equation (2.12.1) becomes

$$u = \begin{cases} y_0 + (t - x_0)/(y_0 - u), & u \neq y_0 \\ y_0, & u = y_0 \end{cases}$$

which has a unique solution $u = y_0$. Thus, in view of Theorem 2.12.2, for the problem (2.12.5), $y(x) \equiv y_0$ is the only solution. ∎

The converse of Theorem 2.12.1 does not hold, i.e., the uniqueness of (1.1.1) does not imply the uniqueness of (2.12.1). This we shall demonstrate by the following :

Example 2.12.2. Consider the initial value problem

$$(2.12.6) \qquad \begin{aligned} y' = f(x,y) &= \begin{cases} |y|^\alpha \sin(1/y); & y \neq 0,\ 0 < \alpha < 2 \\ 0, & y = 0 \end{cases} \\ y(0) &= 0 \end{aligned}$$

for which by elementary methods it can be shown that $y(x) \equiv 0$ is the only solution. For this function $f(x,y)$ the equation (2.12.1) takes the form

$$(2.12.7) \qquad u = \begin{cases} t|u|^\alpha \sin(1/u), & u \neq 0 \\ 0, & u = 0. \end{cases}$$

If $\alpha = 1/2$, then

$$u = \left(\frac{\pi}{2} + 2n\pi\right)^{-1}, \quad t = \left(\frac{\pi}{2} + 2n\pi\right)^{-1/2}; \quad n = 0, 1, 2, \cdots$$

are solutions of (2.12.7).

Further, if $\alpha = 1$, then (2.12.7) reduces to

$$\operatorname{sgn}(u) = t \sin(1/u),$$

which cannot be satisfied if $t < 1$. Thus, Theorem 2.12.2 is applicable even when f is neither Lipschitzian nor decreasing in y. ∎

CHAPTER 3

FIRST ORDER DIFFERENTIAL SYSTEMS

3.1 INTRODUCTION

With a few exceptions throughout in this chapter we shall consider (1.1.1) as an n - dimensional system, i.e., $f = (f_1, \cdots, f_n)$ and $y = (y_1, \cdots, y_n)$. If $f(x, y)$ is continuous in a domain $D \subseteq R^{n+1}$, and $(x_0, y_0) \in D$ then by a n - **solution** of (1.1.1) in an interval J containing x_0, we mean a function $y(x) = (y_1(x), \cdots, y_n(x))$ satisfying :

(i) $y_i(x_0) = y_{0i}$, $1 \leq i \leq n$

(ii) for all $x \in J$, the points $(x, y_1(x), \cdots, y_n(x)) \in D$,

(iii) $y'(x) = (y_1{}'(x), \cdots, y_n{}'(x))$ exists and continuous for all $x \in J$; and

(iv) $y_i{}'(x) = f_i(x, y_1(x), \cdots, y_n(x))$, $1 \leq i \leq n$.

As in the scalar case, if the function $f(x, y)$ is only defined in the domain $D = J \times T \subseteq R^{n+1}$, where J is one of the intervals $[x_0, x_0 + a], [x_0, x_0 +$

$a), (x_0, x_0 + a], (x_0, x_0 + a)$, and $y_0 \in T \subseteq R^n$, then by a n - **classical solution** of (1.1.1) in the interval $\{x_0\} \cup J$, we mean a function $y(x) = (y_1(x), \cdots, y_n(x))$ satisfying the above conditions (i) and (ii), and

(v) $y(x) = (y_1(x), \cdots, y_n(x))$ is continuous in $\{x_0\} \cup J$,

(vi) $y'(x) = ({y_1}'(x), \cdots, {y_n}'(x))$ exists in J^o; and

(vii) for all $x \in J^o$, ${y_i}'(x) = f_i(x, y_1(x), \cdots, y_n(x))$, $1 \le i \le n$.

We shall also consider the case when the function $f(x, y)$ satisfies the Caratheòdory conditions, i.e., $f_i(x, y_1, \cdots, y_n)$, $1 \le i \le n$ are defined in D; measurable in x for each fixed $y = (y_1, \cdots, y_n)$; continuous in $y = (y_1, \cdots, y_n)$ for each fixed x, and there exist (Lebesgue) integrable functions $M_i(x)$, $1 \le i \le n$ in the interval $\bar{J}$ such that

$$|f_i(x, y_1, \cdots, y_n)| \le M_i(x), \ (x, y_1, \cdots, y_n) \in D.$$

In such a case (or otherwise) by a n - **extended solution** of (1.1.1) in the interval $\bar{J}$, we mean a function $y(x) = (y_1(x), \cdots, y_n(x))$ satisfying the above conditions (i) and (ii), and

(viii) $y(x) = (y_1(x), \cdots, y_n(x))$ is absolutely continuous in $\bar{J}$; and

(ix) ${y_i}'(x) = f_i(x, y_1(x), \cdots, y_n(x))$, $1 \le i \le n$ almost everywhere in $\bar{J}$, i.e., for all $x \in \bar{J}$ except on a set of measure zero.

For $y \in R^n$ we shall mostly consider the norm as $|y| = \sum_{i=1}^n |y_i|$, so that the definitions (but, not the meaning) of $\bar{S}, S, \ \bar{S}_+, \ S_+, \ \bar{S}_-, \bar{\Delta}$, and Δ remain the same as in the scalar case. We shall have occasions to use the Euclidean norm and to distinguish it from $|.|$, we shall denote it as $\|.\|$. For $y, \bar{y} \in R^n$, $y \cdot \bar{y}$ represents the usual scalar multiplication. For two vectors $y = (y_1, \cdots, y_n)$ and $\bar{y} = (\bar{y}_1, \cdots, \bar{y}_n)$ we define $y \le \bar{y}$ ($y < \bar{y}$), if and only if, $y_i \le \bar{y}_i$ ($y_i < \bar{y}_i$), $1 \le i \le n$. We say $f(x, y) = (f_1(x, y), \cdots, f_n(x, y))$ is nondecreasing in x if $f(x_1, y) \le f(x_2, y)$ whenever (x_1, y) and (x_2, y) are in D and $x_1 \le x_2$, and $f(x, y)$ is nondecreasing in y if $f(x, y) \le f(x, \bar{y})$ whenever (x, y) and $(x, \bar{y})$ are in D and $y \le \bar{y}$, while $f(x, y)$ is quasi - nondecreasing in y if for each

$i = 1, \cdots, n$, $f_i(x, y) \le f_i(x, \bar{y})$ whenever $y \le \bar{y}$ and $y_i = \bar{y}_i$. Finally, we say that $y(x)$ is strictly increasing function of x if $y(x_1) < y(x_2)$ whenever $x_1 < x_2$.

Many results of Chapters 1 and 2 remain the same for the systems also, and, therefore we shall begin with those results which require some explanation and/or modification. This will be followed by several new results.

3.2 LIPSCHITZ UNIQUENESS THEOREM

We begin with the observation that if for $y \in R^n$ we use the norm $|y| = \sum_{i=1}^n |y_i|$ then the uniform Lipschitz continuity for the function $f(x, y)$ introduced in Definition 1.2.1 remains the same. However, to understand its meaning we consider the function $f(x, y) = (a_{11}y_1 + a_{12}y_2,\ a_{21}y_1 + a_{22}y_2)$ and $D = R^3$. Since

$$\begin{aligned}
&|f(x, y) - f(x, \bar{y})| \\
&= |(a_{11}(y_1 - \bar{y}_1) + a_{12}(y_2 - \bar{y}_2),\ a_{21}(y_1 - \bar{y}_1) + a_{22}(y_2 - \bar{y}_2))| \\
&= |a_{11}(y_1 - \bar{y}_1) + a_{12}(y_2 - \bar{y}_2)| + |a_{21}(y_1 - \bar{y}_1) + a_{22}(y_2 - \bar{y}_2)| \\
&\le |a_{11}||y_1 - \bar{y}_1| + |a_{12}||y_2 - \bar{y}_2| + |a_{21}||y_1 - \bar{y}_1| + |a_{22}||y_2 - \bar{y}_2| \\
&= [|a_{11}| + |a_{21}|]\,|y_1 - \bar{y}_1| + [|a_{12}| + |a_{22}|]\,|y_2 - \bar{y}_2| \\
&\le \max\{|a_{11}| + |a_{21}|, |a_{12}| + |a_{22}|\}\,[|y_1 - \bar{y}_1| + |y_2 - \bar{y}_2|] \\
&\le \max\{|a_{11}| + |a_{21}|, |a_{12}| + |a_{22}|\}|y - \bar{y}|
\end{aligned}$$

this function satisfies the condition (1.2.1) in R^3 with the Lipschitz constant

$$L = \max\{|a_{11}| + |a_{21}|, |a_{12}| + |a_{22}|\}.$$

As in Lemma 1.2.1 the following result gives sufficient conditions for the function $f(x, y)$ to satisfy the Lipschitz condition (1.2.1).

Lemma 3.2.1. Let the domain D be convex and for all (x, y) in D the partial derivatives $\frac{\partial f}{\partial y_i}$; $i = 1, \cdots, n$ exist and $|\frac{\partial f}{\partial y}| \le L$. Then, the function $f(x, y)$ satisfies the Lipschitz condition (1.2.1).

Proof. Let (x,y) and $(x,\bar{y})$ be fixed points in D. Then, since D is convex, for all $0 \le t \le 1$ the points $(x, \bar{y}+t(y-\bar{y}))$ are in D. Thus, the function $F(t) = f(x, \bar{y}+t(y-\bar{y}))$, $0 \le t \le 1$ is well defined, also

$$\begin{aligned} F'(t) &= (y_1-\bar{y}_1)\frac{\partial f}{\partial y_1}(x,\bar{y}+t(y-\bar{y})) + \cdots \\ &+ (y_n-\bar{y}_n)\frac{\partial f}{\partial y_n}(x,\bar{y}+t(y-\bar{y})) \end{aligned}$$

and hence

$$\begin{aligned} |F'(t)| &\le \sum_{i=1}^{n}\left|\frac{\partial f_i}{\partial y_1}(x,\bar{y}+t(y-\bar{y}))\right| |y_1-\bar{y}_1| + \cdots \\ &+ \sum_{i=1}^{n}\left|\frac{\partial f_i}{\partial y_n}(x,\bar{y}+t(y-\bar{y}))\right| |y_n-\bar{y}_n| \\ &\le L\sum_{i=1}^{n}|y_i-\bar{y}_i| \\ &= L\,|y-\bar{y}|. \end{aligned}$$

Now from the relation

$$f(x,y) - f(x,\bar{y}) = F(1) - F(0) = \int_0^1 F'(t)dt$$

we find that

$$|f(x,y) - f(x,\bar{y})| \le \int_0^1 |F'(t)|dt \le L|y-\bar{y}|. \quad \blacksquare$$

As an example once again we consider $f(x,y) = (a_{11}y_1 + a_{12}y_2,\ a_{21}y_1 + a_{22}y_2)$. Since $\frac{\partial f}{\partial y_1} = (a_{11}, a_{21})$, $\frac{\partial f}{\partial y_2} = (a_{12}, a_{22})$, it is clear that $|\frac{\partial f}{\partial y}| = \max\{|a_{11}|+|a_{21}|, |a_{12}|+|a_{22}|\} = L$, as it should be.

Next, we note that like the differentiation if the integration is also performed component - wise, i.e., $\int_\alpha^\beta y(t)dt = (\int_\alpha^\beta y_1(t)dt, \cdots, \int_\alpha^\beta y_n(t)dt)$ then the integral representation (1.2.4) of the n - solutions in Lemma 1.2.2 remains the same.

With these preliminaries it is clear that the statement of the Lipschitz Uniqueness Theorem 1.2.4 and its first proof for n - solutions of (1.1.1) remain the same. However, its second proof needs some modification because for the vector - valued functions there is nothing similar to the mean value theorem. This one can see from the simple example $y(x) = (-1 + \cos x, \sin x)$ in the interval $[0, 2\pi]$.

Let $y(x)$ and $\bar{y}(x)$ be two n - solutions of the initial value problem (1.1.1). Then, as in the second proof of Theorem 1.2.4 for the function $|y(x) - \bar{y}(x)| = \sum_{i=1}^{n} |y_i(x) - \bar{y}_i(x)|$ we arrive at the numbers ε, $\hat{\varepsilon}$ and l. However, since

$$\begin{aligned} 0 < |y(\hat{\varepsilon}) - \bar{y}(\hat{\varepsilon})| &= \sum_{i=1}^{n} |(y_i(\hat{\varepsilon}) - \bar{y}_i(\hat{\varepsilon})) - (y_i(\varepsilon) - \bar{y}_i(\varepsilon))| \\ &= \sum_{i=1}^{n} \delta_i((y_i(\hat{\varepsilon}) - \bar{y}_i(\hat{\varepsilon})) - (y_i(\varepsilon) - \bar{y}_i(\varepsilon))) \\ &= \sum_{i=1}^{n} \delta_i(\hat{\varepsilon} - \varepsilon)(y_i'(\varepsilon^*) - \bar{y}_i'(\varepsilon^*)) \\ &= \sum_{i=1}^{n} \delta_i(\hat{\varepsilon} - \varepsilon)(f_i(\varepsilon^*, y(\varepsilon^*)) - f_i(\varepsilon^*, \bar{y}(\varepsilon^*))) \\ &\leq (\hat{\varepsilon} - \varepsilon)|f(\varepsilon^*, y(\varepsilon^*)) - f(\varepsilon^*, \bar{y}(\varepsilon^*))| \\ &\leq lL|y(\varepsilon^*) - \bar{y}(\varepsilon^*)|, \end{aligned}$$

where each δ_i is 1 or -1, and for each i the number ε^* is the same and $\varepsilon < \varepsilon^* < \hat{\varepsilon}$, we arrive at the same contradiction.

To extend the third proof of Theorem 1.2.4 for systems, we recall that in R^n all the norms are equivalent and therefore there exist positive constants m and M such that for all $y \in R^n$, $m|y| \leq \|y\| \leq M|y|$. Once again, let $y(x)$ and $\bar{y}(x)$ be two n - solutions of the initial value problem (1.1.1). Then, the function $v(x) = \|y(x) - \bar{y}(x)\|^2 = \sum_{i=1}^{n}(y_i(x) - \bar{y}_i(x))^2$ satisfies

$$\begin{aligned} v'(x) &= 2\sum_{i=1}^{n}(y_i(x) - \bar{y}_i(x))(y_i'(x) - \bar{y}_i'(x)) \\ &\leq 2\sum_{i=1}^{n} |y_i(x) - \bar{y}_i(x)||f_i(x, y(x)) - f_i(x, \bar{y}(x))| \\ &\leq 2\left(\sum_{i=1}^{n}(y_i(x) - \bar{y}_i(x))^2\right)^{1/2} \left(\sum_{i=1}^{n}(f_i(x, y(x)) - f_i(x, \bar{y}(x)))^2\right)^{1/2} \end{aligned}$$

$$\begin{aligned}
&\le 2M\|y(x)-\bar{y}(x)\|\ |f(x,y(x))-f(x,\bar{y}(x))| \\
&\le \frac{2LM}{m}\|y(x)-\bar{y}(x)\|^2 \\
&= \frac{2LM}{m}v(x),
\end{aligned}$$

which leads to the same contradiction as in the third proof of Theorem 1.2.4.

The one - sided analog of Lipschitz uniqueness theorem namely Theorem 1.2.5 for systems takes the following form :

Theorem 3.2.2. Let $f(x,y)$ be continuous in $\bar{S}_+$ and for all (x,y), $(x,\bar{y})$ in $\bar{S}_+$ it satisfies one - sided Lipschitz condition

$$(3.2.1) \qquad (f(x,y)-f(x,\bar{y}))\cdot(y-\bar{y}) \le L\|y-\bar{y}\|^2.$$

Then, the initial value problem (1.1.1) has at most one n - solution in $[x_0, x_0 + a]$.

Proof. Let $y(x)$ and $\bar{y}(x)$ be two n - solutions of (1.1.1). Then, for the function $v(x)=\|y(x)-\bar{y}(x)\|^2$, $x\in[x_0,x_0+a]$ as above, we find that

$$\begin{aligned}
v'(x) &= 2\sum_{i=1}^{n}(y_i(x)-\bar{y}_i(x))(f_i(x,y(x))-f_i(x,\bar{y}(x))) \\
&= 2(f(x,y(x))-f(x,\bar{y}(x)))\cdot(y(x)-\bar{y}(x)) \\
&\le 2L\|y(x)-\bar{y}(x)\|^2 \\
&\le 2Lv(x).
\end{aligned}$$

Since $v(x_0)=0$ the above inequality implies that $v(x)\equiv 0$, i.e., $y(x)=\bar{y}(x)$ in $[x_0,x_0+a]$. ∎

3.3 PEANO'S UNIQUENESS THEOREM

Corresponding results of Section 1.3 for n - solutions of (1.1.1) can be stated as follows :

Theorem 3.3.1 (Peano's Uniqueness Theorem). Let $f(x,y)$ be continuous in $\bar{S}_+$ and for all (x,y), $(x,\bar{y})$ in $\bar{S}_+$ it satisfies

$$(f(x,y) - f(x,\bar{y})) \cdot (y - \bar{y}) \leq 0. \tag{3.3.1}$$

Then, the initial value problem (1.1.1) has at most one n - solution in $[x_0, x_0 + a]$. ∎

Theorem 3.3.2. Let $f(x,y)$ be continuous in $\bar{S}_-$ and for all (x,y), $(x,\bar{y})$ in $\bar{S}_-$ it satisfies

$$(f(x,y) - f(x,\bar{y})) \cdot (y - \bar{y}) \geq 0. \tag{3.3.2}$$

Then, the initial value problem (1.1.1) has at most one n - solution in $[x_0 - a, x_0]$. ∎

Theorem 3.3.3. Let $f(x,y)$ be continuous in $\bar{S}$ and let the inequalities (3.3.1) and (3.3.2) in $\bar{S}_+$ and $\bar{S}_-$ respectively, be satisfied. Then, the initial value problem (1.1.1) has at most one n - solution in $|x - x_0| \leq a$. ∎

3.4 HEINS' UNIQUENESS THEOREM

We shall prove a result here which accommodates Lipschitz as well as Osgood's uniqueness theorem. For this, we need the following :

Lemma 3.4.1. Assume that

(i) $\phi(x)$ is continuous in $[x_0, x_0 + a]$, $D_+\phi(x) < \infty$ when $\phi(x) \neq 0$, $x_0 < x < x_0 + a$,

(ii) $L(x)$ is integrable in $[x_0, x_0 + a]$ and assumes values in the extended real line; and

(iii) $g(x)$ is defined for finite reals and assumes finite real values, $g(0) = 0$, $\lim_{x\to 0^-} g(x) = -\infty$, $\lim_{x\to 0^+} g(x) = \infty$, and, $g'(x)$ exists and is nonnegative for all finite $x \neq 0$.

Then, the inequality

$$(3.4.1) \qquad D_+ g \circ \phi(x) \le L(x)$$

for almost all x for which $\phi(x) \neq 0$ implies that each of the sets $\{\phi(x) > 0\}$, $\{\phi(x) = 0\}$, $\{\phi(x) < 0\}$ is connected. Also, $\phi(x) > 0 = \phi(\bar{x})$ implies $x < \bar{x}$, and $0 = \phi(\bar{x}) > f(x^*)$ implies that $\bar{x} < x^*$. Further, on the first (third) set $g \circ \phi(x) - \bar{L}(x)$ is nonincreasing, where $\bar{L}(x) = \int_{x_0}^{x} L(t)dt$.

The above lemma remains valid if D_+ is replaced by D_-.

Proof. First we shall show that $g \circ \phi(x) - \bar{L}(x)$ is nonincreasing on each component of $\{\phi(x) > 0\}$. If $\inf L(x) > -\infty$, we introduce a nonnegative, lower semicontinuous, integrable function $k(x)$, $x_0 \le x \le x_0 + a$ which takes the value ∞ at each point of the set of measure zero consisting those x for which $\phi(x) > 0$ and either (3.4.1) fails to hold or $L(x)$ is not the derivative (finite or not) of $\bar{L}(x)$ at x. Such a function $k(x)$ can be constructed in a well - known manner as the restriction to $[x_0, x_0 + a]$ of the sum of the characteristic functions of open sets O_n; $n = 1, 2, \cdots$ each of which contains the set of zero measure in question and which are such that the sum of their measures is finite. Let $\varepsilon > 0$ and define

$$(3.4.2) \qquad \bar{L}_\varepsilon(x) = \int_{x_0}^{x} (L(t) + \varepsilon k(t))dt.$$

Then, the lower right derivative of $g \circ \phi(x) - \bar{L}_\varepsilon(x)$ is nonpositive at each point of $\{\phi(x) > 0\}$ different from x_0. Consequently, $g \circ \phi(x) - \bar{L}_\varepsilon(x)$ is nonincreasing on each component of $\{\phi(x) > 0\}$. Thus, as $\varepsilon \to 0$, we find that the same conclusion holds for $g \circ \phi(x) - \bar{L}(x)$. The fact that $g \circ \phi(x) - \bar{L}(x)$ is nonincreasing on each component of $\{\phi(x) > 0\}$ when $\inf L(x) = -\infty$ follows on noting that, for each finite constant c, the inequality (3.4.1) holds for almost all x of the set $\{\phi(x) > 0\}$ when $L(x)$ is replaced by $\max\{L(x), c\}$. The same result holds if D_+ is replaced by D_-.

It is now easy to see that $g \circ \phi(x) - \bar{L}(x)$ is nonincreasing on each component of $\{\phi(x) < 0\}$. In fact it suffices to introduce $g_1(x) = -g(-x)$ and $\phi_1(x) = -\phi(2x_0 + a - x)$ and to observe that

(3.4.3) $$D_{-}g_1 \circ \phi_1(x) = D_{+}g \circ \phi(2x_0 + a - x).$$

Now from the fact that the map $x \to g_1 \circ \phi_1(x) + \bar{L}(2x_0 + a - x)$ is nonincreasing on each component of $\{\phi_1(x) > 0\}$ it follows that $g \circ \phi(x) - \bar{L}(x)$ is nonincreasing on each component of $\{\phi(x) < 0\}$.

Next, we shall show that the set $\{\phi(x) = 0\}$ is connected. If this is not true, then there must be a component of $\{\phi(x) > 0\}$ or of $\{\phi(x) < 0\}$ at the end points of which $\phi(x)$ vanishes. However, such a component cannot exist by virtue of the monotone character of $g \circ \phi(x) - \bar{L}(x)$ and the hypothesis (iii). Further, if $\{\phi(x) = 0\} \neq \emptyset$, then we have

$$\sup\{\phi(x) > 0\} \le \min\{\phi(x) = 0\} \le \max\{\phi(x) = 0\} \le \inf\{\phi(x) < 0\}. \quad \blacksquare$$

Theorem 3.4.2 (Heins' Uniqueness Theorem). Let $f(x, y)$ be defined in $\bar{S}_{+}$, and let there exist a nonnegative integrable function $L(x)$, $x_0 \le x \le x_0 + a$ and a function $g(x)$ satisfying the conditions of Lemma 3.4.1. Further, let for any two n - classical solutions $y(x)$ and $\bar{y}(x)$ of (1.1.1) in $[x_0, x_0 + a]$ the following inequality holds

(3.4.4) $$|f(x, y(x)) - f(x, \bar{y}(x))|\, g'(|y(x) - \bar{y}(x)|) \le L(x).$$

Then, $y(x) = \bar{y}(x)$, $x_0 \le x \le x_0 + a$.

Proof. Let $\phi(x) = |y(x) - \bar{y}(x)|$ so that it satisfies almost everywhere the inequality $D_{+}g\circ\phi(x) = g'(\phi(x))\cdot D_{+}\phi(x) \le L(x)$ when $\phi(x) > 0$. Further, as in Lemma 3.4.1 setting $\phi_1(x) = \phi(2x_0 + a - x)$, we find that $\phi_1(x)$ satisfies almost everywhere the inequality $D_{-}g\circ\phi_1(x) \le L(2x_0 + a - x)$ when $\phi(2x_0 + a - x) > 0$. The result now follows from Lemma 3.4.1. $\blacksquare$

Inequality (3.4.4) with $g'(x) = x^{-1}$, $x > 0$ reduces to the Lipschitz condition; and with $g'(x) = (g_1(x))^{-1}$, $x > 0$ where $g_1(x)$ is defined and continuous in $[0, \infty)$, $g_1(0) = 0$, $g_1(x) > 0$, $x > 0$ and $\lim_{\epsilon \to 0^+} \int_{\epsilon} \frac{dt}{g_1(t)} = \infty$ becomes Osgood's condition.

3.5 GIULIANO'S UNIQUENESS THEOREM

An extension of Montel - Tonelli's Uniqueness Theorem 1.5.1 to n - extended solutions is the following :

Theorem 3.5.1 (Giuliano's Uniqueness Theorem). Let $f(x,y)$ satisfy Caratheodory conditions in $\Delta_0 : x_0 < x < x_0 + a,\ \|y\| < \infty$. Further, let for all $(x,y),\ (x,\bar{y}) \in \Delta_0$ it satisfy **Giuliano's condition**

$$(3.5.1) \qquad (f(x,y) - f(x,\bar{y})) \cdot (y - \bar{y}) \le h(x)g(\|y - \bar{y}\|^2),$$

where the function $h(x) \ge 0$ is integrable (Lebesgue) in every interval $[\alpha,\beta] \subset (x_0, x_0 + a)$ and $g(z) > 0$ is continuous for $z > 0,\ g(0) = 0$ and satisfies the condition (1.4.1). Then, the initial value problem (1.1.1) has at most one n - extended solution in $[x_0, x_0 + a]$.

Proof. As usual let $y(x)$ and $\bar{y}(x)$ be two n - extended solutions of (1.1.1) in $[x_0, x_0 + a]$. Then, $\phi(x) = \|y(x) - \bar{y}(x)\|$ satisfies $\phi'(x) = 2(f(x,y(x)) - f(x,\bar{y}(x))) \cdot (y(x) - \bar{y}(x))$, and hence $\phi'(x) \le 2h(x)g(\phi(x))$. Now following as in Theorem 1.5.1 we find that $\phi(x) \equiv 0$, i.e., $y(x) = \bar{y}(x),\ x \in [x_0, x_0 + a]$. ∎

3.6 NAGUMO'S UNIQUENESS THEOREM

While the statement and the first proof of Nagumo's Uniqueness Theorem 1.6.2 remain the same in the case of systems also, its second proof needs a similar modification as we have made in Section 3.2 for the Lipschitz Uniqueness Theorem 1.2.4. Further, its one - sided analog Theorem 1.6.3 takes the following form :

Theorem 3.6.1. Let $f(x,y)$ be continuous in $\bar{S}_+$ and for all $(x,y),\ (x,\bar{y}) \in \bar{S}_+,\ x \ne x_0$ it satisfies

$$(3.6.1) \qquad (f(x,y) - f(x,\bar{y})) \cdot (y - \bar{y}) \le \frac{\|y - \bar{y}\|^2}{x - x_0}.$$

Then, the initial value problem (1.1.1) has at most one n - solution in $[x_0, x_0 + a]$. ∎

The statement of Theorem 2.3.1 can be modified similarly.

3.7 WITTE'S UNIQUENESS THEOREM

Theorem 1.10.3 remains the same for the systems also, however in its one - sided analog inequalities (1.10.5) and (1.10.6) have to be replaced by

(3.7.1) $(f(x,y) - f(x,\bar{y}))\cdot(y-\bar{y}) \le h(x)\|y-\bar{y}\|^2,\ (x,y),(x,\bar{y}) \in \Delta;$ and

(3.7.2) $\|f(x,y) - f(x,\bar{y})\| = o(h(x)\exp(H(x)))$ as $x \to 0^+$ uniformly with respect to $y, \bar{y} \in [-\delta, \delta]$, $\delta > 0$ arbitrary.

3.8 KAMKE'S UNIQUENESS THEOREM

For n - solutions of (1.1.1) we state a result which is a slight modification of Theorem 1.15.1.

Theorem 3.8.1 (Kamke's Uniqueness Theorem). Assume that

(i) the function $g(x,z)$ is nonnegative and defined for $x_0 < x < x_0 + a$, $0 \le z \le 2b$, measurable in x for fixed z and continuous in z for fixed x; there exists a function $M(x)$ defined for $x_0 < x < x_0 + a$ which is integrable (Lebesgue) in $x^* < x < x_0 + a$ for every $x^* > x_0$, such that $g(x,z) \le M(x)$; further, for every x_1, $x_0 < x_1 < x_0 + a$, $z(x) \equiv 0$ is the only absolutely continuous function in $x_0 \le x < x_1$ for which $z'_+(x_0)$ exists and satisfies (1.15.1), (1.15.2) almost everywhere in $x_0 \le x < x_1$; and

(ii) the function $f(x,y)$ is continuous in $\bar{S}_+$ and for all (x,y), $(x,\bar{y}) \in \bar{S}_+$ where $x \neq x_0$ the inequality (1.11.2) is satisfied.

Then, the initial value problem (1.1.1) has at most one n - solution in $[x_0, x_0 + a]$. ∎

If $f(x,y)$ satisfies Caratheodory conditions instead of being continuous in $\bar{S}_+$ then with obvious changes the above theorem holds for n - extended solutions of (1.1.1).

3.9 WEND'S UNIQUENESS THEOREM

The following result is a generalization of Theorem 2.6.1.

Theorem 3.9.1 (Wend's Uniqueness Theorem). Let $\hat{S}_+$ be the subregion of $\bar{S}_+$ where $y \geq y_0$. Suppose the function $f(x,y)$ is defined in $\bar{S}_+$; and nondecreasing in x and quasi - nondecreasing in y for (x,y) in $\hat{S}_+$. Further, suppose that for any n - extended solution $y(x)$ of (1.1.1), $f(x,y(x)) > 0$ for all $x \in (x_0, x_0 + a]$. Then, the initial value problem (1.1.1) has at most one n - extended solution in $[x_0, x_0 + a]$.

Proof. Assume that there are two n - extended solutions $y(x)$ and $\bar{y}(x)$ of (1.1.1). Then, each component of $y(x)$ and $\bar{y}(x)$ is strictly increasing and absolutely continuous in $[x_0, x_0 + a]$. In $[x_0, x_0 + a]$ let the functions $v_i(x)$ be defined by $\bar{y}_i(v_i(x)) = y_i(x)$. Obviously, each $v_i(x)$ is strictly increasing and absolutely continuous, $v_i(x_0) = x_0$ and $v_i'(x)$ exists almost everywhere in $[x_0, x_0 + a]$. Thus, in this interval

$$y_i'(x) = \bar{y}_i'(v_i(x))v_i'(x), \ 1 \leq i \leq n \tag{3.9.1}$$

almost everywhere. Suppose that for some $x_1 \in (x_0, x_0 + a)$, there exists an integer i, $1 \leq i \leq n$ such that $y_i(x_1) < \bar{y}_i(x_1)$. Then, $v_i(x_1) < x_1$, and it follows that there exists $\bar{x}$, $x_0 < \bar{x} < x_1$ and an integer j, $1 \leq j \leq n$ such that $v_j(\bar{x}) < \bar{x}$, $v_j(\bar{x}) \leq v_k(\bar{x})$, $1 \leq k \leq n$, and $v_j'(\bar{x})$, $y_j'(\bar{x})$ and $\bar{y}_j'(v_j(\bar{x}))$ exist.

Thus, from (3.9.1) and the hypotheses, we obtain

$$\begin{aligned} 0 < y_j'(\bar{x}) &= f_j(v_j(\bar{x}), \bar{y}_1(v_j(\bar{x})), \cdots, \bar{y}_n(v_j(\bar{x})))v_j'(\bar{x}) \\ &\le f_j(\bar{x}, \bar{y}_1(v_1(\bar{x})), \cdots, \bar{y}_n(v_n(\bar{x})))v_j'(\bar{x}) \\ &= y_j'(\bar{x})v_j'(\bar{x}). \end{aligned}$$

Hence, it follows that $v_j'(\bar{x}) \ge 1$.

For each $x \in [x_0, x_0 + a]$, let $v(x) = \min\{v_1(x), \cdots, v_n(x)\}$. It is clear that $v(x)$ is absolutely continuous, and if $v'(x)$ and $v_i'(x)$ exist and $v(x) = v_i(x)$, then $v'(x) = v_i'(x)$. Let $x^* = \inf\{x : x_0 \le x \le \bar{x}$ and for each $t, x < t \le \bar{x}$ there exists an integer $i, 1 \le i \le n$ such that $y_i(t) < \bar{y}_i(t)\}$. Then, $x_0 \le x^* < \bar{x}$. By the above argument $v'(x) \ge 1$ almost everywhere in $[x^*, \bar{x}]$, and hence in view of the absolute continuity of $v(x)$ in this interval it follows that $v(x^*) \le v(\bar{x}) + (x^* - \bar{x}) = (v_j(\bar{x}) - \bar{x}) + x^* < x^*$. But this implies that for some i, $v(x^*) = v_i(x^*)$ and $y_i(x^*) = \bar{y}_i(v_i(x^*)) < \bar{y}_i(x^*)$, so $x^* > x_0$ and $y_i(x) < \bar{y}_i(x)$ immediately to the left of x^*. This contradicts the definition of x^*. ∎

3.10 BOWNDS' UNIQUENESS THEOREM

In the following result the ideas of Theorem 2.9.1 are used to obtain an analogous result for n - classical solutions of (1.1.1). However, for $n = 1$ it reduces only to a weaker form of Theorem 2.9.1.

Theorem 3.10.1 (Bownds' Uniqueness Theorem). Suppose that it is possible to find three functions g, h, k in $\bar{S}_+$ such that for all $(x, y) \in \bar{S}_+$

$$f(x, y) = \operatorname{diag}[g_1(x, y_1), \cdots, g_n(x, y_n)]\, h(x, y) + k(x, y), \tag{3.10.1}$$

where g is continuous and strictly positive in $\bar{S}_+$, and $\frac{\partial}{\partial x}g(x, y)$ exists, is non-negative, and is dominated by an (Lebesgue) integrable function of y in $\bar{S}_+$. Consider the function q in $\bar{S}_+$ which is defined component - wise as follows

$$q_i(x, y) = \frac{k_i(x, y)}{g_i(x, y_i)}, \tag{3.10.2}$$

and suppose that h and q satisfy a generalized Lipschitz condition in $\bar{S}_+$, i.e., there exists a nonnegative (Lebesgue) integrable function $\lambda(x)$, $x \in (x_0, x_0+a)$ such that for all (x,y), $(x,\bar{y})$ in $\bar{S}_+$

$$|h(x,y) - h(x,\bar{y})| + |q(x,y) - q(x,\bar{y})| \le \lambda(x)|y - \bar{y}|. \tag{3.10.3}$$

Then, the initial value problem (1.1.1) has at most one n - classical solution in $[x_0, x_0 + a]$.

Proof. Suppose there are two n - classical solutions $y(x)$ and $\bar{y}(x)$ of (1.1.1). In the interval $[x_0, x_0 + a]$ we define a scalar function $\phi(x)$ as follows

$$\phi(x) = \frac{1}{2}\sum_{i=1}^{n}\left(\int_{\bar{y}_i(x)}^{y_i(x)} \frac{1}{g_i(x,z)}dz\right)^2. \tag{3.10.4}$$

Then, $\phi(x_0) = 0$, and

$$\phi(x) \ge 0, \ \ x \in [x_0, x_0 + a]. \tag{3.10.5}$$

By the hypotheses on g there exists an (Lebesgue) integrable function $\mu(y)$ in $|y - y_0| \le b$ so that

$$0 \le \frac{\partial}{\partial x} g_i(x,y) \le \mu(y), \ 1 \le i \le n \tag{3.10.6}$$

where y lies between $y_i(x)$ and $\bar{y}_i(x)$, $x \in [x_0, x_0+a]$. This implies that (3.10.4) can be differentiated to obtain

$$\phi'(x) = \sum_{i=1}^{n}\left(\int_{\bar{y}_i(x)}^{y_i(x)} \frac{dz}{g_i(x,z)}\right) \times \left(\int_{\bar{y}_i(x)}^{y_i(x)} \frac{\partial}{\partial x}\frac{dz}{g_i(x,z)} + \frac{y_i'(x)}{g_i(x,y_i(x))} - \frac{\bar{y}_i'(x)}{g_i(x,\bar{y}_i(x))}\right). \tag{3.10.7}$$

Now because of the positivity of g_i and the nonnegativity of $\frac{\partial}{\partial x}g_i$, it follows that

$$\left(\int_{\bar{y}_i(x)}^{y_i(x)} \frac{dz}{g_i(x,z)}\right)\left(\int_{\bar{y}_i(x)}^{y_i(x)} \frac{\partial}{\partial x}\frac{dz}{g_i(x,z)}\right) \le 0, \ 1 \le i \le n. \tag{3.10.8}$$

Therefore, we find that

$$\phi'(x) \le \sum_{i=1}^{n}\left(\int_{\bar{y}_i(x)}^{y_i(x)} \frac{dz}{g_i(x,z)}\right)\left(\frac{y_i'(x)}{g_i(x,y_i(x))} - \frac{\bar{y}_i'(x)}{g_i(x,\bar{y}_i(x))}\right). \tag{3.10.9}$$

Next, in view of (3.10.1) and (3.10.2), we have

$$(3.10.10)\qquad \frac{y_i'(x)}{g_i(x, y_i(x))} = \frac{g_i(x, y_i(x))h_i(x, y(x)) + k_i(x, y(x))}{g_i(x, y_i(x))}$$

$$= h_i(x, y(x)) + q_i(x, y(x))$$

with a similar expression involving $\bar{y}_i(x)$. Also, since g is bounded, there exists a $m > 0$ such that

$$(3.10.11)\qquad \left|\int_{\bar{y}_i(x)}^{y_i(x)} \frac{dz}{g_i(x, z)}\right| \le \frac{|y_i(x) - \bar{y}_i(x)|}{m}, \quad x \in [x_0, x_0 + a].$$

Using (3.10.10) and (3.10.11) in (3.10.9), we get

$$(3.10.12)\qquad \phi'(x) \le \frac{1}{m}\sum_{i=1}^{n} |y_i(x) - \bar{y}_i(x)|$$

$$\times \left[|h_i(x, y(x)) - h_i(x, \bar{y}(x))| + |q_i(x, y(x)) - q_i(x, \bar{y}(x))|\right].$$

Now from the condition (3.10.3) it is clear that

$$(3.10.13)\qquad |h_i(x, y(x)) - h_i(x, \bar{y}(x))| + |q_i(x, y(x)) - q_i(x, \bar{y}(x))|$$

$$\le \bar{\lambda}(x) \sum_{i=1}^{n} |y_i(x) - \bar{y}_i(x)|,$$

where $\bar{\lambda}(x)$ is a nonnegative (Lebesgue) integrable function in $(x_0, x_0 + a)$. Therefore, we find that

$$\phi'(x) \le \frac{\bar{\lambda}(x)}{m}\left(\sum_{i=1}^{n} |y_i(x) - \bar{y}_i(x)|\right)^2$$

$$(3.10.14)\qquad \le \frac{n\bar{\lambda}(x)}{m}\sum_{i=1}^{n} |y_i(x) - \bar{y}_i(x)|^2.$$

If $M = \max_{1\le i\le n}\{g_i(x, y) : x \in [x_0, x_0 + a],\ |y - y_0| \le b\}$, then from (3.10.4) it follows that

$$(3.10.15)\qquad \phi(x) \ge \frac{1}{2M^2}\sum_{i=1}^{n} (y_i(x) - \bar{y}_i(x))^2.$$

Now combining (3.10.14) and (3.10.15), we obtain

$$\phi'(x) \le \frac{n\bar{\lambda}(x)}{m} 2M^2 \phi(x), \quad x \in (x_0, x_0 + a)$$

and hence

$$\psi'(x) \geq 0 \tag{3.10.16}$$

almost everywhere in $(x_0, x_0 + a)$, where

$$\psi(x) = -\phi(x) \exp\left(-\frac{2nM^2}{m}\int_{x_0}^{x} \bar{\lambda}(t)dt\right).$$

Also it is not difficult to see that

$$D^+\psi(x) \geq -\exp\left(-\frac{2nM^2}{m}\int_{x_0}^{x} \bar{\lambda}(t)dt\right)\phi'(x) \tag{3.10.17}$$

everywhere in (x_0, x_0+a), and so, from the fact that the right side of (3.10.12) is finite, it follows that

$$D^+\psi(x) > -\infty \tag{3.10.18}$$

everywhere in $(x_0, x_0 + a)$. Therefore, (3.10.16) together with (3.10.18) imply that

$$0 \leq \int_{x_0}^{x} \psi'(t)dt \leq \psi(x) - \psi(x_0).$$

Consequently, since $\psi(x_0) = 0$, it follows that $\psi(x) \geq 0$, or $\phi(x) \leq 0$, which together with (3.10.4) and (3.10.5) imply that $\phi(x) \equiv 0$, i.e., $y(x) = \bar{y}(x)$, $x \in [x_0, x_0 + a]$. ∎

Example 3.10.1. For the initial value problem

$$y' = f(x,y) = \left([1+|y_1|]y_2x^{-1/2}\ ,\ (1-x)[1+|y_2|^{1/2}]x^{-1/2}\right),\ 0 < x < 1$$

$$y_0 = (1\ ,\ 0)$$

we can choose $g(x,y) = (1+|y|\ ,\ 1+|y|^{1/2})$, $h(x,y) = (y_2x^{-1/2}\ ,\ x^{-1/2})$, and $k(x,y) = (0\ ,\ -x^{1/2}[1+|y_2|^{1/2}])$, so that the conditions of Theorem 3.10.1 are satisfied, and hence it has at most one solution. ∎

Example 3.10.2. For the initial value problem

$$y' = G(x,y) = \left([1+|y_1|]y_2x^{-1/2}\ ,\ (1-x)|y_2|^{1/2}x^{-1/2}\right),\ 0 < x < 1$$

$$y_0 = (1\ ,\ 0)$$

there are at least two solutions, one is given by $y(x) = (1\ ,\ 0)$, while the second is

$$\bar{y}(x) = \left(2\exp\left\{\frac{2}{3}x^{3/2}\left[1 - \frac{2}{5}x + \frac{1}{21}x^2\right]\right\} - 1\ ,\ \left[x^{1/2} - \frac{1}{3}x^{3/2}\right]^2\right). \quad \blacksquare$$

As in Section 2.9 here also the above examples are so chosen that

$$G(x,y) - G(x,\bar{y}) = f(x,y) - f(x,\bar{y})$$

and hence Lipschitz type uniqueness theorems cannot be applied.

3.11 GARD'S UNIQUENESS THEOREM

Theorem 3.11.1 (Gard's Uniqueness Theorem). Let in Theorem 2.11.1 the condition (i) be replaced by
(i)′ for all $(x,y),\ (x,\bar{y}) \in S_+$ the following inequality holds

$$(f(x,y) - f(x,\bar{y})) \cdot (y - \bar{y}) \le \frac{\phi'(x)}{\phi(x)}\|y - \bar{y}\|^2. \tag{3.11.1}$$

Then, the initial value problem (1.1.1) has at most one n - classical solution in $[x_0, x_0 + a)$.

Proof. Let $y(x)$ and $\bar{y}(x)$ be two n - classical solutions of (1.1.1). Then, as in Theorem 2.11.1 we define a function

$$v(x) = \begin{cases} \frac{1}{2}\left[\|y(x) - \bar{y}(x)\|/\phi(x)\right]^2, & x \in (x_0, x_0 + a) \\ 0, & x = x_0. \end{cases}$$

Using (ii) and l'Hospital's rule, we obtain

$$\lim_{x\to x_0^+} \frac{y_i(x) - \bar{y}_i(x)}{\phi(x)} = \lim_{x\to x_0^+} \frac{f_i(x,y(x)) - f_i(x,\bar{y}(x))}{\phi'(x)} = 0,\ 1 \le i \le n.$$

Hence, the function $v(x)$ is continuous in $[x_0, x_0 + a)$. Now for $x > x_0$, using

(3.11.1), we obtain

$$\begin{aligned} v'(x) &= -\frac{\|y(x)-\bar{y}(x)\|^2}{\phi^3(x)}\phi'(x) \\ &+ \sum_{i=1}^{n}\frac{(y_i(x)-\bar{y}_i(x))(f_i(x,y(x))-f_i(x,\bar{y}(x)))}{\phi^2(x)} \\ &\le 0. \end{aligned}$$

Thus, $v(x)$ is continuous and nondecreasing in $[x_0, x_0+a)$. Since, $v(x)$ is nonnegative, we find that $v(x)\equiv 0$, i.e., $y(x)=\bar{y}(x)$ in $[x_0, x_0+a)$.

3.12 DRIVER - NORRIS' UNIQUENESS THEOREM

The result proved here provides a weaker criterion than the Lipschitz Uniqueness Theorem 1.2.4.

Theorem 3.12.1 (Driver - Norris' Uniqueness Theorem). Let $f(x,y)$ be continuous in $\bar{S}$; and each point in $\bar{S}$ has an open neighborhood U, a constant $L>0$, an integer $m\ge 0$, and functions h_i and g_i for $i=1,\cdots,m$ such that

$$(3.12.1)\qquad |f(x,y)-f(x,\bar{y})|\le L|y-\bar{y}| + L\sum_{i=1}^{m}|g_i(h_i(x,y))-g_i(h_i(x,\bar{y}))|$$

in $U\cap\bar{S}$, where $h_i : U\to R$ is continuously differentiable with

$$(3.12.2)\qquad \frac{\partial h_i(x,y)}{\partial x} + \sum_{j=1}^{n}\frac{\partial h_i(x,y)}{\partial y_j}f_j(x,y)\ne 0, \quad \text{in } U\cap\bar{S}$$

and each $g_i : R\to R$ is continuous and of bounded variation in bounded subintervals. Then, the initial value problem (1.1.1) has at most one n - solution in $|x-x_0|\le a$.

Proof. Let $y(x)$ and $\bar{y}(x)$ be two n - solutions of (1.1.1). We shall show that $y(x)=\bar{y}(x)$ in $[x_0, x_0+a]$. The proof in the interval $[x_0-a, x_0]$ is similar. Let $c=\inf\{x\in(x_0,x_0+a] : y(x)\ne\bar{y}(x)\}$. Then, $y(c)=\bar{y}(c)$.

For the point $(c, y(c))\in\bar{S}$ let U, L, m, h_i and g_i be as in the hypotheses of the theorem. Without loss of generality, we assume that for each i the expression in (3.12.2) is positive at $(c, y(c))$. Then, reducing U if necessary, the continuity of the derivatives of h_i assures that there exist positive constants p and M such that, for $i = 1, \cdots, m$

$$\frac{\partial h_i(x,y)}{\partial x} + \sum_{j=1}^{n} \frac{\partial h_i(x,y)}{\partial y_j} f_j(x,y) \geq p \quad \text{in } U \cap \bar{S} \tag{3.12.3}$$

and

$$|h_i(x,y) - h_i(x,\bar{y})| \leq M|y - \bar{y}| \quad \text{in } U. \tag{3.12.4}$$

Let $[\alpha_i, \beta_i]$ be a bounded interval which contains $h_i(U \cap \bar{S})$. This can always be done by reducing U. Then, g_i is the difference of two continuous nondecreasing functions in $[\alpha_i, \beta_i]$, and each of the latter can be extended to a continuous nondecreasing function in R by defining it to be constant in $(-\infty, \alpha_i]$ and constant in $[\beta_i, \infty)$. Without loss of generality, we shall assume that each g_i is itself nondecreasing in R and that

$$g_i(h_i(c, y(c))) = 0. \tag{3.12.5}$$

We define

$$\phi(x) = \int_c^x |y'(t) - \bar{y}'(t)|dt, \qquad \text{for } c \leq x \leq x_0 + a.$$

Then, $\phi(c) = 0$, $\phi'(c) = 0$, $\phi(x)$ and $\phi'(x)$ are continuous, $\phi'(x) \geq 0$ and $|y(x) - \bar{y}(x)| \leq \phi(x)$ in $[c, x_0 + a]$.

Let $d \in (c, x_0 + a]$ be so small that $(x, y(x))$ and $(x, \bar{y}(x))$ remain in U for $x \in [c, d)$. Then, from (3.12.4)

$$h_i(x, y(x)) - M\phi(x) \leq h_i(x, y(x)) \leq h_i(x, y(x)) + M\phi(x),$$

and, from (3.12.3)

$$\frac{d}{dx} h_i(x, y(x)) \geq p \quad \text{for } c \leq x < d \quad \text{and } i = 1, \cdots, m.$$

Thus, for $c \leq x < d$, using (3.12.1) and the monotonicity of each g_i gives

$$\phi(x) \leq L \int_c^x \left\{ |y(t) - \bar{y}(t)| + \sum_{i=1}^{m} |g_i(h_i(t, y(t))) - g_i(h_i(t, \bar{y}(t)))| \right\} dt$$

$$\leq L(x-c)\phi(x) \ + \ \frac{L}{p}\sum_{i=1}^{m}\int_c^x (g_i(h_i(t,y(t))+M\phi(t))$$

$$-g_i(h_i(t,\bar{y}(t))-M\phi(t)))\,\frac{d}{dt}h_i(t,y(t))dt$$

$$= L(x-c)\phi(x) \ + \ \frac{L}{p}\sum_{i=1}^{m}\int_{h_i(x,y(x))-M\phi(x)}^{h_i(x,y(x))+M\phi(x)} g_i(s)ds$$

$$-\frac{L}{p}\sum_{i=1}^{m}\int_c^x (g_i(h_i(t,y(t))+M\phi(t))-g_i(h_i(t,y(t))-M\phi(t)))\,M\phi'(t)dt.$$

Choose $\delta_1 > 0$ such that for each i

$$|g_i(s)| < \frac{p}{6mLM} \quad \text{when} \quad |s-h_i(c,y(c))| < \delta_1.$$

Then, let $\delta \in (0, \frac{1}{6L})$ be such that $c+\delta \leq d$, and for each i

$$|h_i(x,y(x))-h_i(c,y(c))| \ + \ M\phi(x) < \delta_1 \quad \text{when} \quad c \leq x < c+\delta.$$

Now for $c < x < c+\delta$ one finds $\phi(x) \leq \frac{5}{6}\phi(x)$. This contradiction completes the proof. ∎

While Driver and Norris [46,118] designed the above theorem to resolve a problem in classical electrodynamics, we shall illustrate its importance by considering the following :

Example 3.12.1. In Theorem 3.12.1 we can choose $h_1(x,y) = x + y^{5/3}$ and $g_1(y) = y^{1/3}$ so that the initial value problem

$$y' = (x+y^{5/3})^{1/3}, \ y(0) = 0 \tag{3.12.6}$$

for $x \geq 0$ has at most one solution. ∎

Example 3.12.2. In Theorem 3.12.1 we can choose $h_1(x,y) = y$ and $h_2(x,y) = x+y$ so that the initial value problem

$$y' = g_1(y) + g_2(x+y), \ y(0) = 0 \tag{3.12.7}$$

for $x \geq 0$ has at most one solution. ∎

3.13 GOEBEL - ZUR'S UNIQUENESS THEOREM

The generalized Lipschitz condition

$$|f(x,y) - f(x,\bar{y})| \le \lambda(x)|y - \bar{y}| \tag{3.13.1}$$

has already appeared in several uniqueness theorems. Here, we shall assume that the function $\lambda(x)$ in (3.13.1) is measurable and finite almost everywhere in $[x_0, x_0 + a]$.

In the following result we shall use the fact that the continuity of the function $f(x,y)$ in $\bar{S}_+$ with $(x_0, y_0) = (0,0)$ implies that the function $\omega(h)$ defined by

$$\omega(h) = \sup\{|f(x,y) - f(t,y)| \ : \ |y| \le b, \ x,t \in [0,a], \ |x-t| \le h\}$$

tends to zero as $h \to 0$.

Theorem 3.13.1 (Goebel - Zur's Uniqueness Theorem). Let $f(x,y)$ be continuous in $\bar{S}_+$ with $(x_0, y_0) = (0,0)$ and satisfy the generalized Lipschitz condition (3.13.1). Further, let

$$\inf\left\{\mu(A_N)\omega\left(\frac{\mu(A_N)}{2}\right)\exp\left(\int_0^a \lambda_N(t)dt\right) \ : \ N \ge 0\right\} = 0, \tag{3.13.2}$$

where $A_N = [x \ : \ \lambda(x) \ge N]$,

$$\lambda_N(x) = \begin{cases} \lambda(x), & x \notin A_N \\ N, & x \in A_N \end{cases}$$

and $\mu(\ )$ denotes the Lebesgue measure in $[0,a]$. Then, the initial value problem (1.1.1) with $(x_0, y_0) = (0,0)$ has at most one n - solution in $[0,a]$.

Proof. We cover the set A_N by a sequence of open intervals $G_i = (a_i, b_i)$ in such a way that

$$\mu(G) = \mu\left(\bigcup_{i=1}^{\infty} G_i\right) \le \mu(A_N) + \varepsilon.$$

Then, construct the function

$$f_N(x,y) = \begin{cases} f(x,y) & \text{for} \quad x \notin G \\ f(a_i,y) & \text{for} \quad x \in (a_i, \frac{1}{2}(a_i+b_i)) \\ f(b_i,y) & \text{for} \quad x \in [\frac{1}{2}(a_i+b_i), b_i). \end{cases}$$

It is obvious that $f_N(x,y)$ satisfies the Carathèodory conditions, it is bounded by the same constant as $f(x,y)$ and, moreover,

$$|f_N(x,y) - f_N(x,\bar{y})| \le \lambda_N(x)|y-\bar{y}| \le N|y-\bar{y}|$$

holds, so the initial value problem

$$y' = f_N(x,y), \; y(0) = 0 \tag{3.13.3}$$

has at most one n - extended solution $y_N(x)$ in $[0,a]$.

Let $y(x)$ be an arbitrary n - solution of (1.1.1) with $(x_0,y_0) = (0,0)$. Then, we have

$$\begin{aligned} &|y(x) - y_N(x)| \\ &\le \int_0^x |f(t,y(t)) - f_N(t,y_N(t))|dt \\ &\le \int_0^x |f(t,y(t)) - f_N(t,y(t))|dt \; + \; \int_0^x |f_N(t,y(t)) - f_N(t,y_N(t))|dt \\ &\le \int_G |f(t,y(t)) - f_N(t,y(t))|dt \; + \; \int_0^x \lambda_N(t)|y(t) - y_N(t)|dt \\ &\le \int_G \omega\left(\frac{\sup(b_i - a_i)}{2}\right) dt \; + \; \int_0^x \lambda_N(t)|y(t) - y_N(t)|dt \\ &\le (\mu(A_N)+\varepsilon)\omega\left(\frac{\mu(A_N)+\varepsilon}{2}\right) \; + \; \int_0^x \lambda_N(t)|y(t) - y_N(t)|dt. \end{aligned}$$

Hence, in view of the Gronwall inequality (1.12.13) it follows that

$$|y(x) - y_N(x)| \le (\mu(A_N)+\varepsilon)\;\omega\left(\frac{\mu(A_N)+\varepsilon}{2}\right)\exp\left(\int_0^a \lambda_N(t)dt\right).$$

Therefore, if $y(x)$ and $\bar{y}(x)$ are two n - solutions of (1.1.1) with $(x_0,y_0) = (0,0)$ then we have

$$|y(x) - \bar{y}(x)| \le 2(\mu(A_N)+\varepsilon)\;\omega\left(\frac{\mu(A_N)+\varepsilon}{2}\right)\exp\left(\int_0^a \lambda_N(t)dt\right).$$

However, since ε can be chosen arbitrarily small the condition (3.13.2) implies that $|y(x) - \bar{y}(x)| \equiv 0$, i.e., $y(x) = \bar{y}(x)$ in $[0, a]$. ∎

Theorem 3.13.2. Let $f(x, y)$ satisfy the Caratheodory conditions in $\bar{S}_+$ with $(x_0, y_0) = (0, 0)$ and $M(x) = \sup\{M_1(x), \cdots, M_n(x)\}$. Further, let $f(x, y)$ in $\bar{S}_+$ satisfy the generalized Lipschitz condition (3.13.1), and

$$\inf\left\{\int_{A_N} M(t)dt \ \exp\left(\int_0^a \bar{\lambda}_N(t)dt\right) \ : \ N \geq 0\right\} = 0, \tag{3.13.4}$$

where

$$\bar{\lambda}_N(x) = \begin{cases} \lambda(x), & x \notin A_N \\ 0, & x \in A_N. \end{cases}$$

Then, the initial value problem (1.1.1) with $(x_0, y_0) = (0, 0)$ has at most one n - extended solution in $[0, a]$.

Proof. We set

$$f_N(x, y) = \begin{cases} f(x, y), & x \notin A_N \\ 0, & x \in A_N. \end{cases}$$

It is obvious that $f_N(x, y)$ satisfies the Caratheodory conditions in $\bar{S}_+$ with $(x_0, y_0) = (0, 0)$ and

$$|f_N(x, y) - f_N(x, \bar{y})| \leq \bar{\lambda}_N(x)|y - \bar{y}| \leq N \ |y - \bar{y}|$$

holds. Hence, the initial value problem (3.13.3) has at most one n - extended solution $y_N(x)$ in $[0, a]$. Now let $y(x)$ and $\bar{y}(x)$ be two arbitrary n - extended solutions of (1.1.1) with $(x_0, y_0) = (0, 0)$. Then, as in Theorem 3.13.1 it follows that

$$|y(x) - \bar{y}(x)| \leq 2 \int_{A_N} M(t)dt \ \exp\left(\int_0^a \bar{\lambda}_N(t)dt\right)$$

for $N \geq 0$. The above inequality in view of (3.13.4) implies that $y(x) = \bar{y}(x)$ in $[0, a]$. ∎

Example 3.13.1. Suppose in Theorem 3.13.1, $\lambda(x) = \frac{k}{x}$. In this case $A_N = [0, \frac{k}{N}]$ and the expression considered in (3.13.2) has the form

$$\frac{k}{N}\omega\left(\frac{k}{2N}\right) \exp\left(\int_0^{\frac{k}{N}} N dt \ + \ \int_{\frac{k}{N}}^a \frac{k}{t} dt\right) = CN^{k-1}\omega\left(\frac{k}{2N}\right),$$

where C is a constant. Hence, the condition (3.13.2) holds if $\lim_{N\to\infty} N^{k-1}\,\omega\left(\frac{k}{2N}\right) = 0$, which is always true if $k \leq 1$.

Similarly, if in Theorem 3.13.2, $\lambda(x) = \frac{k}{x}$, then the expression considered in (3.13.4) has the form

$$CN^k \int_0^{\frac{k}{N}} M(t)dt,$$

where again C is a constant. Thus, if $M(x)$ is bounded then, the condition (3.13.4) holds if $k < 1$. ∎

3.14 BOWNDS - DIAZ'S UNIQUENESS THEOREM

In what follows we shall assume that $f(x,y)$ is continuous for $x_0 \leq x \leq x_0 + a$, $y \in \bar{T} \subseteq R^n$ and continuously differentiable for $x_0 < x < x_0 + a$, $y \in T$, where T is an open set in R^n, and $y_0 \in T$. We set $F(x,y) = (1, f(x,y))$ which is obviously defined for $x_0 < x < x_0 + a$, $y \in T$, and let $\phi(x,y) = (< \nabla f_1 \cdot F >, \cdots, < \nabla f_n \cdot F >)$, where ∇ denotes the gradient in R^{n+1} and $< \cdot >$ denotes the scalar multiplication in R^{n+1}.

Theorem 3.14.1 (Bownds - Diaz's Uniqueness Theorem). Let for all (x,y), $(x,\bar{y})$, $x_0 < x < x_0 + a$, y, $\bar{y} \in T$

$$(f(x,y) - f(x,\bar{y})) \cdot (\phi(x,y) - \phi(x,\bar{y})) \leq 0, \tag{3.14.1}$$

i.e.,

$$\sum_{i=1}^{n}(f_i(x,y) - f_i(x,\bar{y}))\left[\left(\frac{\partial f_i(x,y)}{\partial x} + \sum_{j=1}^{n}\frac{\partial f_i(x,y)}{\partial y_j}f_j(x,y)\right)\right.$$
$$\left. - \left(\frac{\partial f_i(x,\bar{y})}{\partial x} + \sum_{j=1}^{n}\frac{\partial f_i(x,\bar{y})}{\partial \bar{y}_j}f_j(x,\bar{y})\right)\right] \leq 0.$$

Then, the initial value problem (1.1.1) has at most one n - solution $y(x)$ in $[x_0, x_0 + a]$, such that for $x \in (x_0, x_0 + a)$, $y(x) \in T$.

Proof. Define $V(x) = \frac{1}{2}\|f(x,y(x)) - f(x,\bar{y}(x))\|^2$, $x_0 \leq x \leq x_0 + a$, where $y(x)$ and $\bar{y}(x)$ are two solutions of (1.1.1). Obviously, $V(x_0) = 0$, $V(x) \geq$

$0,\ x_0 \le x \le x_0 + a$, and

$$V'(x) = (f(x,y(x)) - f(x,\bar{y}(x)))\cdot\frac{d}{dx}(f(x,y(x)) - f(x,\bar{y}(x))),\ x_0 < x < x_0 + a.$$

However, since

$$\frac{d}{dx}(f(x,y(x)) - f(x,\bar{y}(x))) = (\phi(x,y(x)) - \phi(x,\bar{y}(x)))$$

in view of (3.14.1) it follows that $V'(x) \le 0,\ x_0 < x < x_0 + a$. Thus, $V(x) = 0,\ x_0 \le x \le x_0 + a$, i.e., $f(x,y(x)) = f(x,\bar{y}(x)),\ x_0 < x < x_0 + a$. But, then $y'(x) = \bar{y}'(x),\ x_0 < x < x_0 + a$, which in view of $y(x_0) = \bar{y}(x_0)$ leads to $y(x) = \bar{y}(x),\ x_0 \le x \le x_0 + a$. ∎

It is interesting to note that in the above theorem the choice of the open set T plays an important role. For example, the initial value problem $y' = \sqrt{|y|},\ 0 < x < 1,\ y(0) = 0$ has no solution $y(x)$ in $[0,1]$ if $T = \{y : \frac{1}{4} < y < 1\}$, which satisfies the condition that for $x \in (0,1),\ y(x) \in T$. However, it has a unique solution $y(x) = \frac{1}{4}x^2$ in $[0,1]$ if $T = \{y : 0 < y < 1\}$, since $\frac{1}{4}x^2 \in T$ for $0 < x < 1$.

Example 3.14.1. Consider the initial value problem

$$(3.14.2)\qquad y' = f(x,y) = \begin{cases} k(x) + \sqrt{y}, & 0 < x < 1,\ y \ge 0 \\ k(x), & 0 < x < 1,\ y < 0 \end{cases}$$
$$y(0) = 0$$

where $k(x)$ is continuous in $[0,1]$, continuously differentiable in $(0,1)$, and $k(x) \ge 0$ for $x \ge 0$. For $(x,y) \in (0,1) \times T$, where $T = \{y : y > 0\}$ we have

$$\phi(x,y) = \ <\nabla f\cdot(1,f)> \ = \ k'(x) + \frac{1}{2\sqrt{y}}[k(x) + \sqrt{y}].$$

Hence, for $0 < x < 1,\ y,\bar{y} \in T$ it follows that

$$(f(x,y) - f(x,\bar{y}))\cdot(\phi(x,y) - \phi(x,\bar{y})) = \frac{1}{2}\,k(x)[\sqrt{y} - \sqrt{\bar{y}}]\left[\frac{1}{\sqrt{y}} - \frac{1}{\sqrt{\bar{y}}}\right]$$

$$= -\frac{1}{2\sqrt{yz}}k(x)[\sqrt{y} - \sqrt{\bar{y}}]^2 \le 0.$$

Thus, the conditions of Theorem 3.14.1 are satisfied, and in conclusion the problem (3.14.2) has at most one solution $y(x)$, $0 \leq x \leq 1$ such that $y(x) > 0$, $0 < x < 1$. ∎

3.15 OKAMURA'S UNIQUENESS THEOREM

In the year 1892 Lyapunov introduced a method, which is called the second or direct method, to establish stability results. However, this method has been recognized to be very general and powerful in the qualitative theory of ordinary differential equations. The success of this method depends on the construction of a function called **Lyapunov function.** Here, we shall use this powerful method to establish necessary and sufficient conditions for uniqueness of n - solutions of the initial value problem (1.1.1).

Definition 3.15.1. A scalar function $V(x,y)$ is said to satisfy a local Lipschitz condition with respect to y in $D \subseteq R^{n+1}$ if for each compact subset D_1 of D there is a constant $L = L(D_1)$ such that for all (x,y), $(x,\bar{y})$ in D_1

$$(3.15.1) \qquad |V(x,y) - V(x,\bar{y})| \leq L\|y - \bar{y}\|.$$

Let $V(x,y)$ be a continuous function defined in D and let it satisfy a local Lipschitz condition (3.15.1) with respect to y. Corresponding to $V(x,y)$, we define the function

$$(3.15.2) \qquad D^+V_f(x,y) = \limsup_{h\to 0+} \frac{1}{h}[V(x+h, y+hf(x,y)) - V(x,y)].$$

Let $y(x)$ be a n - solution of $y' = f(x,y)$ which stays in D, and denote by

$$(3.15.3) \qquad D^+V(x,y(x)) = \limsup_{h\to 0+} \frac{1}{h}[V(x+h, y(x+h)) - V(x,y(x))].$$

For a point $(x,y) \in D$ and small h, there exists a neighborhood U of (x,y) such that $\bar{U} \subset D$, $(x+h, y+hf(x,y)) \in U$ and $(x+h, y(x+h)) \in U$. Thus, in view of (3.15.1) it follows that

$$(3.15.4) \quad V(x+h, y(x+h)) - V(x, y(x))$$

$$= V(x+h, y+hf(x,y)+h\varepsilon) - V(x,y)$$

$$\leq V(x+h, y+hf(x,y)) + Lh\|\varepsilon\| - V(x,y),$$

where ε tends to zero as $h \to 0$. From (3.15.4) it follows that

$$(3.15.5) \qquad D^+V(x, y(x)) \leq D^+V_f(x,y).$$

On the other hand, we have

$$V(x+h, y(x+h)) - V(x, y(x)) \geq V(x+h, y+hf(x,y)) - Lh\|\varepsilon\| - V(x,y),$$

which implies that

$$(3.15.6) \qquad D^+V(x, y(x)) \geq D^+V_f(x,y).$$

Thus, from (3.15.5) and (3.15.6) it follows that

$$(3.15.7) \qquad D^+V(x, y(x)) = D^+V_f(x,y).$$

In a similar manner, we also obtain the relation

$$(3.15.8) \qquad D_+V(x, y(x)) = D_+V_f(x,y).$$

In case $V(x,y)$ has continuous partial derivatives of the first order, then it is clear that

$$(3.15.9) \qquad D^+V_f(x,y) = \frac{\partial V}{\partial x} + \frac{\partial V}{\partial y} \cdot f(x,y).$$

If $D^+V_f(x,y) \leq 0$ and consequently $D^+V(x, y(x)) \leq 0$, then $V(x, y(x))$ is a nonincreasing function of x, which implies that $V(x,y)$ is nonincreasing along a n - solution of $y' = f(x,y)$. Conversely, if $V(x,y)$ is nonincreasing along a n - solution of $y' = f(x,y)$ then we have $D^+V_f(x,y) \leq 0$. Further, if $D_+V_f(x,y) \geq 0$ then $V(x,y)$ is nondecreasing along a n - solution of $y' = f(x,y)$, and conversely.

Theorem 3.15.1 (Okamura's Uniqueness Theorem). Suppose that $f(x,y)$ is continuous in $\bar{S}_+$ with $(x_0, y_0) = (0,0)$, and that $f(x,0) \equiv 0$. Further, suppose that there exists a Lyapunov function $V(x,y)$ defined in $\bar{S}_+$ which satisfies the following conditions

(i) $V(x, 0) \equiv 0$,

(ii) $V(x, y) > 0$ for $\|y\| \neq 0$; and

(iii) $V(x, y)$ satisfies the locally Lipschitz condition (3.15.1) with respect to y, and $D^+V_f(x, y) \leq 0$ in the interior of $\bar{S}_+$.

Then, for the initial value problem (1.1.1) with $(x_0, y_0) = (0, 0)$, $y(x) \equiv 0$ is the only n - solution in $[0, a]$.

Proof. Suppose that the initial value problem (1.1.1) with $(x_0, y_0) = (0, 0)$, has a n - solution $y(x)$ which is different from the zero solution. Then, there are x_1, x_2, $0 \leq x_1 < x_2 \leq a$, such that $y(x_1) = 0$, $y(x_2) \neq 0$, and that $\|y(x)\| > 0$ for $x \in (x_1, x_2)$. However, since by (ii) $V(x_2, y(x_2)) > 0$, by (iii) $V(x_2, y(x_2)) \leq V(x_1, y(x_1))$, and by (i) $V(x_1, y(x_1)) = 0$ we arrive at a contradiction. ∎

Theorem 3.15.2. Suppose that $f(x, y)$ is continuous in $\bar{S}_-$ with $(x_0, y_0) = (0, 0)$, and that $f(x, 0) \equiv 0$. Further, suppose that there exists a Lyapunov function $V(x, y)$ defined in $\bar{S}_-$ which satisfies the conditions (i) and (ii) of Theorem 3.15.1 and

(iv) $V(x, y)$ satisfies the locally Lipschitz condition (3.15.1) with respect to y, and $D_+V_f(x, y) \geq 0$ in the interior of $\bar{S}_-$.

Then, for the initial value problem (1.1.1) with $(x_0, y_0) = (0, 0)$, $y(x) \equiv 0$ is the only n - solution in $[-a, 0]$.

Proof. The proof is similar to that of Theorem 3.15.1. ∎

On combining Theorems 3.15.1 and 3.15.2, we obtain

Theorem 3.15.3. Suppose that $f(x, y)$ is continuous in $\bar{S}$ with $(x_0, y_0) = (0, 0)$, and that $f(x, 0) \equiv 0$. Further, suppose that there exists a Lyapunov function $V(x, y)$ defined in $\bar{S}$ which satisfies the conditions of Theorems 3.15.1 and 3.15.2 in $\bar{S}_+$ and $\bar{S}_-$, respectively. Then, for the initial value problem (1.1.1) with $(x_0, y_0) = (0, 0)$, $y(x) \equiv 0$ is the only n - solution in $[-a, a]$. ∎

Now we shall show that the existence of a Lyapunov function for the uniqueness of the zero solution of the initial value problem (1.1.1) in Theorem 3.15.1 is a necessary condition. For this, let $P_1(x_1, y_1)$ and $P_2(x_2, y_2)$ with $x_1 < x_2$ be two points in $\bar{S}_+$. Let $Y_{P_1P_2}$ denote the family of all functions $y(x)$ which are continuous in $[x_1, x_2]$, with their derivatives continuous except for at most finite values of x and that $y(x_1) = y_1$, $y(x_2) = y_2$ and $(x, y(x)) \in \bar{S}_+$ for $x \in [x_1, x_2]$. Let $v(P_1, P_2)$ for $x_1 < x_2$ be defined as follows

$$v(P_1, P_2) = \inf_{y(x) \in Y_{P_1 P_2}} \int_{x_1}^{x_2} \|y'(t) - f(t, y(t))\| dt. \tag{3.15.10}$$

If $x_1 = x_2$, then we set $v(P_1, P_2) = \|y_1 - y_2\|$ and if $x_1 > x_2$ then $v(P_1, P_2) = v(P_2, P_1)$.

Lemma 3.15.4. The points P_1 and P_2 are on a n - solution of $y' = f(x, y)$ if and only if $v(P_1, P_2) = 0$.

Proof. The necessary part is immediate. To show the sufficient part we assume that $v(P_1, P_2) = 0$, and there exists a sequence of functions $\{y_n(x)\}$, $y_n(x) \in Y_{P_1P_2}$, such that

$$\lim_{m \to \infty} \int_{x_1}^{x_2} \|y'_m(t) - f(t, y_m(t))\| dt = 0.$$

Let $\phi_m(x) = y_m(x) - y_1 - \int_{x_1}^{x} f(t, y_m(t)) dt$, so that $\lim_{m\to\infty} \phi_m(x) = 0$. For $x_1 \le x' \le x'' \le x_2$ and $z_m(x) = y_m(x) - \phi_m(x)$, we have $z_m(x'') - z_m(x') = \int_{x'}^{x''} f(t, y_m(t)) dt$, and hence $\|z_m(x'') - z_m(x')\| \le M(x'' - x')$, where M is a constant such that $\|f(x, y)\| \le M$ in $\bar{S}_+$. Therefore, $\{z_m(x)\}$ is equicontinuous and by Arzela's theorem, there exists a uniformly convergent subsequence, which we denote be $\{z_m(x)\}$ again. Let $\bar{y}(x)$ be the limit function. Then, clearly $\bar{y}(x_1) = y_1$, $\bar{y}(x_2) = y_2$ and in view of $y_m(x) \to \bar{y}(x)$ as $\phi_m(x) \to 0$ we have

$$\bar{y}(x) = y_1 + \int_{x_1}^{x_2} f(t, \bar{y}(t)) dt.$$

Thus, there exists a n - solution of $y' = f(x, y)$ passing through P_1 and P_2.

Lemma 3.15.5. For two points P_1 and P_2 with $x_1 < x_2$ we have

$$|v(P_1, P_2) - \|y_1 - y_2\|| \le M(x_2 - x_1). \tag{3.15.11}$$

Proof. For any $y(x) \in Y_{P_1P_2}$, we have

$$\int_{x_1}^{x_2} \|y'(t) - f(t,y(t))\|dt \geq \int_{x_1}^{x_2} \|y'(t)\|dt - \int_{x_1}^{x_2} \|f(t,y(t))\|dt$$

$$\geq \|y_2 - y_1\| - M(x_2 - x_1),$$

which implies that $v(P_1,P_2) \geq \|y_2 - y_1\| - M(x_2 - x_1)$. Next, let $y(x)$ be the function which represents the segment $\overline{P_1P_2}$. Obviously, $y(x) \in Y_{P_1P_2}$. Since

$$\int_{x_1}^{x_2} \|y'(t)\|dt = \left\|\int_{x_1}^{x_2} y'(t)dt\right\| = \|y_2 - y_1\|,$$

we have

$$\int_{x_1}^{x_2} \|y'(t) - f(t,y(t))\|dt \leq \|y_2 - y_1\| + M(x_2 - x_1).$$

Thus, $v(P_1,P_2) \leq \|y_2 - y_1\| + M(x_2 - x_1)$. Hence, (3.15.11) follows. ∎

Lemma 3.15.6. For the points $P_i(x_i,y_i)$, $i = 1,2,3$ where $x_1 \leq x_2 \leq x_3$ the following inequality holds

$$v(P_1,P_3) \leq v(P_1,P_2) + v(P_2,P_3). \tag{3.15.12}$$

Proof. If $x_1 < x_2 < x_3$ then the inequality (3.15.12) is obvious. Thus, we assume that $x_1 < x_2 = x_3$. Let $\{y_m(x)\}$ be a sequence such that

$$v(P_1,P_2) = \lim_{m\to\infty} \int_{x_1}^{x_2} \|y_m'(t) - f(t,y_m(t))\|dt$$

and take the points $Q_m = (x', y_m(x'))$, $x_1 < x' < x_2 = x_3$. Then, for P_1, Q_m and P_3 we have $v(P_1,P_3) \leq v(P_1,Q_m) + v(Q_m,P_3)$, and hence

$$\int_{x_1}^{x_2} \|y_m'(t) - f(t,y_m(t))\|dt = \int_{x_1}^{x'} \|y_m'(t) - f(t,y_m(t))\|dt$$

$$+ \int_{x'}^{x_2} \|y_m'(t) - f(t,y_m(t))\|dt$$

$$\geq v(P_1,Q_m) + v(Q_m,P_2)$$

$$\geq v(P_1,P_3) - v(Q_m,P_3) + v(Q_m,P_2).$$

From the above inequality and (3.15.11) it follows that

$$\int_{x_1}^{x_2} \|y_m'(t) - f(t,y_m(t))\|dt \geq v(P_1,P_3) - \|y_2 - y_3\| - 2M(x_2 - x')$$

and hence as $m \to \infty$ and $x' \to x_2$, we find that $v(P_1, P_2) \geq v(P_1, P_3) - \|y_2 - y_3\|$. But, in view of $v(y_2, y_3) = \|y_2 - y_3\|$ this is the same as (3.15.12). ∎

Lemma 3.15.7. For the points $P_i(x_i, y_i)$, $i = 1, 2, 3$ where $x_1 \leq x_2 \leq x_3$ the following inequalities hold

$$v(P_1, P_3) \geq v(P_1, P_2) - \|y_3 - y_2\| - M(x_3 - x_2) \tag{3.15.13}$$

and

$$v(P_1, P_3) \geq v(P_2, P_3) - \|y_2 - y_1\| - M(x_2 - x_1). \tag{3.15.14}$$

Proof. In case $x_1 < x_2 < x_3$, let $\{y_m(x)\}$ be a sequence such that $v(P_1, P_3) = \lim_{m\to\infty} \int_{x_1}^{x_3} \|y_m'(t) - f(t, y_m(t))\| dt$. For the points $Q_m = (x_2, y_m(x_2))$, in view of (3.15.12) and (3.15.11) we have

$$\begin{aligned}
&\int_{x_1}^{x_3} \|y_m'(t) - f(t, y_m(t))\| dt \\
&\geq v(P_1, Q_m) + v(Q_m, P_3) \\
&\geq v(P_1, P_2) - v(Q_m, P_2) + v(Q_m, P_3) \\
&\geq v(P_1, P_2) - \|y_m(x_2) - y_2\| + \|y_3 - y_m(x_2)\| - M(x_3 - x_2) \\
&\geq v(P_1, P_2) - \|y_3 - y_2\| - M(x_3 - x_2),
\end{aligned}$$

which implies that $v(P_1, P_3) \geq v(P_1, P_2) - \|y_3 - y_2\| - M(x_3 - x_2)$. The proof of (3.15.14) is similar. ∎

From the above lemmas it is clear that $v(P_1, P_2)$ is a nonnegative continuous function of (P_1, P_2) and that

$$|v(P_1, P_2) - v(P_1, P_3)| \leq \|y_2 - y_3\| + M|x_3 - x_2|. \tag{3.15.15}$$

Theorem 3.15.8. Suppose that $f(x, y)$ is continuous in $\bar{S}_+$ with $(x_0, y_0) = (0, 0)$, and that $f(x, 0) \equiv 0$. Further, suppose that $y(x) \equiv 0$ is the only n - solution of the initial value problem (1.1.1) with $(x_0, y_0) = (0, 0)$. Then, there exists a Lyapunov function $V(x, y)$ which is defined in $\bar{S}_+$ and satisfies the conditions (i), (ii) and (iii) of Theorem 3.15.1.

Proof. For $P_1 = (0,0)$ and $P_2 = (x,y) \in \bar{S}_+$, we set $V(x,y) = v(P_1,P_2)$. Since, $y(x) \equiv 0$ is a n - solution of the initial value problem (1.1.1) it is clear that $V(x,0) \equiv 0$. Further, since for any point $(x,y) \in \bar{S}_+$ such that $y \neq 0$, there is no n - solution of (1.1.1) passing through (x,y), from Lemma 3.15.4 it follows that $V(x,y) > 0$. From (3.15.15), we also have

$$|V(x,y) - V(\bar{x},\bar{y})| \leq \|y - \bar{y}\| + M|x - \bar{x}|,$$

i.e., $V(x,y)$ satisfies the locally Lipschitz condition (3.15.1). Finally, Lemmas 3.15.4 and 3.15.6 ensure that the function $V(x,y)$ is nondecreasing along the n - solutions of $y' = f(x,y)$. Therefore, $D^+V_f(x,y) \leq 0$, as required. ∎

Theorem 3.15.9. Suppose that $f(x,y)$ is continuous in $\bar{S}_-$ with $(x_0,y_0) = (0,0)$, and that $f(x,0) \equiv 0$. Further, suppose that $y(x) \equiv 0$ is the only n - solution of the initial value problem (1.1.1) with $(x_0,y_0) = (0,0)$. Then, there exists a Lyapunov function $V(x,y)$ which is defined in $\bar{S}_-$ and satisfies the conditions of Theorem 3.15.2.

Proof. The proof is similar to that of Theorem 3.15.8. ∎

Theorems 3.15.8 and 3.15.9 lead to

Theorem 3.15.10. Suppose that $f(x,y)$ is continuous in $\bar{S}$ with $(x_0,y_0) = (0,0)$, and that $f(x,0) \equiv 0$. Further, suppose that $y(x) \equiv 0$ is the only n - solution of the initial value problem (1.1.1) with $(x_0,y_0) = (0,0)$. Then, there exists a Lyapunov function $V(x,y)$ which is defined in $\bar{S}$ and satisfies the conditions of Theorem 3.15.3. ∎

Finally, we shall prove a generalization of this theorem.

Theorem 3.15.11. Suppose that $f(x,y)$ is continuous in $S_1 : 0 \leq x < a$, $\|y\| < b$. In order that every n - solution of $y' = f(x,y)$ starting from a point in S_1 is unique to the right it is necessary and sufficient that for any point $(x_1,y_1) \in S_1$ there exists a neighborhood U of (x_1,y_1) which has the following property : Let W be a set of $(x,y,\bar{y})$ such that (x,y), $(x,\bar{y}) \in U$. Then, there exists a Lyapunov function $V(x,y,\bar{y})$ defined in W which satisfies the following conditions

(i) $V(x, y, \bar{y}) \equiv 0$ if $y = \bar{y}$,

(ii) $V(x, y, \bar{y}) > 0$ if $y \neq \bar{y}$; and

(iii) $V(x, y, \bar{y})$ satisfies the locally Lipschitz condition (3.15.1) with respect to $(y, \bar{y})$, and

(3.15.16) $D^+V_f(x, y, \bar{y})$

$$\leq \limsup_{h\to 0^+} \frac{1}{h}[V(x+h, y+hf(x,y), \bar{y}+hf(x,\bar{y})) - V(x, y, \bar{y})]$$

$$\leq 0.$$

Proof. The sufficient part is obvious, and therefore we shall prove only the necessary part. Associated with $y' = f(x, y)$ we consider the system

(3.15.17) $$y' = f(x, y),\ \bar{y}' = f(x, \bar{y}).$$

For each point $(x_1, y_1) \in S_1$, let U be a neighborhood of (x_1, y_1) which is contained in the set $\hat{S} = \{(x, y) : |x - x_1| \leq \alpha,\ \|y - y_1\| \leq \beta\} \subset S_1$. For two points $P(x, y, \bar{y})$ and $Q(\tau, \xi, \eta)$, $(x, y) \in \hat{S}, (x, \bar{y}) \in \hat{S},\ (\tau, \xi) \in \hat{S},\ (\tau, \eta) \in \hat{S}$, we define the function $v(P, Q)$. Let

$$v^*(P) = \min_{\substack{0 \leq \tau \leq x \\ \xi = \eta}} v(P, Q).$$

Since $v(P, Q)$ is continuous in Q, and Q varies on a compact set, the $\min v(P, Q) = v^*(P)$ exists and is continuous in P, also $v^*(P) = v(P, Q^*)$ for some Q^*. Let $V(x, y, \bar{y}) = v^*(P)$. If P is on $y = \bar{y}$ then $v^*(P) = 0$ or $V(x, y, \bar{y}) = 0$, and if $y \neq \bar{y}$ then $v^*(P) > 0$ or $V(x, y, \bar{y}) > 0$, because of the uniqueness of the solutions. For two points $P(x, y, \bar{y})$, $\bar{P}(x, w, \bar{w})$, for which we can assume that $v^*(\bar{P}) \geq v^*(P)$, $0 \leq v^*(\bar{P}) - v^*(P) \leq v(\bar{P}, Q) - v(P, Q) \leq \overline{\bar{P}P}$, where $v^*(P) = v(P, Q)$. From this, it follows that $|V(x, y, \bar{y}) - V(x, w, \bar{w})| \leq \|y - w\| + \|\bar{y} - \bar{w}\|$.

Finally, suppose that $P(x, y, \bar{y})$, $R(\bar{x}, w, \bar{w})$, $\bar{x} < x$, are on a solution of (3.15.17). Then there exists a $Q(x_1, y_1, \bar{y}_1)$ on $y = \bar{y}$ such that $v^*(R) = v(R, Q)$. By Lemmas 3.15.4 and 3.15.6, $v(P, Q) \leq v(P, R) + v(R, Q) = v(R, Q)$. As we have $v^*(P) \leq v(P,Q)$, $v^*(P) \leq v^*(R)$, which means that $V(x,y,\bar{y})$ is nonincreasing along the solution of (3.15.17). ∎

Example 3.15.1. If the function $f(x,y)$ satisfies the Lipschitz condition (1.2.1) with the norm $\|.\|$ then the function $V(x,y,\bar{y}) = e^{-2Lx}\|y-\bar{y}\|^2$ satisfies

$$D^+V_f(x,y,\bar{y})$$

$$\leq e^{-2Lx}[-2L\|y-\bar{y}\|^2 + 2\|y-\bar{y}\|\|f(x,y)-f(x,\bar{y})\|] \leq 0.$$

Thus, it is easily seen that this function $V(x,y,\bar{y})$ satisfies the conditions of Theorem 3.15.11 and hence every solution is unique to the right. ∎

3.16 CHU - DIAZ'S UNIQUENESS THEOREM

For the scalar problem (1.1.1) we shall provide one - sided analog of Theorem 3.15.1. The importance of the obtained result is in the fact that the proof requires only the mean value theorem.

Theorem 3.16.1 (Chu - Diaz's Uniqueness Theorem). Assume that the function $f(x,y)$ is defined in S_+ with $x_0 = 0$, and there exists a Lyapunov function $V(x,y)$ which is defined for $0 < x < a,\ 0 < y < 2b$ and satisfies the following conditions

(i) $V(x,y)$ is continuously differentiable for $0 < x < a,\ 0 < y < 2b$,

(ii) $V(x,y) > 0$ for $0 < x < a,\ 0 < y < 2b$ and $\lim_{y\to 0^+} V(x,y) = V(x,0^+) = 0$ for $0 < x < a$,

(iii) $\lim_{(x,y)\to(0^+,0^+)} V(x,y) = 0$; and

(iv) for all $(x,y),\ (x,\bar{y}) \in S_+,\ \bar{y} > y$

$$V_x(x,\bar{y}-y) + V_y(x,\bar{y}-y)(f(x,\bar{y}) - f(x,y)) \leq 0. \tag{3.16.1}$$

Then, the initial value problem (1.1.1) with $x_0 = 0$ has at most one classical solution in $[0, a)$.

Proof. Suppose $y(x)$ and $\bar{y}(x)$ are two classical solutions of (1.1.1) with $x_0 = 0$ in $[0, a)$. Then, there are numbers x_1, x_2 such that $0 \le x_1 < x_2 < a$ and $\bar{y}(x) > y(x)$, $x_1 < x \le x_2$ and $\bar{y}(x_1) = y(x_1)$. Therefore, $V(x_2, \bar{y}(x_2) - y(x_2)) > 0$, and $V(x_1, \bar{y}(x_1) - y(x_1)) = 0$, provided that $x_1 \neq 0$. However, if $x_1 = 0$, then in view of (iii) we have $\lim_{(x,y)\to(0^+,0^+)} V(x, y) = 0$. Thus, by the mean value theorem there exists a number $\bar{x}$ with $x_1 < \bar{x} < x_2$, such that

$$\begin{aligned} 0 < V(x_2, \bar{y}(x_2) - y(x_2)) &= V(x_2, \bar{y}(x_2) - y(x_2)) - V(x_1, \bar{y}(x_1) - y(x_1)) \\ &= (x_2 - x_1)\,[V_x(\bar{x}, \bar{y}(\bar{x}) - y(\bar{x})) \\ &\quad + V_y(\bar{x}, \bar{y}(\bar{x}) - y(\bar{x}))(f(\bar{x}, \bar{y}(\bar{x})) - f(\bar{x}, y(\bar{x})))]\,. \end{aligned}$$

Thus, from (3.16.1) it follows that $0 < V(x_2, \bar{y}(x_2) - y(x_2)) \le 0$. This contradiction completes the proof. ∎

Corollary 3.16.2. In Theorem 3.16.1 the condition (iii) can be replaced by

(iii)′ $\lim_{(x,y)\to(0^+,0^+)} V(x, y)$ exists, and $\lim_{x\to 0^+} V(x, xh(x)) = 0$, where $h(x)$ is any nonnegative function, defined in $(0, a)$ such that $\lim_{x\to 0^+} h(x) = 0$.

Proof. It suffices to note that for any two classical solutions $y(x)$ and $\bar{y}(x)$ of (1.1.1) such that $\bar{y}(x) \ge y(x)$, $0 \le x \le x_1 < a$

$$\begin{aligned} V(x, \bar{y}(x) - y(x)) &= V(x, (\bar{y}(x) - y(x)) - (\bar{y}(0) - y(0))) \\ &= V(x, x(f(\bar{x}, \bar{y}(\bar{x})) - f(\bar{x}, y(\bar{x})))),\ 0 < \bar{x} < x_1. \end{aligned}$$ ∎

3.17 SATO'S UNIQUENESS THEOREM

For the scalar problem (1.9.4) where $f(x, y)$ is continuous for $0 \le x \le a$, $0 \le y \le b$ and $f(x, 0) \equiv 0$ Theorems 3.15.1 and 3.15.8 can be combined to obtain the following :

Theorem 3.17.1 (Okamura's Uniqueness Theorem). For the initial value problem (1.9.4), $y(x) \equiv 0$ is the only solution if and only if there exists a continuous and bounded function $g(x,z)$, $0 \le x \le a$, $0 \le z < \infty$ for which the continuous partial derivatives $g_x(x,z)$ and $g_z(x,z)$ exist for $0 \le x \le a$, $0 < z < \infty$, further $g(x,0) \equiv 0$, $g(x,y) \ge f(x,y)$, $0 \le x \le a$, $0 < y \le b$; and the only solution of the problem

$$z' = g(x,z), \ z(0) = 0 \tag{3.17.1}$$

is the trivial solution $z(x) \equiv 0$.

Here we shall use this fundamental result to obtain several different forms of necessary and sufficient conditions for the uniqueness of the solution $y(x) \equiv 0$ of the problem (1.9.4). For this we need the following :

Lemma 3.17.2. Assume that

(i) $\phi(x,z)$ is defined and continuously differentiable for $0 \le x \le a$, $0 \le z < \infty$, $\phi(x,0) \equiv 0$, and $\phi(x,z) > 0$ for $0 \le x \le a$, $0 < z < \infty$,

(ii) $\xi(x)$ is a continuous function for which the right derivative $D^+\xi(x)$ exists in $0 \le x < a$; and

(iii) $\phi(0,0) \le \xi(0)$, and for a solution $y(x)$ of (1.9.4)

$$\phi_x(x,y(x)) + \phi_z(x,y(x))f(x,y(x)) < D^+\xi(x), \ 0 \le x < a. \tag{3.17.2}$$

Then, it follows that

$$\phi(x,y(x)) \le \xi(x), \ 0 \le x < a. \tag{3.17.3}$$

Proof. Assume that there exists a $x_1 \in (0,a)$ such that $\phi(x_1,y(x_1)) > \xi(x_1)$. Let $\bar{x}$ be the lower bound of the numbers t such that $\phi(x,y(x)) > \xi(x)$, $t \le x \le x_1$. Then, in view of the continuity of the functions it follows that $\phi(\bar{x},y(\bar{x})) = \xi(\bar{x})$ and for sufficiently small $h > 0$, $\phi(\bar{x}+h,y(\bar{x}+h)) > \xi(\bar{x}+h)$, and hence

$$\frac{\phi(\bar{x}+h,y(\bar{x}+h)) - \phi(\bar{x},y(\bar{x}))}{h} > \frac{\xi(\bar{x}+h) - \xi(\bar{x})}{h},$$

i.e.,

$$\phi_x(\bar{x}, y(\bar{x})) + \phi_z(\bar{x}, y(\bar{x}))f(\bar{x}, y(\bar{x})) \geq D^+\xi(\bar{x}),$$

which contradicts (3.17.2). ∎

Theorem 3.17.3 (Sato's Uniqueness Theorem). For the initial value problem (1.9.4), $y(x) \equiv 0$ is the only solution if and only if

(i) there exists a function $\phi(x, z)$ which satisfies the condition (i) of Lemma 3.17.2,

(ii) there exists a function $h(x, z)$ which is continuous and bounded for $0 \leq x \leq a,\ 0 \leq z < \infty,\ h(x, 0) \equiv 0$ and the only solution of the problem

$$z' = h(x, z),\ z(0) = 0 \tag{3.17.4}$$

is the trivial solution $z(x) \equiv 0$; and

(iii) the following inequality holds

$$\phi_x(x, y) + \phi_z(x, y)f(x, y) \leq h(x, \phi(x, y)),\ 0 \leq x \leq a,\ 0 \leq y \leq b. \tag{3.17.5}$$

Proof. Let $\{\varepsilon_n\}$ be a sequence of positive numbers converging to 0 and $z_n(x)$ be a solution of the initial value problem $z' = h(x, z) + \varepsilon_n,\ z(0) = 0$. Since the sequence $\{z_n(x)\}$ is uniformly bonded and equicontinuous, and for the problem (3.17.4), $z(x) \equiv 0$ is the only solution it follows that $\lim_{n\to\infty} z_n(x) = 0$.

Now from the hypotheses we have

$$\phi_x(x, y) + \phi_z(x, y)f(x, y) < h(x, \phi(x, y)) + \varepsilon_n$$

and hence in Lemma 3.17.2 we can take $\xi(x)$ to be $z_n(x)$, to obtain

$$\phi(x, y(x)) \leq z_n(x),$$

where $y(x)$ is a solution of (1.9.4). But, now from the properties of $\phi(x, z)$ and $\{z_n(x)\}$ it is clear that $y(x) \equiv 0$.

To show the necessary part of the theorem it suffices to choose in Theorem 3.17.1, $\phi(x, z) \equiv z$, and $h(x, z) \equiv g(x, z)$. ∎

Corollary 3.17.4. Let $f(x, y)$ be continuous in $\bar{S}_+$ with $x_0 = 0$, and

(i) there exists a function $h(x,z)$ which satisfies the condition (ii) of Theorem 3.17.3,

(ii) there exists a function $\psi(x,y,z)$ which is defined and continuously differentiable for $0 \le x \le a$, $|y-y_0| \le b$, $0 \le z < \infty$, $\psi(x,y,0) \equiv 0$, and $\psi(x,y,z) > 0$ for $0 \le x \le a$, $|y-y_0| \le b$, $0 < z < \infty$; and

$$(3.17.6)\quad \psi_x(x,y,z)+\psi_y(x,y,z)f(x,y)+\psi_z(x,y,z)(f(x,y+z)-f(x,y)) \le h(x,\psi(x,y,z)).$$

Then, the initial value problem (1.1.1) with $x_0 = 0$ has at most one solution in $[0,a]$.

Proof. Let $y = \rho(x)$ be the minimal solution and $\rho(x) + z(x)$ ($z(x) \ge 0$) be any other solution of (1.1.1) with $x_0 = 0$. Then, it follows that $z' = F(x,z) = f(x,\rho(x)+z) - f(x,\rho(x))$, $z(0) = 0$. Thus, in Theorem 3.17.3 it suffices to choose $\phi(x,z) \equiv \psi(x,\rho(x),z)$ and $f(x,z) \equiv F(x,z)$. ∎

Corollary 3.17.5. Let $\Psi(y,z)$ be a function defined and continuously differentiable for $|y-y_0| \le b$, $0 \le z < \infty$, $\Psi(y,0) \equiv 0$, $\Psi(y,z) > 0$ for $|y-y_0| \le b$, $0 < z < \infty$; and

$$(3.17.7)\qquad \Psi_z(y,z)f(x,y+z) + (\Psi_y(y,z) - \Psi_z(y,z))f(x,y) \le 0.$$

Then, the initial value problem (1.1.1) with $x_0 = 0$ has at most one solution in $[0,a]$.

Proof. Setting $h(x,z) \equiv 0$ and $\psi(x,y,z) \equiv \Psi(y,z)$ in Corollary 3.17.4 the result follows immediately. ∎

Corollary 3.17.6. Let there exist a function $\eta(y)$ defined and continuous for $|y| < \infty$ and positive almost everywhere. Further, let for all (x,y), $(x,\bar{y})$, $0 \le x \le a$, $\bar{y} > y$

$$(3.17.8)\qquad \eta(\bar{y})f(x,\bar{y}) \le \eta(y)f(x,y).$$

Then, the initial value problem (1.1.1) with $x_0 = 0$ has at most one solution in $[0,a]$.

Proof. Setting $\Psi(y,z) = \int_y^{y+z} \eta(t)dt$ in Corollary 3.17.5 the result follows immediately. ∎

For $\eta(y) \equiv 1$ Corollary 3.17.6 reduces to Peano's Uniqueness Theorem 1.3.1.

3.18 BRAUER - STERNBERG'S UNIQUENESS THEOREM

By employing Lyapunov functions it is possible to give very general conditions for the uniqueness of n - solutions of (1.1.1). We begin with a result of Perron's type.

Theorem 3.18.1. Assume that

(i) the function $g(x,z)$ is continuous for $x_0 \le x \le x_0 + a$, $z \ge 0$ and for every x_1, $x_0 < x_1 < x_0 + a$, $z(x) \equiv 0$ is the only differentiable function in $[x_0, x_1)$, which satisfies (1.11.1),

(ii) the function $f(x,y)$ is continuous in $\bar{S}_+$; and

(iii) there exists a Lyapunov function $V(x, y - \bar{y})$ defined and continuously differentiable in $\bar{S}_0 = \{(x, y-\bar{y}) : (x,y) \in \bar{S}_+,\ (x,\bar{y}) \in \bar{S}_+\}$ such that $V(x,0) \equiv 0$, $V(x,y-\bar{y}) > 0$ if $\|y - \bar{y}\| \ne 0$, and for all $(x, y-\bar{y}) \in \bar{S}_0$

$$V'(x,y-\bar{y}) = \frac{\partial V(x,y-\bar{y})}{\partial x} + \frac{\partial V(x,y-\bar{y})}{\partial y} \cdot (f(x,y) - f(x,\bar{y}))$$

$$\le g(x, V(x,y-\bar{y})). \tag{3.18.1}$$

Then, the initial value problem (1.1.1) has at most one n - solution in $[x_0, x_0 + a]$.

Proof. Let us suppose that there are two n - solutions $y(x)$ and $\bar{y}(x)$ of the initial value problem (1.1.1) in $[x_0, x_0 + a]$. Consider the function $\phi(x) = V(x, y(x) - \bar{y}(x))$. We have in view of (3.18.1) and the continuous differentiability of $V(x, y - \bar{y})$ the inequality $\phi'(x) \le g(x, \phi(x))$. Also, $\phi(x_0) = 0$. For any x_1 such that $x_0 < x_1 < x_0 + a$, we obtain, by Theorem 1.11.2 the estimate $\phi(x) \le r(x)$, $x_0 \le x \le x_1$, where $r(x)$ is the maximal solution of (1.11.1). The assumptions on the Lyapunov function V now ensure that $\phi(x) \equiv 0$, i.e., $y(x) = \bar{y}(x)$ in $[x_0, x_0 + a]$. ∎

Corollary 3.18.2. Peano's Uniqueness Theorem 1.3.1 is a particular case of Theorem 3.18.1.

Proof. The function $V(x, y - \bar{y}) = (y - \bar{y})^2$ satisfies all the conditions of Theorem 3.18.1 and for this function the inequality (3.18.1) reduces to

$$2(y - \bar{y}) \cdot (f(x, y) - f(x, \bar{y})) \le g(x, V(x, y - \bar{y})),$$

which is also satisfied if $g(x, z) \equiv 0$. ∎

The next theorem is analogous to Kamke's result.

Theorem 3.18.3 (Brauer - Sternberg's Uniqueness Theorem). Assume that

(i) the function $g(x, z)$ is continuous for $x_0 < x \le x_0 + a$, $z \ge 0$ and for every x_1, $x_0 < x_1 < x_0 + a$, $z(x) \equiv 0$ is the only differentiable function in $x_0 < x < x_1$ and continuous in $x_0 \le x \le x_1$ for which $z'_+(x_0) = \lim_{x \to x_0^+} \frac{z(x) - z(x_0)}{x - x_0}$ exists and (1.15.1) and (1.15.2) hold; and

(ii) hypotheses (ii) and (iii) of Theorem 3.18.1 hold except that the condition (3.18.1) is satisfied for all $(x, y - \bar{y}) \in \bar{S}_0$ with $x \ne x_0$.

Then, the initial value problem (1.1.1) has at most one n - solution in $[x_0, x_0 + a]$.

Proof. Define the function $g_f(x, z) = \sup_{V(x, y - \bar{y}) = z} V'(x, y - \bar{y})$ for $x_0 \le x \le x_0 + a$ and $z \ge 0$. Since $f(x, y)$, $\frac{\partial V}{\partial x}$, $\frac{\partial V}{\partial y}$ are all continuous, it follows that $g_f(x, z)$ is continuous for $x_0 \le x \le x_0 + a$, $z \ge 0$. Rest of the proof is similar to that of Theorem 1.15.1. ∎

Our final result here is analogous to Brauer's Uniqueness Theorem 1.18.1.

Theorem 3.18.4. Assume that

(i) functions $g_1(x,z)$, $g_2(x,z)$ are continuous for $x_0 \le x \le x_0 + a$, $z \ge 0$,

(ii) hypotheses (i), (ii) and (iv) of Theorem 1.18.1 hold,

(iii) hypothesis (ii) of Theorem 3.18.1 holds; and

(iv) there exists a Lyapunov function $V(x, y-\bar{y})$ defined and continuously differentiable in $\bar{S}_0$ such that $V(x,0) \equiv 0$, $V(x,y-\bar{y}) > 0$ if $\|y-\bar{y}\| \ne 0$, and for all $(x, y-\bar{y}) \in \bar{S}_0$

$$V'(x,y-\bar{y}) \le g_i(x, V(x,y-\bar{y})); \; i = 1,2.$$

Then, the initial value problem (1.1.1) has at most one n - solution in $[x_0, x_0 + a]$. ∎

In Theorems 3.18.1, 3.18.3 and 3.18.4 it is easily seen that the continuous differentiability of the Lyapunov function $V(x,y-\bar{y})$ in $\bar{S}_0$ can be replaced by its existence of one - sided derivatives. Further, with such a replacement the function $V(x,y-\bar{y}) = |y-\bar{y}| = \sum_{i=1}^{n} |y_i - \bar{y}_i|$ satisfies all the conditions of Theorem 3.18.1 and for this function the inequality (3.18.1) reduces to

$$|f(x,y) - f(x,\bar{y})| \le g(x, V(x, |y-\bar{y}|).$$

3.19 MOYER'S UNIQUENESS THEOREM

Here we shall prove a result which unifies various uniqueness theorems.

Theorem 3.19.1 (Moyer's Uniqueness Theorem). Assume that $f(x,y)$ is continuous in $\bar{S}_+$ with $x_0 = 0$, and $W(x,z)$ is continuously differentiable for $0 < x \le a$, $z > 0$, and

(i) $\dfrac{\partial W(x,z)}{\partial x} \le 0$, $\dfrac{\partial W(x,z)}{\partial z} \ge 0$, $0 < x \le a$, $z > 0$,

(ii) $\dfrac{\partial W(x,\|y-\bar{y}\|)}{\partial z}\|f(x,y) - f(x,\bar{y})\| \le -\dfrac{\partial W(x,\|y-\bar{y}\|)}{\partial x}$ for all (x,y), $(x,\bar{y}) \in \bar{S}_+, x \ne x_0 = 0$, $y \ne \bar{y}$; and

(iii) $W(x, 0^+) = -\infty, \ 0 < x \leq a.$

Further, assume that $y(x)$ and $\bar{y}(x)$ are two n - solutions of (1.1.1) with $x_0 = 0$ in $[0, a]$. Then, $y(x) = \bar{y}(x), \ 0 \leq x \leq a$ if and only if

$$\liminf_{x \to 0^+} W(x, \|y(x) - \bar{y}(x)\|) = -\infty. \tag{3.19.1}$$

Proof. Let $\phi(x) = \|y(x) - \bar{y}(x)\|$ and let $P = \{x \in (0, a) : \phi(x) > 0\}$. Since, $\|D^+\phi(x)\| \leq \|y'(x) - \bar{y}'(x)\|, \ 0 < x < a$ from (i) and (ii) it follows that

$$\begin{aligned} D^+W(x, \phi(x)) &\leq \frac{\partial W(x, \phi(x))}{\partial x} + \frac{\partial W(x, \phi(x))}{\partial z} D^+\phi(x) \\ &\leq \frac{\partial W(x, \phi(x))}{\partial x} + \frac{\partial W(x, \phi(x))}{\partial z} \|y'(x) - \bar{y}'(x)\| \\ &\leq \frac{\partial W(x, \phi(x))}{\partial x} + \frac{\partial W(x, \phi(x))}{\partial z} \|f(x, y(x)) - f(x, \bar{y}(x))\| \\ &\leq 0. \end{aligned}$$

Hence, $W(x, \phi(x))$ is nonincreasing in P. But, $W(x, \phi(x))$ is continuous for $0 < x < a$ and $W(x, \phi(x)) = -\infty$ when $x \notin P$. Therefore, $W(x, \phi(x))$ is nonincreasing for $0 < x < a$. Thus, either $P = \emptyset$ or $P = (0, \alpha)$ for some $\alpha > 0$. Since in the latter case $W(x, \phi(x))$ will be bounded below near $x = 0$, it follows that $\phi(x) \equiv 0$ if and only if $\liminf_{x \to 0^+} W(x, \phi(x)) = -\infty$. ∎

Corollary 3.19.2. Assume that $f(x, y)$ and $W(x, z)$ are as in Theorem 3.19.1. Then, the initial value problem (1.1.1) with $x_0 = 0$ has at most one n - solution if any one of the following hold :

(a) There is an $z_0 > 0$ such that $W(0^+, z) < \infty, \ 0 < z \leq z_0$ and $\liminf_{z \to 0^+} W(0^+, z) = -\infty$.

(b) There exists a differentiable function $w_1(z), \ 0 \leq z < \infty$, and an absolutely continuous function $w_2(x), \ 0 \leq x \leq a$, such that

(1) $w_1'(z) > 0, \ z > 0, \ w_2'(x) \geq 0$, almost everywhere in $(0, a)$,

(2) $w_1(0) = w_2(0) = 0$,

(3) $w_1'(\|y - \bar{y}\|)\|f(x, y) - f(x, \bar{y})\| \leq w_2'(x)$, almost everywhere in $(0, a)$, (x, y), $(x, \bar{y}) \in \bar{S}_+, x \neq x_0 = 0, \ y \neq \bar{y}$; and

(4) $\liminf_{x\to 0^+} W(x, w_1^{-1}(w_2(x))) = -\infty.$

(c) $f(x,y)$ is continuous in $\bar{S}_+$ and $\liminf_{x\to 0^+} W(x,x) = -\infty$.

(d) $f(x,y)$ is continuous in $\bar{S}_+$, $W(x,x)$ is bounded in a neighborhood of zero, $\frac{\partial W}{\partial x}$ is a nonincreasing function of z and there is an $x_1 > 0$ such that $\liminf_{x\to 0^+}[W(x_1, xh(x)) - W(x_1, x)] = -\infty$ for every $h(x)$ such that $h(x)\to 0$ as $x \to 0^+$.

(e) $f(x,y)$ is continuous in $\bar{S}_+$ and $\liminf_{x\to 0^+} W(x, xh(x)) = -\infty$ for every nonnegative $h(x)$ such that $h(x)\to 0$ as $x\to 0^+$.

Proof. As usual let $y(x)$ and $\bar{y}(x)$ be two n - solutions of (1.1.1) in $[0,a]$ and $\phi(x) = \|y(x) - \bar{y}(x)\|$.

(a) By the hypothesis (i) of Theorem 3.19.1 the function $W(x,z)$ is nonincreasing in x. Thus, $W(x,\phi(x)) \le W(0^+, \phi(x))$, $x \in (0,a]$. But, since $\phi(x)\to 0$ as $x\to 0^+$, it follows that $\liminf_{x\to 0^+} W(x,\phi(x)) \le -\infty$.

(b) From Lemma 5.2.1 it is clear that $w_1(\phi(x)) - w_2(x)$ is a nonincreasing function of $x \in (0,a)$. Since $\phi(0) = 0$ in view of (2) we have $w_1(\phi(0)) - w_2(0) = 0$ and thus $w_1(\phi(x)) \le w_2(x)$. But, $w_1(z)$ is increasing and thus w_1^{-1} exists. Therefore, we find that $\phi(x) \le w_1^{-1}(w_2(x))$, $x\in[0,a]$. Since $W(x,z)$ is nondecreasing function of z, $W(x,\phi(x)) \le W(x, w_1^{-1}(w_2(x)))$, $x\in(0,a]$, and hence (4) implies that $\liminf_{x\to 0^+} W(x,\phi(x)) \le -\infty$.

(c) If $f(x,y)$ is continuous, then $y(x)$ and $\bar{y}(x)$ are continuously differentiable and hence $\|y'(0) - \bar{y}'(0)\| = \|f(0,y_0) - f(0,y_0)\| = 0$. Thus, $\liminf_{x\to 0^+} \frac{\phi(x)}{x} = 0$. But, then $\phi(x) \le x$ for all sufficiently small x. Therefore, it follows that $\liminf_{x\to 0^+} W(x,\phi(x)) \le \liminf_{x\to 0^+} W(x,x) \le -\infty$.

(d) Again we have $\liminf_{x\to 0^+} \frac{\phi(x)}{x} = 0$, and hence for sufficiently small t we find that $\phi(t) \le t$ and thus $\frac{\partial W(x,\phi(t))}{\partial x} \ge \frac{\partial W(x,t)}{\partial x}$. Integrating this inequality from t to x_1 where $t \le x_1$, we obtain

$$W(x_1,\phi(t)) - W(t,\phi(t)) \ge W(x_1,t) - W(t,t)$$

and hence

$$W(t,\phi(t)) \le W\left(x_1, t\frac{\phi(t)}{t}\right) - W(x_1,t) + W(t,t).$$

But, $W(t,t)$ is bounded and $\liminf_{x\to 0^+} \frac{\phi(x)}{x} = 0$, therefore $\liminf_{t\to 0^+} W(t, \phi(t)) \le -\infty$.

(e) Again, since $\liminf_{x\to 0^+} \frac{\phi(x)}{x} = 0$ it follows that

$$\liminf_{x\to 0^+} W(x,\phi(x)) = \liminf_{x\to 0^+} W\left(x, x\frac{\phi(x)}{x}\right) = -\infty. \qquad \blacksquare$$

In the following table we list various uniqueness conditions which can be deduced from Corollary 3.19.2.

$W(x,z)$	Assumptions	Part	Condition
1. $lnz - Lx$		(a)	**Lipschitz (1.2.1)**
2. $w_1(z) - Lx$	$w_1(0^+) = -\infty$	(a)	**Osgood (1.4.3)**
3. $w_1(z) - w_2(x)$	$w_1(0^+) = -\infty,\ \lvert w_2(0^+)\rvert < \infty$	(a)	**Montel - Tonelli (1.5.1)** **Heins (3.4.4)**
4. $lnz - klnx$	$0 < k < 1$	(c)	**Rosenblatt (1.6.1)**
5. $lnz - lnx$		(c)	**Nagumo (1.6.1)**
6. $lnz - klnx$	$w_1(z) = z^{1-\alpha},\ w_2(x) = cx,$ $0 < k(1-\alpha) < 1$	(b)	**Krasnosel'skii -Krein (1.7.1),(1.7.2)**
7. $w_1(z) - w_2(x)$	$w_1(x) - w_2(x) \to -\infty$ **as** $x \to 0^+$	(c)	**Filippov - LaSalle (1.15.19)**
8. $w_1(z) - w_2(x)$	$w_1(x) - w_2(x)$ **bounded and** $w_1(xh(x))$ $-w_2(x) \to -\infty$ **for those** $h(x) \ge 0$ **such that** $h(x) \to 0$ **as** $x \to 0^+$	(d)	**Filippov - LaSalle (1.15.20)**
9. $w_1(z/x) - w_2(x)$	$w_1(0^+) = -\infty,\ \lvert w_2(0^+)\rvert < \infty$	(c)	**Walter [164]**

Theorem 3.19.1 as well as Corollary 3.19.2 carry over for n - extended solutions of (1.1.1). In fact, we shall prove

Theorem 3.19.3. Assume that $f(x,y)$ satisfies the Carathèodory conditions in $\bar{S}_+$; $W(x,z)$ is continuous and for each $z > 0$ it is absolutely continuous function of x on compact subsets of $(0,a]$, $\frac{\partial W(x,z)}{\partial z}$ exists and is continuous for $0 < x \le a,\ z > 0$ and that $W(x,z)$ satisfies conditions (i) - (iii) of Theorem 3.19.1. Further, assume that $y(x)$ and $\bar{y}(x)$ are two n - extended solutions of (1.1.1) with $x_0 = 0$ in $[0,a]$. Then, $y(x) = \bar{y}(x),\ 0 \le x \le a$ if and only if (3.19.1) holds.

Proof. As in Theorem 3.19.1 let $\phi(x) = \|y(x) - \bar{y}(x)\|$ and $P = \{x \in (0,a) \ ; \ \phi(x) > 0\}$ to deduce that $D^+W(x,\phi(x)) \leq 0$ almost everywhere in P. Further, since $W(x,\phi(x))$ is an absolutely continuous function on compact subsets of $(0,a]$, it follows that $W(x,\phi(x))$ is a nonincreasing function of x. The result now follows as in Theorem 3.19.1. ∎

Corollary 3.19.4. Assume that $f(x,y)$ and $W(x,y)$ are as in Theorem 3.19.3 and any one of the conditions of Corollary 3.19.2 is satisfied. Then, n - extended solutions of (1.1.1) with $x_0 = 0$ are unique. ∎

Corollary 3.19.5. Assume that $f(x,y)$ satisfies the Caratheodory conditions in $\bar{S}_+$ and let $M(x) = \max(M_1(x), \cdots, M_n(x))$. Further, assume that $W(x,z)$ is as in Theorem 3.19.3, and $\lim_{x\to 0^+} W(x, 2\bar{M}(x)) = -\infty$, where $\bar{M}(x) = \int_0^x M(t)dt$. Then, the initial value problem (1.1.1) has at most one n - extended solution in $[0,a]$.

Proof. Let $w_1(z) = z$ and $w_2(x) = 2\bar{M}(x)$. Then,

$$\begin{aligned} w_1'(\|y - \bar{y}\|)\|f(x,y) - f(x,\bar{y})\| &= \|f(x,y) - f(x,\bar{y})\| \\ &\leq 2M(x) \\ &= w_1'(x). \end{aligned}$$

The result now follows from the part (b) of Corollary 3.19.2. ∎

Remark 3.19.1. Moyer's uniqueness theorem is a particular case of Theorems 3.15.1 and 3.15.8. Indeed, if $W(x,z)$ is as in Theorem 3.19.1 and $\bar{y}(x)$ is a n - solution of (1.1.1) with $x_0 = 0$ in $[0,a]$ then the function

$$V(x,y) = V(x, \bar{y}(x) - y) = e^{W(x, \|\bar{y}(x) - y\|)} \tag{3.19.2}$$

satisfies the hypotheses of Theorem 3.15.1 in S_o. Therefore, Theorem 3.19.1 can be stated as follows :

Theorem 3.19.6. Assume that $f(x,y)$ is continuous in $\bar{S}_+$ with $x_0 = 0$, and there exists a function $V(x,y)$ in S_o which satisfies the conditions of Theorem 3.15.1. Further, assume that $y(x)$ and $\bar{y}(x)$ are two n - solutions of (1.1.1) with $x_0 = 0$ in $[0, a]$. Then, $y(x) = \bar{y}(x),\ 0 \le x \le a$ if and only if

$$\liminf_{x \to 0^+} V(x, y(x) - \bar{y}(x)) = 0. \tag{3.19.3}$$ ∎

3.20 BERNFELD - DRIVER - LAKSHMIKANTHAM'S UNIQUENESS THEOREM

The following result extends various earlier uniqueness results.

Theorem 3.20.1 (Bernfeld - Driver - Lakshmikantham's Uniqueness Theorem). Let $f(x, y)$ be defined in S_+, $x_1 \in (x_0, x_0 + a)$, and assume that

(i) the function $g(x, z)$ is continuous for $x_0 < x < x_0 + a,\ z \ge 0$ with $g(x, 0) \equiv 0$ and such that $z(x) \equiv 0$ is the only solution of (1.13.1) which satisfies the condition that $\lim_{x \to x_0^+} \frac{z(x)}{B(x)} = 0$, where $B(x)$ is continuous in $[x_0, x_0 + a)$ with $B(x) > 0$ for $x > x_0$; and

(ii) there exists a Lyapunov function $V(x, y, \bar{y}) \ge 0$ defined and continuous for $x_0 < x < x_0 + a,\ \|y\| < \infty,\ \|\bar{y}\| < \infty$ such that $V(x, y, \bar{y}) = 0$ if and only if $y = \bar{y}$; for any two n - classical solutions $y(x),\ \bar{y}(x)$ of (1.1.1) in $[x_0, x_1)$, $\lim_{x \to x_0^+} \frac{V(x, y(x), \bar{y}(x))}{B(x)} = 0$, and $D_+V(x, y(x), \bar{y}(x)) \le g(x, V(x, y(x), \bar{y}(x)))$ whenever $x_0 < x < x_1$, and $V(x, y(x), \bar{y}(x)) > 0$.

Then, the initial value problem (1.1.1) has at most one n - classical solution in $[x_0, x_1)$.

Proof. Let $y(x)$ and $\bar{y}(x)$ be two n - classical solutions of (1.1.1) in $[x_0, x_1)$. Define $\phi(x) = V(x, y(x), \bar{y}(x))$ for $x_0 \le x < x_1$ and suppose that $\phi(x_2) > 0$ for some $x_2 \in (x_0, x_1)$. Then, as long as $\phi(x)$ remains positive to the left of x_2 we have $D_+\phi(x) \le g(x, \phi(x))$. Let $\rho(x)$ be the minimal solution of (1.13.1) for $x \le x_2$ with $\rho(x_2) = \phi(x_2)$. Then, as in Theorem 2.8.3, $\rho(x)$ can be continued up to $x = x_0$ and $0 \le \rho(x) \le \phi(x),\ 0 \le x \le x_2$. Further,

$$0 \le \lim_{x \to x_0^+} \frac{\rho(x)}{B(x)} \le \lim_{x \to x_0^+} \frac{\phi(x)}{B(x)} = 0,$$

which, by hypotheses, implies that $\rho(x) \equiv 0$. This contradiction forces that $\phi(x) = V(x, y(x), \bar{y}(x)) \equiv 0$, which means that $y(x) = \bar{y}(x)$, $x_0 \le x < x_1$. ∎

Corollary 3.20.2. Let $f(x,y)$, $g(x,z)$ and $B(x)$ be as in Theorem 3.20.1 and assume that for any two n - classical solutions $y(x)$, $\bar{y}(x)$ of (1.1.1) in $[x_0, x_1)$, $\lim_{x \to x_0^+} \frac{\|y(x)-\bar{y}(x)\|}{B(x)} = 0$, and the following inequality is satisfied

$$\|f(x,y(x)) - f(x,\bar{y}(x))\| \le g(x, \|y(x) - \bar{y}(x)\|), \ x_0 < x < x_1. \tag{3.20.1}$$

Then, the initial value problem (1.1.1) has at most one n - classical solution in $[x_0, x_1)$.

Proof. The proof is immediate by letting $V(x, y(x), \bar{y}(x)) = \|y(x) - \bar{y}(x)\|$ in Theorem 3.20.1. ∎

A comparison of Corollary 3.20.2 with Kamke's Uniqueness Theorem 1.15.1 as well as with Laksmikantham's Uniqueness Theorem 2.8.3 suggests that in both of these theorems it suffices for the inequality (1.11.2) to hold in a subset of $\bar{S}_+$. While it is imposed here in terms of the unknown solutions of (1.1.1), the importance of Corollary 3.20.2 is illustrated in the following :

Example 3.20.1. Consider the initial value problem

$$\begin{aligned} &y' = x^\beta + y^\alpha, \ 0 \le x < \infty, \ |y| < \infty \\ &y(0) = 0, \end{aligned} \tag{3.20.2}$$

where $0 < \alpha < 1$ and $\beta > 0$ are constants.

To show the uniqueness of the solutions of (3.20.2) for $\beta > \alpha(1-\alpha)^{-1}$ we let $B(x) = ((1-\alpha)x)^{(1-\alpha)^{-1}}$. Then, $B(x)$ is the maximal solution of the initial value problem $w' = w^\alpha$, $w(0) = 0$. Let $y(x)$ be any solution of (3.20.2). Then, it follows that $y' > y^\alpha$ for $x > 0$ so that

$$y(x) > B(x), \ x > 0. \tag{3.20.3}$$

Now let $\varepsilon > 0$ and consider $w_\varepsilon(x) = (1+\varepsilon)B(x)$ so that $w_\varepsilon(0) = 0$ and for $x > 0$

$$w'_\varepsilon(x) = (1+\varepsilon)\left(\frac{w_\varepsilon(x)}{1+\varepsilon}\right)^\alpha.$$

But, since $w_\varepsilon^\alpha(x) = (1+\varepsilon)^\alpha((1-\alpha)x)^{\alpha(1-\alpha)^{-1}}$ and $\beta > \alpha(1-\alpha)^{-1}$, it follows that $w_\varepsilon'(x) > x^\beta + w_\varepsilon^\alpha(x)$ for all sufficiently small $x > 0$. Thus, there exists some $\delta(\varepsilon) > 0$ such that

$$w_\varepsilon(x) = (1+\varepsilon)B(x) > y(x),\ 0 < x < \delta(\varepsilon). \tag{3.20.4}$$

On combining (3.20.3) and (3.20.4) we find that for each $\varepsilon > 0$

$$1 < \frac{y(x)}{B(x)} < 1+\varepsilon,\ 0 < x < \delta(\varepsilon).$$

From this we conclude that $\lim_{x\to 0^+} y(x)/B(x) = 1$ for each solution of (3.20.2). Thus, for any two solutions $y(x)$ and $\bar{y}(x)$ of (3.20.2) it is clear that $\lim_{x\to 0^+} |y(x) - \bar{y}(x)|/B(x) = 0$.

Next, if $x_0 < x < x_0 + a$ and $s,\ t > B(x)$, then

$$|s^\alpha - t^\alpha| = \alpha\xi^{\alpha-1}|s-t| < \frac{\alpha}{(1-\alpha)x}\,|s-t|.$$

Thus, in Corollary 3.20.2 the inequality (3.20.1) is satisfied with $g(x,z) = \frac{\alpha}{(1-\alpha)x}z$. (It is important to note that this condition cannot be satisfied in the set of the form S_+ and therefore Theorems 1.15.1 and 2.8.3 are powerless for the problem (3.20.2).) Further, since the solutions of $z' = \frac{\alpha}{(1-\alpha)x}z$ are of the form $z(x) = cx^{\alpha(1-\alpha)^{-1}}$ the condition $\lim_{x\to x_0^+}\frac{z(x)}{B(x)} = 0$ implies that $c = 0$, and hence $z(x) \equiv 0$. Thus, the hypotheses of Corollary 3.20.2 are satisfied and in conclusion the problem (3.20.2) has at most one solution in a sufficiently small interval $[0, x_1)$. The uniqueness to the right of this interval can be deduced from the Lipschitz Uniqueness Theorem 1.2.4.

For the case $\beta \le \alpha(1-\alpha)^{-1}$ the uniqueness of (3.20.2) can be established similarly by taking $B(x) = x^{\beta+1}/(\beta+1)$. ∎

Corollary 3.20.3. Consider the scalar initial value problem

$$y' = p(x,y)q(x,y),\ y(x_0) = y_0 \tag{3.20.5}$$

where the functions $p(x,y)$ and $q(x,y)$ are defined in S_+, and

(i) $0 < p(x,y) \le K$ for all $(x,y) \in S_+$,

(ii) $\lim \int_{y_0}^{y} dt/p(x,t) = 0$ as $(x,y) \to (x_0, y_0)$ in S_+,

(iii) $p(x,y)$ is continuous in S_+ and nondecreasing in x for each fixed y; and

(iv) for all (x,y), $(x,\bar{y}) \in S_+$ with $y \geq \bar{y}$, $q(x,y) - q(x,\bar{y}) \leq L(x)(y-\bar{y})$, where $L(x)$ is continuous and integrable in $(x_0, x_0 + a)$.

Then, the initial value problem (3.20.5) has at most one classical solution in $[x_0, x_1)$.

Proof. Let in Theorem 3.20.1, $B(x) \equiv 1$ and $V(x,y,\bar{y}) = |\int_y^{\bar{y}} dt/p(x,t)|$ for (x,y), $(x,\bar{y}) \in S_+$. Then, for any two solutions $y(x)$ and $\bar{y}(x)$ of (3.20.5) in $[x_0, x_1)$, we have

$$\begin{aligned} D_+ \int_{y(x)}^{\bar{y}(x)} \frac{dt}{p(x,t)} &= \liminf_{\Delta x \to 0^+} \frac{1}{\Delta x} \left\{ \int_{\bar{y}(x)}^{\bar{y}(x+\Delta x)} \frac{dt}{p(x+\Delta x, t)} \right. \\ &\quad \left. - \int_{y(x)}^{y(x+\Delta x)} \frac{dt}{p(x+\Delta x, t)} + \int_{y(x)}^{\bar{y}(x)} \left[\frac{1}{p(x+\Delta x, t)} - \frac{1}{p(x,t)} \right] dt \right\} \\ &= \frac{\bar{y}'(x)}{p(x,\bar{y}(x))} - \frac{y'(x)}{p(x,y(x))} \\ &\quad + \liminf_{\Delta x \to 0^+} \frac{1}{\Delta x} \int_{y(x)}^{\bar{y}(x)} \left[\frac{1}{p(x+\Delta x, t)} - \frac{1}{p(x,t)} \right] dt. \end{aligned}$$

If $V(x, y(x), \bar{y}(x)) > 0$ we might as well assume that $\bar{y}(x) > y(x)$. Then, the above inequality and the fact that $p(x + \Delta x, t) \geq p(x,t)$, gives

$$\begin{aligned} D_+ V(x, y(x), \bar{y}(x)) &= D_+ \int_{y(x)}^{\bar{y}(x)} \frac{dt}{p(x,t)} \\ &< q(x, \bar{y}(x)) - q(x, y(x)) \\ &\leq L(x)(\bar{y}(x) - y(x)) \leq K L(x) V(x, y(x), \bar{y}(x)). \end{aligned}$$

Thus, the conditions of Theorem 3.20.1 are satisfied and the conclusion follows. ∎

Setting $f_1 = p$, $f_2 = q$ and assuming without loss of generality that $f_3 = 0$, we note that Corollary 3.20.3 generalizes Bownds' Uniqueness Theorem 2.9.1 except for the assumption that $L(x)$ is continuous.

Example 3.20.2. Consider the initial value problem

$$y' = k(x) + g(y), \; y(0) = 0 \tag{3.20.6}$$

where k and g are nonnegative and continuous in $[0, \infty)$. If k and g both are positive in $(0, \infty)$ with $\int_0^b \frac{dt}{g(t)} < \infty$ for any $b > 0$ and k nondecreasing, then (3.20.6) has a unique solution. This follows from Corollary 3.20.3 with $g(x, y) \equiv 1$ and $q(x, y) \equiv k(x) + g(y)$. Obviously, (3.20.6) includes the initial value problem (3.20.2). ∎

It is interesting to note the following :

(i) If g is nonincreasing then for (3.20.6) the uniqueness follows from Peano's Uniqueness Theorem 1.3.1.

(ii) The condition $\int_0^b \frac{dt}{g(t)} < \infty$ used in the above example is the same which assures nonuniqueness (cf. Theorem 1.4.3) for the problem $y' = g(y), \; y(0) = 0$.

(iii) If k and g both are positive and k is nondecreasing then the uniqueness of (3.20.6) follows from Wend's Uniqueness Theorem 1.19.1.

(iv) If k and g both are positive and g in nondecreasing then the uniqueness of (3.20.6) can be deduced from Bownds' Uniqueness Theorem 2.9.1 for $y' = f_1(x, y) f_2(x, y), \; y(0) = 0$ by setting

$$f_1(x, y) = \inf_{0 \le s \le x} k(s) + g(y), \; f_2(x, y) = \frac{k(x) + g(y)}{f_1(x, y)}.$$

(v) Example 3.20.2 permits $k(0) = 0$ as well as $g(0) = 0$.

3.21 KAPER - KWONG'S UNIQUENESS THEOREM

Once again we shall consider the initial value problem (3.20.6), where $g(y)$ is defined and differentiable in $(0, a)$, a may be infinite. Further, it vanishes with a vertical tangent at the origin and monotonically increasing and concave (i.e., g' is decreasing) in $(0, a)$. The function $k(x)$ is defined and continuous in $[0, \infty)$ and may change sign near the origin. Let $K(x) = \int_0^x k(t)dt, \; x \ge 0$.

Theorem 3.21.1 (Kaper - Kwong's Uniqueness Theorem). If $K(x) \geq 0$ for $x \geq 0$ and there is a sequence $\{x_n\}$ of points $x_n > 0$ that converges to 0 such that $K(x_n) > 0$ for all n, then the initial value problem (3.20.6) has at most one nonnegative (nontrivial) solution.

To prove this theorem we need the following three lemmas.

Lemma 3.21.2. If $\int_0^b \frac{dt}{g(t)} = \infty$ for any $b > 0$, then (3.20.6) has at most one solution.

Proof. In view of the concavity of the function $g(y)$ the function $f(x,y) = k(x)+g(y)$ satisfies the inequality $|f(x,y)-f(x,\bar{y})| = |g(y)-g(\bar{y})| \leq g(|y-\bar{y}|)$ for all (x,y), $(x,\bar{y})$ with $x > 0$. Further, since $f(x,y)$ is continuous, from Theorem 1.4.4 the conclusion follows. ∎

If $\int_0^b \frac{dt}{g(t)} < \infty$, then by Theorem 1.4.3 the initial value problem

$$w' = g(w),\ x > 0,\ w(0) = 0 \tag{3.21.1}$$

has an infinite number of solutions. In what follows we shall denote by $w_M(x)$ the maximal solution of (3.21.1). It is clear that this solution can be written as

$$\int_0^{w_M(x)} \frac{dt}{g(t)} = x,\ x \geq 0.$$

We shall assume that $w_M(x)$ exists in $[0, X)$.

Lemma 3.21.3. Every solution $y(x)$ of (3.20.6) satisfies the inequality $y(x) \geq w_M(x)$.

Proof. Every solution $y(x)$ of (3.20.6) satisfies the integral equation $y(x) = T(y)(x)$, where T is given by

$$T(y)(x) = K(x) + \int_0^x g(y(t))dt,\ x \geq 0. \tag{3.21.2}$$

For $0 \leq y(x) \leq z(x)$ it is clear that $T(y)(x) \leq T(z)(x)$, i.e., T is monotone. We define a sequence of functions $\{y_n(x)\}$ as follows

$$y_{n+1}(x) = T(y_n)(x),\ y_1(x) = K(x);\ n = 1, 2, \cdots . \tag{3.21.3}$$

Since $y_1(x) = K(x) \geq 0$ it is clear that $y_2(x) \geq y_1(x)$ and by induction $y_{n+1}(x) \geq y_n(x)$; $n = 1, 2, \cdots$. Hence, the sequence $\{y_n(x)\}$ is monotonically increasing. Since it is also bounded above, it converges as $n \to \infty$ to an element $y_*(x)$. This limit satisfies the integral equation (3.21.2) and hence is a solution of (3.20.6).

If $z(x)$ is also a solution of (3.20.6), then $z(x) = T(z)(x) \geq y_1(x)$. Because $T(z)(x) = z(x)$ and T is monotone, it follows that $z(x) \geq y_n(x)$, $n = 1, 2, \cdots$. Hence, $z(x) \geq y_*(x)$, therefore $y_*(x)$ is the minimal solution of (3.20.6). Thus, it suffices to establish the lemma for $y_*(x)$.

The initial value problem (3.21.1) is equivalent to $w(x) = L(w)(x)$, where

$$L(w)(x) = \int_0^x g(w(t))dt, \; x \geq 0. \tag{3.21.4}$$

Like T the operator L is monotone.

Let ε be any of the points x_n of the sequence mentioned in the theorem. Let $w_{M,\varepsilon}(x)$ be the maximal solution of the problem

$$w' = g(w), \; x > \varepsilon, \; w(\varepsilon) = 0$$

and let this function be trivially extended backwards to $x = 0$. Because $K(\varepsilon) > 0$ and $w_{M,\varepsilon}(\varepsilon) = 0$, there exists a $\tau > \varepsilon$ such that $K(x) > w_{M,\varepsilon}(x)$, $x \in [\varepsilon, \tau]$. Taking such a τ we define the function $w_1(x)$ by

$$w_1(x) = w_{M,\varepsilon}(x), \; x \in [0, \tau], \quad w_1(x) = 0, \; x \in [\tau, X].$$

With this $w_1(x)$ as the initial function we define a sequence of functions $\{w_n(x)\}$ as follows

$$w_{n+1}(x) = L(w_n)(x); \; n = 1, 2, \cdots .$$

This sequence converges as $n \to \infty$ to a limit function $w_*(x)$ which is a solution of (3.21.1).

A direct computation shows that each $w_n(x)$ coincides with $w_{M,\varepsilon}(x)$ in $[0, \tau]$, so the same is true for $w_*(x)$. However, since $w_{M,\varepsilon}(x)$ is the unique maximal solution of (3.21.1) beyond ε, $w_*(x)$ and $w_{M,\varepsilon}(x)$ are one and the same function.

Clearly, $y_1(x) \geq w_1(x) \geq 0$. Further, since T and L satisfy the ordering relation $T(y)(x) \geq L(w)(x)$ whenever $y(x) \geq w(x) \geq 0$, it follows that $y_n(x) \geq w_n(x)$; $n = 1, 2, \cdots$. In the limit $n \to \infty$, we therefore have $y_*(x) \geq w_*(x)$.

The proof now follows by taking ε successively equal to each of the points of the sequence $\{x_n\}$. This sequence converges to 0, and $w_{M,\varepsilon}(x)$ converges pointwise to $w_M(x)$, as ε tends to 0 through the points of the sequence. ∎

Lemma 3.21.4. If $y_m(x)$ is the minimal solution of (3.20.6), then any solution $y(x)$ of (3.20.6) satisfies the inequality $y(x) \leq y_m(x) + w_M(x)$.

Proof. Since $y(x) = T(y)(x)$ and $y_m(x) = T(y_m)(x)$ it follows that

$$y(x) - y_m(x) = \int_0^x (g(y(t)) - g(y_m(t)))dt \leq \int_0^x g(y(t) - y_m(t))dt.$$

Thus, $w(x) = y(x) - y_m(x)$ satisfies the integral inequality $w(x) \leq L(w)(x)$. The conclusion $w(x) = y(x) - y_m(x) \leq w_M(x)$ is now obvious. ∎

Proof of Theorem 3.21.1. Suppose $y(x)$ and $\bar{y}(x)$ are two solutions of (3.20.6) in $[0, X)$; we assume that $\bar{y}(x) > y(x)$. The function $\phi(x) = \bar{y}(x) - y(x)$ satisfies the initial value problem

$$\phi'(x) = g(\bar{y}(x)) - g(y(x)), \; x > 0, \; \phi(0) = 0.$$

From the mean value theorem, we have $g(\bar{y}(x)) - g(y(x)) = g'(\xi(x))\phi(x)$ where $\xi(x) \in (y(x), \bar{y}(x))$. The derivative g' is decreasing, so estimating $\xi(x)$ from below by $y(x)$, and $y(x)$ in view of Lemma 3.21.3 by $w_M(x)$ we obtain the differential inequality $\phi'(x) \leq g'(w_M(x))\phi(x)$. However, since $g'(w_M(x)) = [\ln g(w_M(x))]'$ it follows that $[\ln \phi(x)]' \leq [\ln g(w_M(x))]'$. Thus, for any two points $0 < x_1 < x_2 < X$ an integration gives

$$\phi(x_2) \leq \frac{g(w_M(x_2))}{g(w_M(x_1))}\phi(x_1).$$

From Lemma 3.21.4 we can estimate $\phi(x_1)$ from above by $w_M(x_1)$. Thus, in view of $g(w_M(x_1)) = g'(\eta)w_M(x_1)$ for some $\eta \in (0, w_M(x_1))$ it follows that

$$\phi(x_2) \leq \frac{1}{g'(\eta)}\phi(w_M(x_2)).$$

Since $g(y)$ has a vertical tangent at the origin, the coefficient $1/g'(\eta)$ tends to 0 as $t_1 \to 0$. Therefore, we can make the right side of the above inequality arbitrarily small by choosing x_1 sufficiently close to 0. Hence, it follows that $\phi(x_2) = 0$. But, since x_2 is arbitrary, we conclude that $\phi(x) \equiv 0$, i.e., $y(x) = \bar{y}(x)$. ∎

Theorem 3.21.1 is different from any of the criteria discussed in Section 3.20. We also note that this theorem can be trivially generalized to the initial value problem

$$y' = k(x) + p(x)g(y), \; x > 0, \; y(0) = 0$$

where $p(x)$ is a positive continuous function. Indeed, a transformation of the independent variable from x to $\int_0^x p(t)dt$ reduces to a problem of the type (3.20.6). Finally, one can construct examples to show that in Theorem 3.21.1 the condition $K(x) \geq 0$ cannot be relaxed.

3.22 TESAROVA - DOSLY'S UNIQUENESS THEOREM

The following result as well as its proof is a slight variation of Theorem 3.20.1.

Theorem 3.22.1 (Tesarova - Dosly's Uniqueness Theorem). Assume that $f(x,y)$ is continuous for $x_0 < x \leq x_0 + a$, $\|y - y_0\| \leq b$, and

(i) the function $B(x) > 0$ is continuous in $(x_0, x_0 + a)$ and the function $g(x,z)$ is continuous for $x_0 < x \leq x_0 + a$, $z \geq 0$, and that for every $x_1 \in (x_0, x_0 + a)$, $z(x) \equiv 0$ is the only differentiable function satisfying (1.15.1) and

(3.22.1) $$z(x) = o(B(x)) \quad \text{as} \quad x \to x_0^+,$$

(ii) there exists a Lyapunov function $V(x, y, \bar{y}) \geq 0$ defined and continuous in

$$S_\varepsilon = \{(x, y, \bar{y}) : x_0 < x \leq x_0 + a + \varepsilon,$$

$$\|y - y_0\| \le b + \varepsilon, \ \|\bar{y} - y_0\| \le b + \varepsilon, \ \varepsilon > 0\}$$

and satisfies the locally Lipschitz condition (3.15.1) with respect to $(y, \bar{y})$; for any two n - solutions $y(x)$ and $\bar{y}(x)$ of (1.1.1), $V(x, y(x), \bar{y}(x)) \equiv 0$ if and only if $y(x) = \bar{y}(x)$, $x \in [x_0, x_0+a]$, and that for all (x, y), $(x, \bar{y})$, $x_0 < x < x_0 + a$, $\|y - y_0\| \le b$, $\|\bar{y} - y_0\| \le b$, $y \neq \bar{y}$, $D_+V_f(x, y, \bar{y}) \le g(x, V(x, y, \bar{y}))$; and

(iii) for any two n - solutions $y(x)$ and $\bar{y}(x)$ of (1.1.1)

$$V(x, y(x), \bar{y}(x)) = o(B(x)), \quad \text{as } x \to x_0^+. \tag{3.22.2}$$

Then, the initial value problem (1.1.1) has at most one n - solution in $[x_0, x_0 + a]$. ∎

It is interesting to note that the results of Section 1.10 can be deduced from Theorem 3.22.1. For this, let $V(x, y, \bar{y}) = |y - \bar{y}|$ and $g(x, z) = h(x)z$ so that for all (x, y), $(x, \bar{y}) \in \Delta$, $y \neq \bar{y}$ the condition (1.10.1) leads to

$$\begin{aligned} D_+V_f(x, y, \bar{y}) &= \frac{1}{|y - \bar{y}|}(y - \bar{y})(f(x, y) - f(x, \bar{y})) \\ &\le h(x)|y - \bar{y}| = g(x, |y - \bar{y}|). \end{aligned}$$

Further, the general solution of the equation $z' = h(x)z$ is $z(x) = z(x_1)\times \exp\left(-\int_x^{x_1} h(t)dt\right)$, where $x_1 \in (0, a)$ is fixed. Thus, to obtain uniqueness criterion as a corollary of Theorem 3.22.1 we need to find a function $B(x)$ such that

(i) $z(x) \equiv 0$ is the only solution of $z' = h(x)z$ satisfying (3.22.1); and

(ii) $|y(x) - \bar{y}(x)| = o(B(x))$ for any two solutions $y(x)$ and $\bar{y}(x)$ of (1.9.4).

Corollary 3.22.2. Theorem 1.10.1 with the condition (1.10.2) replaced by Lemmert's condition (1.10.7) is a particular case of Theorem 3.22.1.

Proof. Let $B(x) = \exp\left(\int_a^x h(t)dt\right)$, so that

$$\lim_{x\to 0+} \frac{z(x)}{B(x)} = z(x_1) \lim_{x\to 0+} \frac{\exp\left(-\int_x^{x_1} h(t)dt\right)}{\exp\left(\int_a^x h(t)dt\right)} = z(x_1)\exp\left(\int_{x_1}^a h(t)dt\right)$$

and $z(x) = o(B(x))$ if and only if $z(x_1) = 0$. So $z(x) \equiv 0$ because $x_1 \in (0, a)$ is arbitrary.

Further, in view of (1.10.7) for every $\varepsilon > 0$ there exists a $\delta(\varepsilon) > 0$ such that for all (x_1, y), $(x_2, \bar{y}) \in \Delta$, $|x_1 - x_2| + |y - \bar{y}| < \delta(\varepsilon)$ it holds that

$$\left| \frac{f(x_1, y)}{h(x_1) \exp\left(\int_a^{x_1} h(t)dt\right)} - \frac{f(x_2, \bar{y})}{h(x_2) \exp\left(\int_a^{x_2} h(t)dt\right)} \right| < \varepsilon. \tag{3.22.3}$$

Let $y(x)$ and $\bar{y}(x)$ be two solutions of (1.9.4). Since $y(x_0) = \bar{y}(x_0) = 0$ there exists a $\tau \in (0, a)$ such that $|y(x) - \bar{y}(x)| < \delta$ for $0 < x < \tau$. Thus, from (3.22.3) it follows that

$$\lim_{x \to 0^+} \frac{|f(x, y(x)) - f(x, \bar{y}(x))|}{h(x) \exp\left(\int_a^x h(t)dt\right)} = 0. \tag{3.22.4}$$

If $\lim_{x \to 0^+} \int_x^a h(t)dt = C$ is finite then $\lim_{x \to 0^+} B(x) = e^{-C}$ and $|y(x) - \bar{y}(x)| = o(B(x))$ as $x \to 0^+$.

If $\lim_{x \to 0^+} \int_x^a h(t)dt = \infty$ then $\lim_{x \to 0^+} B(x) = 0$. Therefore, in this case an application of l'Hospital's rule and (3.22.4) leads to $|y(x) - \bar{y}(x)| = o(B(x))$ as $x \to 0^+$. ∎

Corollary 3.22.3. Theorem 1.10.1 with the condition (1.10.2) replaced by Došlà - Došlý's condition (1.10.8) is a particular case of Theorem 3.22.1.

Proof. Let $B(x) = \int_0^x H(t)dt$. In view of (1.10.8) there exists a sequence $\{s_n\}$, $s_n \in (0, x_1)$ and $\inf s_n = 0$ such that

$$\lim_{n \to \infty} \int_0^{s_n} H(t)dt \ \exp\left(\int_{s_n}^{x_1} h(t)dt\right) = C > 0.$$

If a solution $z(x)$ of $z' = h(x)z$ satisfies $z(x) = o(B(x))$ as $x \to 0^+$ then

$$\lim_{n \to \infty} \frac{z(s_n)}{B(s_n)} = \lim_{n \to \infty} z(x_1) \frac{\exp\left(-\int_{s_n}^{x_1} h(t)dt\right)}{\int_0^{s_n} H(t)dt} = \frac{z(x_1)}{C} = 0.$$

As $C > 0$ we get $z(x_1) = 0$ and thus $z(x) = 0$, $x \in (0, x_1)$.

Further, since in view of (1.10.8) we have $\lim_{x \to 0^+} \int_0^x |f(t, y(t))|dt / \int_0^x H(t)dt = 0$ for any two solutions $y(x)$ and $\bar{y}(x)$ of (1.9.4) we find that

$$\lim_{x \to 0^+} \frac{|y(x) - \bar{y}(x)|}{B(x)} \le \lim_{x \to 0^+} \frac{|y(x)|}{B(x)} + \lim_{x \to 0^+} \frac{|\bar{y}(x)|}{B(x)} = 0. \qquad ∎$$

Corollary 3.22.4. Theorem 1.10.3 is a particular case of Theorem 3.22.1.

Proof. Let $B(x) = e^{H(x)}$ so that in view of $H'(x) = h(x)$ almost everywhere

$$\lim_{x\to 0+} \frac{z(x)}{B(x)} = z(x_1) \lim_{x\to 0+} \exp\left(\int_{x_1}^{x} h(t)dt - H(x)\right) = z(x_1)e^C.$$

Thus, the condition (3.22.1) holds if and only if $z(x_1) = 0$, i. e., $z(x) \equiv 0$. Rest of the proof is similar to that of Corollary 3.22.2. ∎

Next, we shall deduce Kamke's Uniqueness Theorem 1.15.1 for n - solutions of (1.1.1) from Theorem 3.22.1.

Corollary 3.22.5. Let the functions $f(x,y)$ and $g(x,z)$ be as in Theorem 1.15.1 except that for all (x,y), $(x,\bar{y}) \in \bar{S}_+$, $x \neq x_0$, $y \neq \bar{y}$ the inequality (1.11.2) is replaced by

$$(3.22.5) \qquad \frac{1}{\|y-\bar{y}\|}(f(x,y) - f(x,\bar{y}))\cdot(y-\bar{y}) \le g(x, \|y-\bar{y}\|).$$

Then, the initial value problem (1.1.1) has at most one n - solution in $[x_0, x_0 + a]$.

Proof. Without loss of generality we may assume that $g(x,z)$ is continuous for $x_0 < x \le x_0 + a$, $z \ge 0$. Setting $V(x,y,\bar{y}) = \|y-\bar{y}\|$ and $B(x) = x - x_0$ in Theorem 3.22.1 we find that for all (x,y), $(x,\bar{y}) \in \bar{S}_+$, $x \neq x_0$, $y \neq \bar{y}$

$$D_+V_f(x,y,\bar{y}) = \frac{1}{\|y-\bar{y}\|}(f(x,y) - f(x,\bar{y}))\cdot(y-\bar{y}) \le g(x, \|y-\bar{y}\|).$$

Further, since for any two n - solutions $y(x)$ and $\bar{y}(x)$ of (1.1.1) $\lim_{x\to x_0^+} \|y(x) - \bar{y}(x)\|/(x-x_0) = 0$ the conclusion follows immediately. ∎

Corollary 3.22.6. In Theorem 3.22.1 the condition (i) can be replaced by

(i)′ the function $B(x) > 0$ is continuous in $(x_0, x_0 + a)$ and the function $g(x,y,z)$ is continuous for $x_0 < x \le x_0 + a$, $\|y - y_0\| \le b$, $z \ge 0$, and that for every n - solution $y(x)$ of (1.1.1) and $x_1 \in (x_0, x_0 + a)$, $z(x) \equiv 0$ is the only differentiable function satisfying

$$(3.22.6) \qquad z'(x) = g(x, y(x), z(x)), \; x_0 < x < x_1$$

and (3.22.1). ∎

In view of the above corollary it is clear that Borùvka's Uniqueness Theorem 2.5.1 is a particular case of Theorem 3.22.1.

Corollary 3.22.7 (Perov's Uniqueness Theorem). Assume that $f(x,y)$ is continuous in $\bar{S}_+$ with $x_0 = 0$. Further, let for all $(x,y),\ (x,\bar{y}) \in \bar{S}_+,\ x \neq x_0 = 0,\ y \neq \bar{y},\ \|y-\bar{y}\| \leq \gamma x^k\ (\gamma > 0)$

$$\text{(3.22.7)} \qquad \frac{1}{\|y-\bar{y}\|}(f(x,y) - f(x,\bar{y}))\cdot(y-\bar{y}) \leq k\frac{\|y-\bar{y}\|}{x} + x^k h(x) g\left(\frac{\|y-\bar{y}\|}{x^k}\right),$$

where $0 \leq k \leq 1$, $h(x)$ is such that $\liminf_{x\to 0^+} \int_x^a h(t)dt < \infty$, and $g(z)$ is continuous for $0 \leq z \leq \gamma,\ g(z) > 0,\ z > 0$ and $\int_0^\gamma \frac{dt}{g(t)} = \infty$. Then, the initial value problem (1.1.1) has at most one n - solution in $[0,a]$.

Proof. In Theorem 3.22.1 setting $B(x) = 1,\ g(x,z) = h(x)g(z)$ and $V(x,y,\bar{y}) = \|y-\bar{y}\|/x^k$ the conclusion follows immediately. ∎

3.23 ANTOSIEWICZ'S UNIQUENESS THEOREM

Using Lyapunov functions Antosiewicz in the year 1962 proved the following general uniqueness result.

Theorem 3.23.1 (Antosiewicz's Uniqueness Theorem). Assume that $f(x,y)$ is continuous in $\bar{S}_+$ and

(i) there exist functions $g_i(x,z)$ defined and continuous for $x_0 < x \leq x_0 + a,\ |z| < \infty$, and that each equation $z_i' = (-1)^i g_i(x,(-1)^i z_i)$ has $r_i(x) \equiv 0$ as the maximal solution in $[x_0, x_0+a]$, where $i = 1,2$; and

(ii) there exists a Lyapunov function $V(x, y-\bar{y})$ defined and continuous in S_ε and satisfies the locally Lipschitz condition (3.15.1) with respect to the second argument, $V(x,0) \equiv 0$, $V(x, y-\bar{y}) > 0$ for $\|y-\bar{y}\| > 0$ and that for all (x,y), $(x,\bar{y})$, $x_0 < x < x_0+a$, $\|y-y_0\| \le b$, $\|\bar{y}-y_0\| \le b$,

$$\begin{aligned}(-1)^i D_+ V_{if}(x, y-\bar{y}) \\ = \liminf_{h\to 0+} \frac{(-1)^i}{h}[V(x+h, y-\bar{y}+h(f(x,y)-f(x,\bar{y}))) - V(x,y-\bar{y})] \\ \le (-1)^i g_i(x, V(x,y-\bar{y})),\ i=1,2.\end{aligned}$$

Then, the initial value problem (1.1.1) has at most one n - solution in $[x_0, x_0+a]$.

A generalization of the above result is proved in the following :

Theorem 3.23.2. Assume that $f(x,y)$ is continuous for $x_0 < x \le x_0+a$, $\|y-y_0\| \le b$, and

(i) functions $B_i(x) > 0$ are continuous in (x_0, x_0+a) and the functions $g_1(x,z)$, $g_2(x,z)$ are continuous for $x_0 < x \le x_0+a$, $z \ge 0$ and $x_0 < x \le x_0+a$, $z \le 0$ respectively, and that for every $x_1 \in (x_0, x_0+a)$, $z_i(x) \equiv 0$ are the only differentiable functions satisfying

$$(-1)^{i-1} z_i(x) \ge 0$$

$$z_i'(x) = g_i(x, z_i(x)),\ x_0 < x < x_1$$

$$z_i(x) = o(B_i(x)) \quad \text{as} \quad x \to x_0^+$$

where $i = 1,2$,

(ii) there exists a Lyapunov function $V(x,y,\bar{y})$ defined and continuous in S_ϵ and satisfies the locally Lipschitz condition (3.15.1) with respect to $(y,\bar{y})$; for any two n - solutions $y(x)$ and $\bar{y}(x)$ of (1.1.1), $V(x,y(x),\bar{y}(x)) \equiv 0$ if and only if $y(x) = \bar{y}(x)$, $x \in [x_0, x_0+a]$, and that for all (x,y), $(x,\bar{y})$, $x_0 < x < x_0+a$, $\|y-y_0\| \le b$, $\|\bar{y}-y_0\| \le b$, $y \ne \bar{y}$, $D_+V_f(x,y,\bar{y}) \le g_1(x, V(x,y,\bar{y}))$ and $D^+V_f(x,y,\bar{y}) \ge g_2(x, V(x,y,\bar{y}))$; and

(iii) for any two n - solutions $y(x)$ and $\bar{y}(x)$ of (1.1.1) the relation (3.22.2) with $B(x) = \min\{B_1(x), B_2(x)\}$ holds.

Then, the initial value problem (1.1.1) has at most one n - solution in $[x_0, x_0+a]$.

Proof. Let $y(x)$ and $\bar{y}(x)$ be two n - solutions of (1.1.1) in $[x_0, x_0+a]$. Define $\phi(x) = V(x, y(x), \bar{y}(x))$ for $x_0 \le x \le x_1$. Then, there exists some $\bar{x} \in (x_0, x_1)$ such that $\phi(\bar{x}) \ne 0$. From the hypothesis (ii) it follows that

$$D_+\phi(x) \le g_1(x, \phi(x))$$

and

$$D^+\phi(x) \ge g_2(x, \phi(x))$$

for every $x \in (x_0, x_1)$ such that $\phi(x) \ne 0$. Define $\phi(x_0) = 0$.

If $\phi(\bar{x}) < 0$ then for the left maximal solution $r_2(x)$ of the problem $z' = g_2(x, z)$, $z(\bar{x}) = \phi(\bar{x})$ it follows that $r_2(x) \ge \phi(x)$, $x \in (x_2, \bar{x})$, where $x_2 = \sup\{x \in [x_0, \bar{x}) : \phi(x) = 0\}$. If $x_2 > x_0$ then, we can take $r_2(x) = 0$ in (x_0, x_2). Thus, we obtain $|r_2(x)| \le |\phi(x)|$ in $(x_0, \bar{x})$. However, since $\phi(x) = o(B(x))$ as $x \to x_0^+$, we have $\phi(x) = o(B_2(x))$ as $x \to x_0^+$. But, then in view of (i), $r_2(x) \equiv 0$, which contradicts $r_2(\bar{x}) < 0$. In a similar manner the case $\phi(\bar{x}) > 0$ contradicts $\rho_1(\bar{x}) > 0$, where $\rho_1(x)$ is the left minimal solution of $z' = g_1(x, z)$, $z(\bar{x}) = \phi(\bar{x})$. ∎

3.24 LAKSHMIKANTHAM - SAMIMI'S UNIQUENESS THEOREM

We begin with the following result which generalizes Theorem 1.17.1.

Lemma 3.24.1. Assume that

(i) the function $B(x)$ is continuous in $[0, a]$, $B(0) = 0$ and $B(x) > 0$, $0 < x \le a$,

(ii) hypotheses (i) and (ii) of Theorem 1.17.1 hold; and

(iii) for each continuous function $z(x) \ge 0$ in $(0, a]$ satisfying $D_-z(x) \le h(x, z(x))$, $0 < x \le a$ and $\lim_{x\to 0^+} \frac{z(x)}{B(x)} = 0$, there exists a function

$\mu(x, z(\cdot)) \geq 0$ which is continuous in $[0, a]$ and a number δ, $0 < \delta \leq a$ such that $\mu(0, z(\cdot)) \equiv 0$, and

$$B(x)D_{-}z(x) < D_{-}B(x)z(x) + h(x, \rho(x, x_1, \mu(x, z(x))))B^2(x), \quad 0 < x \leq \delta. \tag{3.24.1}$$

Then, each $z(x) = 0$, $x \in (0, a]$.

Proof. Suppose on the contrary that there exists a function $z_1(x)$ satisfying the hypotheses of lemma and $z_1(x) \not\equiv 0$. Then, there exists a $\sigma \in (0, a]$ such that $z_1(\sigma) > 0$. Let $\rho(x, \sigma, z_1(\sigma))$ be the left minimal solution of $z' = h(x, z)$, $z(\sigma) = z_1(\sigma)$. Then, as earlier we have

$$0 \leq \rho(x, \sigma, z_1(\sigma)) \leq z_1(x), \quad 0 < x \leq \sigma. \tag{3.24.2}$$

Now in view of the conditions on $\mu(x, z_1(\cdot))$ for each $\eta \in (0, z_1(\sigma))$ there exists a $x_2 \in (0, \sigma)$ such that $0 < \mu(x, z_1(x)) \leq \eta$, $0 < x \leq x_2$. However, since $\rho(x, \sigma, c)$ is nondecreasing in c, we find that

$$\rho(x, \sigma, \mu(x, z_1(x))) \leq \rho(x, \sigma, \eta) \leq \rho(x, \sigma, z_1(\sigma)), \quad 0 < x \leq x_2. \tag{3.24.3}$$

Now we shall show that there exists a $x_1 \in (0, x_2]$ such that

$$\frac{z_1(x_1)}{B(x_1)} - \rho(x_1, \sigma, \eta) = \rho_0 > 0. \tag{3.24.4}$$

For this, since $B(0) = 0$ and $B(x)$ is continuous in $[0, a]$ we can assume that $0 \leq B(x) \leq 1$, $0 \leq x \leq x_1$. Thus, from (3.24.2) and (3.24.3) it follows that $z_1(x)/B(x) \geq \rho(x, \sigma, \eta)$, $0 < x \leq x_2$. If (3.24.4) is not true for any $x_1 \in (0, x_2]$, then we have $z_1(x)/B(x) = \rho(x, \sigma, \eta)$, $0 < x \leq x_2$. But, then

$$\frac{D_{-}z_1(x)B(x) - z_1(x)D_{-}B(x)}{B^2(x)} = \rho'(x, \sigma, \eta), \quad 0 < x \leq x_2$$

which implies that for $0 < x \leq x_2$

$$\begin{aligned} B(x)D_{-}z_1(x) &= D_{-}B(x)z_1(x) + h(x, \rho(x, \sigma, \eta))B^2(x) \\ &\geq D_{-}B(x)z(x) + h(x, \rho(x, \sigma, \mu(x, z_1(x))))B^2(x). \end{aligned}$$

This contradicts (3.24.1) and therefore (3.24.4) holds.

Now setting $w(x) = z_1(x)/B(x)$ in (3.24.1) and using (3.24.3), we obtain

$$D_{-}w(x) < h(x, \rho(x, \sigma, \eta)),\ 0 < x \leq x_1 < x_2. \tag{3.24.5}$$

Let $q(x) = \rho(x, \sigma, \eta) + \rho_0$. Since $w(x_1) = q(x_1)$, it follows that $w(x) \geq q(x)$ for $0 < x \leq x_1$. Thus, in view of $\rho(x, \sigma, \eta) \geq 0$, we find that $\frac{z_1(x)}{B(x)} \geq \rho_0 > 0,\ 0 < x \leq x_1$. Hence, $\lim_{x\to 0+} z_1(x)/B(x) = \rho_0 > 0$. But, this contradicts our assumption that $\lim_{x\to 0+} z_1(x)/B(x) = 0$. ∎

As we have remarked in Section 1.17, here also, if the left solutions of the equation $z' = h(x, z)$ are unique for $0 < x \leq a,\ z \geq 0$ then in (3.24.1) the strict inequality can be replaced by the weaker inequality. For, in this case we directly have $0 < \rho(x_1, \sigma, \eta) < \rho(x_1, \sigma, z_1(\sigma)) \leq z_1(x_1)$, and hence

$$\frac{z_1(x_1)}{B(x_1)} - \rho(x_1, \sigma, \eta) = \rho_0 > 0 \quad \text{for each} \quad x_1 \in (0, x_2].$$

Also, (3.24.5) becomes $w'(x) \leq h(x, \rho(x, \sigma, \eta)),\ 0 < x \leq x_1$, and hence

$$\int_x^{x_1} w'(t)dt \leq \int_x^{x_1} \rho'(t, \sigma, \eta)dt,\ 0 < x \leq x_1.$$

Therefore, we find that

$$w(x) \geq \rho(x, \sigma, \eta) + w(x_1) - \rho(x_1, \sigma, \eta) = \rho(x, \sigma, \eta) + \rho_0 \geq \rho_0 > 0.$$

Lemma 3.24.2. Assume that

(i) hypothesis (i) of Theorem 1.18.1 holds; and

(ii) the function $g(x, z)$ is defined and continuous for $0 < x \leq a,\ z \geq 0,\ g(x, 0) \equiv 0$, and all solutions $z(x)$ of $z' = g(x, z),\ z(0) = 0$ are defined and satisfy the relation $z(x) \leq A(x),\ 0 < x \leq a$.

Then, for any continuous function $z_1(x) \geq 0,\ 0 \leq x \leq a,\ z_1(0) = 0$ satisfying $D_{-}z_1(x) \leq g(x, z_1(x)),\ 0 < x \leq a$ it is necessary that $\lim_{x\to 0+} z_1(x)/B(x) = 0$.

Proof. Suppose that $\lim_{x\to 0+} z_1(x)/B(x) \neq 0$. Then, in view of (1.18.1) there exists a $x_2 \in (0, a]$ such that $z_1(x_2) > A(x_2)$. Let $\rho(x)$ be the left minimal solution of $z' = g(x,z)$, $z(x_2) = z_1(x_2)$. Then, as earlier we find that $0 \le \rho(x) \le z_1(x)$, $0 < x \le x_2$ and $\rho(0) = 0$, since $z_1(0) = 0$. Consequently, by (ii) we have $0 \le \rho(x) \le A(x)$, $0 < x \le a$. But, this leads to a contradiction that $z_1(x_2) = \rho(x_2) \le A(x_2) < z_1(x_2)$. Hence, we must have $0 \le z_1(x) \le A(x)$, $0 < x \le a$. The conclusion of the lemma is now immediate. ∎

Theorem 3.24.3 (Lakshmikantham - Samimi's Uniqueness Theorem). Assume that $f(x,y)$ is continuous for $0 < x \le a$, $\|y - y_0\| \le b$, and

(i) hypotheses of Lemma 3.24.1 hold,

(ii) hypothesis (ii) of Theorem 3.22.1 holds with $x_0 = 0$ and the inequality $D_+V_f(x,y,\bar{y}) \le g(x, V(x,y,\bar{y}))$ replaced by $D_-V_f(x,y,\bar{y}) \le h(x, V(x,y,\bar{y}))$; and

(iii) hypothesis (iii) of Theorem 3.22.1 holds with $x_0 = 0$.

Then, the initial value problem (1.1.1) has at most one n - solution in $[0,a]$.

Proof. The proof uses Lemma 3.24.1 and is similar to that of earlier results. ∎

Corollary 3.24.4. Assume that $f(x,y)$ is continuous for $0 < x \le a$, $\|y - y_0\| \le b$, and

(i) hypotheses of Lemmas 3.24.1 and 3.24.2 hold,

(ii) hypothesis (ii) of Theorem 3.22.1 holds with $x_0 = 0$ and S_ϵ replaced by $\bar{S}_\epsilon = (0,y,\bar{y}) \bigcup S_\epsilon$; and

(iii) hypothesis (ii) of Theorem 3.24.3 holds with S_ϵ replaced by $\bar{S}_\epsilon$.

Then, the initial value problem (1.1.1) has at most one n - solution in $[0,a]$.

Proof. As usual assume that $y(x)$ and $\bar{y}(x)$ are two n - solutions of (1.1.1) in $[0,a]$. Then, for the function $\phi(x) = V(x, y(x), \bar{y}(x))$ the inequalities $D_-\phi(x) \le g(x,\phi(x))$, $0 < x \le a$ and $D_-\phi(x) \le h(x,\phi(x))$, $0 < x \le a$ hold. Rest of the proof is similar to that of earlier results, except now we use Lemmas 3.24.1 and 3.24.2. ∎

3.25 KALAS' NONUNIQUENESS THEOREM

Here by employing Lyapunov functions we shall prove a result which provides sufficient conditions for the nonuniqueness of the n - solutions of (1.1.1).

Theorem 3.25.1 (Kalas' Nonuniqueness Theorem). Let $f(x,y)$ be continuous in $\bar{S}_+$, $\|f(x,y)\| \le M$, $\xi \in (0,1)$, $\alpha = \min(a, \xi b/M)$, $x_1 \in (x_0, x_0+\alpha)$ and assume that

(i) the functions $g_1(x,z)$, $g_2(x,z)$ are continuous for $x_0 < x \le x_1$, $|z| < \infty$ nondecreasing in z and such that in $(x_0, x_1]$ there are solutions $z(x)$ and $w(x)$ of (1.18.2) and (1.18.3), respectively, which satisfy the conditions that $w(x_1) < z(x_1)$, $\lim_{x \to x_0^+} z(x) = \lim_{x \to x_0^+} w(x) = 0$; and

(ii) there exists a continuous positive function $\varepsilon(x)$ in (x_0, x_1) and a Lyapunov function $V(x,y)$ defined and continuous in $x_0 \le x \le x_1$, $\|y - y_0\| \le b$ such that $V(x_0, y) = 0$ if and only if $y = y_0$, $w(x_1) < V(x_1, u_1) < z(x_1)$ for some u_1, $\|u_1 - y_0\| < (1-\xi)b$, $V(x,y)$ satisfies the locally Lipschitz condition (3.15.1) with respect to y in $\Omega_z \cup \Omega_w$, where

$$\Omega_z = \{(x,y) : z(x) < V(x,y) < z(x) + \varepsilon(x),\ x_0 < x < x_1,\ \|y - y_0\| < b\}$$

and

$$\Omega_w = \{(x,y) : w(x) - \varepsilon(x) < V(x,y) < w(x),\ x_0 < x < x_1,\ \|y - y_0\| < b\};$$

and

$$D^+V_f(x,y) \ge g_1(x, V(x,y)) \ \text{ in } \ \Omega_z, \quad \text{if } \ \Omega_z \ne \emptyset, \tag{3.25.1}$$

$$D^+V_f(x,y) \le g_2(x, V(x,y)) \ \text{ in } \ \Omega_w, \quad \text{if } \ \Omega_w \ne \emptyset. \tag{3.25.2}$$

Then, the initial value problem (1.1.1) has at least two n - solutions in $[x_0, x_1]$.

Proof. Since $V(x,y)$ is continuous and $w(x_1) < V(x_1, u_1) < z(x_1)$ for some u_1, $\|u_1 - y_0\| < (1-\xi)b$, there exists a δ, $0 < \delta < (1-\xi)b - \|u_1 - y_0\|$

satisfying the condition $w(x_1) < V(x_1, y) < z(x_1)$ for $\|y - y_0\| < \delta$. Choose $y_1,\ y_2 \in \{y : \|y - y_0\| < \delta\},\ y_1 \neq y_2$ and consider n - solutions $y_i(x)$ of

$$(3.25.3)_i \qquad y' = f(x, y),\ y_i(x_1) = y_i;\ i = 1, 2.$$

Since $\|y_i - y_0\| < (1 - \xi)b$ for $i = 1, 2$ the solutions $y_1(x),\ y_2(x)$ are defined in $[x_0, x_1]$. Thus, to complete the proof it suffices to check that $y_1(x_0) = y_2(x_0) = y_0$. For this, we set $y(x) = y_i(x),\ \phi(x) = V(x, y_i(x))$ for any $i \in \{1, 2\}$. Clearly,

$$(3.25.4) \qquad w(x_1) < \phi(x_1) < z(x_1).$$

We shall show that $\{x \in (x_0, x_1) : (x, y(x)) \in \Omega_z\} = \emptyset$. If this is not true, then there exists some $t \in (x_0, x_1)$ such that $(t, y(t)) \in \Omega_z$. Since $V(x, y)$ satisfies the locally Lipschitz condition (3.15.1) with respect to y in $\Omega_z \bigcup \Omega_w$, as in Section 3.15 we find that

$$(3.25.5) \qquad D^+\phi(t) = D^+V_f(t, y(t)).$$

Further, there exists an interval (x_2, x_3) where $x_0 < x_2 < x_3 < x_1$ such that

$$(3.25.6) \qquad \phi(x_3) = z(x_3)$$

and

$$(3.25.7) \qquad z(t) < \phi(t) < z(t) + \varepsilon(t),\ \ t \in (x_2, x_3).$$

Thus, $(t, y(t)) \in \Omega_z,\ t \in (x_2, x_3)$. Using (3.25.1) and (3.25.5) we deduce that

$$\begin{aligned} D^+[\phi(t) - z(t)] &= D^+\phi(t) - z'(t) \\ &\geq g_1(t, \phi(t)) - z'(t),\quad t \in (x_2, x_3). \end{aligned}$$

Hence, the nondecreasing nature of $g_1(x, z)$ implies that

$$D^+[\phi(t) - z(t)] \geq g_1(t, z(t)) - z'(t) = 0,\quad t \in (x_2, x_3).$$

Consequently, the function $\phi(t) - z(t)$ is nondecreasing in (x_2, x_3) and this leads to a contradiction with (3.25.6) and (3.25.7). Therefore, $\{x \in (x_0, x_1) : (x, y(x)) \in \Omega_z\} = \emptyset$.

In a similar manner it follows that $\{x \in (x_0, x_1) : (x, y(x)) \in \Omega_w\} = \emptyset$. Thus, from (3.25.4) we find that $w(x) \le \phi(x) \le z(x)$, $x \in (x_0, x_1]$. Further, since $V(x_0, y(x_0)) = \phi(x_0) = \lim_{x \to x_0^+} z(x) = \lim_{x \to x_0^+} w(x) = 0$, in view of $V(x_0, y) = 0$ if and only if $y = y_0$, we deduce that $y(x_0) = y_0$. Therefore, $y_1(x_0) = y_2(x_0) = y_0$. ∎

Corollary 3.25.2. Let $f(x,y)$, ξ, α and x_1 be as in Theorem 3.25.1 and assume that

(i) the function $h(x,z)$ is continuous for $x_0 < x \le x_1, 0 \le z < \infty$ nondecreasing in z and such that in $(x_0, x_1]$ there is a solution $z(x)$ of $z' = h(x,z)$ which satisfies the conditions that $0 < z(x_1)$, $\lim_{x \to x_0^+} z(x) = 0$; and

(ii) there exists a continuous positive function $\varepsilon(x)$ in (x_0, x_1) and a Lyapunov function $V(x,y) \ge 0$ defined and continuous in $x_0 \le x \le x_1$, $\|y - y_0\| \le b$ such that $V(x_0, y) = 0$ if and only if $y = y_0$, $V(x_1, u_1) < z(x_1)$ for some u_1, $\|u_1 - y_0\| < (1-\xi)b$, $V(x,y)$ satisfies the locally Lipschitz condition (3.15.1) with respect to y in Ω_z, where Ω_z is the same as in Theorem 3.25.1, and

$$D^+ V_f(x,y) \ge h(x, V(x,y)) \ \text{ in } \ \Omega_z.$$

Then, the initial value problem (1.1.1) has at least two n - solutions in $[x_0, x_1]$.

Proof. Define

$$g_1(x,z) = \begin{cases} h(x,z), & x_0 < x \le x_1, \ 0 \le z < \infty \\ h(x,0), & x_0 < x \le x_1, \ -\infty < z \le 0 \end{cases}$$

and $g_2(x,z) = -1$, $x_0 < x \le x_1$, $|z| < \infty$ so that the assumptions of Theorem 3.25.1 are satisfied with $w(x) = x_0 - x$, $\Omega_w = \emptyset$. ∎

Corollary 3.25.3. Let $f(x,y)$ be continuous in $\bar{S}_+$ and assume that

(i) the function $h_1(x,z)$ is continuous for $x_0 < x \le x_0 + a$, $0 \le z < \infty$ nondecreasing in z and such that in $(x_0, x_0 + a]$ there is a solution $z(x)$ of $z' = h_1(x,z)$ which satisfies the conditions that $0 < z(x)$, $x \in (x_0 . x_0 + a)$, $\lim_{x \to x_0^+} z(x) = 0$; and

(ii) there exists a continuous positive function $\varepsilon(x)$ in (x_0, x_0+a), a differentiable function $\phi(x)$ in $[x_0, x_0+a)$ such that $\phi(x_0) = y_0$ and the inequality

$$(3.25.8) \qquad (f(x,y) - \phi'(x)) \cdot (y - \phi(x)) \geq \|y - \phi(x)\| \; h_1(x, \|y - \phi(x)\|)$$

holds in $\bar{\Omega} = \{(x,y) : z(x) < \|y - \phi(x)\| < z(x) + \varepsilon(x), \; x_0 < x < x_0 + a, \; \|y - y_0\| < b\}$.

Then, there exists a $x^* \in (x_0, x_0+a]$ such that for any $x_2 \in (x_0, x^*]$ the initial value problem (1.1.1) has at least two n - solutions in $[x_0, x_2]$.

Proof. Let $\xi = 1/2$ and $\alpha = \min(a, b/2M)$. Then, there exists a $x^* \in (x_0, x_0 + \alpha)$ such that $\|\phi(x) - y_0\| < b/2, \; x \in (x_0, x^*]$. Let $x_2 \in (x_0, x^*]$ and define

$$V(x,y) = \|y - \phi(x)\|, \; x_0 \leq x \leq x_2, \; \|y - y_0\| \leq b.$$

Since for $x_0 \leq x < x_2, \; \|y - y_0\| < b, \; y \neq \phi(x)$

$$D^+V_f(x,y) = \frac{1}{\|y - \phi(x)\|}(f(x,y) - \phi'(x)) \cdot (y - \phi(x))$$

in view of (3.25.8) it follows that

$$D^+V_f(x,y) \geq h_1(x, \|y - \phi(x)\|), \; (x,y) \in \bar{\Omega}.$$

Thus, Corollary 3.25.2 is applicable and the initial value problem (1.1.1) has at least two n - solutions in $[x_0, x_2]$. ∎

Example 3.25.1. For the initial value problem $y' = y^{1/3}, \; y(0) = 0$ the conditions of Corollary 3.25.3 are satisfied with $h_1(x,z) = \frac{2}{3}\sqrt{x}, \; z(x) = \frac{4}{9}x\sqrt{x}, \; \varepsilon(x) = 1$ and $\phi(x) = 0$. ∎

Example 3.25.2. Consider the initial value problem

$$(3.25.9) \quad y' = f(x,y) = \begin{cases} \dfrac{\gamma x^{m-1} y}{\alpha x^m + \beta y^2}, & 0 \leq x \leq a, \; |y| < \infty, \; x^2 + y^2 \neq 0 \\ 0, & x = y = 0, \end{cases}$$

where $\alpha > 0,\ \beta > 0,\ m > 2$ and $\gamma > \alpha m/2$. Since

$$|f(x,y)| \le \frac{\gamma x^{m/2-1}}{2\sqrt{\alpha\beta}},\ 0 \le x \le a,\ |y| < \infty$$

this function is continuous. We shall show that the conditions of Corollary 3.25.3 for the problem (3.25.9) are satisfied. For this, we set $c = (2\gamma - m\alpha)^{1/4}(4\beta m)^{-1/4}$, $q_1(x,z) = \frac{1}{2}cmx^{m/4-1}\sqrt{z},\ 0 < x \le a,\ 0 \le z < \infty$, $z(x) = c^2x^{m/2},\ 0 < x \le a,\ \varepsilon(x) = z(x),\ 0 < x < a,\ \phi(x) = 0,\ 0 \le x < a$. Then, for $(x,y) \in \bar{\Omega} = \{(x,y) : z(x) < |y| < z(x) + \varepsilon(x),\ 0 < x < a,\ |y| < b\}$ we have

$$yf(x,y) = \frac{\gamma x^{m-1}y^2}{\alpha x^m + \beta y^2} \ge \frac{\gamma c x^{m+m/4-1}|y|\sqrt{|y|}}{\alpha x^m + 4\beta c^4 x^m} = \frac{\gamma c x^{m/4-1}|y|\sqrt{|y|}}{\alpha + 4\beta c^4} = |y|q_1(x,|y|).$$

Thus, the conditions of Corollary 3.25.3 with $x_0 = 0,\ y_0 = 0$ are satisfied, and hence for the initial value problem (3.25.9) there exists a $x^* \in (0,a]$ such that for any $x_2 \in (0,x^*]$ it has at least two solutions in $[0,x_2]$. ∎

3.26 SAMIMI'S NONUNIQUENESS THEOREM

In the following result we shall show that if certain conditions of Theorem 3.20.1 are violated then the initial value problem (1.1.1) has more than one n - solutions.

Theorem 3.26.1 (Samimi's Nonuniqueness Theorem). Let $f(x,y)$ be continuous for $0 < x \le a,\ \|y - y_0\| \le b,\ \|f(x,y)\| \le M$, and $y_0(x)$ is a n - solution of (1.1.1) in $[0,\alpha]$, where $\alpha = \min(a, b/2M),\ x_1 \in (0,\alpha]$ and assume that

(i) the function $g(x,z)$ is continuous for $0 < x \le a,\ 0 \le z \le 2b$ nondecreasing in z and such that in $(0,x_1]$ there is a solution of (1.13.1) which satisfies the conditions that $z(0) = 0,\ z(x_1) > 0,\ \lim_{x\to 0^+} \frac{z(x)}{B(x)} = 0$, where $B(x)$ is continuous in $[0,x_1]$ and $B(x) > 0$ for $x > 0$; and

(ii) there exists a Lyapunov function $V(x,y) \ge 0$ defined and continuous for $0 \le x \le x_1,\ \|y\| < \infty$ such that $V(0,y) = 0$ if and only if $y = y_0$,

$V(x_1, u_1 - y_0(x_1)) < z(x_1)$ for some u_1, $\|u_1 - y_0\| < b/4$, $V(x,y)$ satisfies the locally Lipschitz condition (3.15.1) with respect to y, and

$$\begin{aligned} D^+V(x, y - y_0(x)) &= \limsup_{h\to 0+} \frac{1}{h}[V(x+h, y - y_0(x) + h(f(x,y) \\ &\quad -f(x, y_0(x)))) - V(x, y - y_0(x))] \\ &\geq g(x, V(x, y - y_0(x))) \quad \text{in} \quad \Omega \end{aligned}$$

where $\Omega = \{(x,y) : z(x) < V(x, y - y_0(x)),\ 0 < x < x_1,\ \|y - y_0\| \leq b\}$.

Then, the initial value problem (1.1.1) has a n - solution $y_1(x) \neq y_0(x)$ in $[x_0, x_1]$ such that $\lim_{x\to 0+} \frac{V(x, y_1(x) - y_0(x))}{B(x)} = 0$.

Proof. Let $y_1(x)$ be a n - solution of the initial value problem $y' = f(x,y)$, $y(x_1) = u_1$ to the left of x_1. It is clear that $y_1(x)$ exists in $(0, x_1]$. Indeed, if for some $x_2 \in (0, x_1)$, $\|y_1(x_2) - y_0\| = b$, then

$$\begin{aligned} b &= \|y_1(x_2) - y_0\| \leq \|u_1 - y_0\| + \int_{x_2}^{x_1} \|f(t, y_1(t))\| dt \\ &< \frac{b}{4} + (x_1 - x_2)M \leq \frac{b}{4} + \frac{b}{2} < b, \end{aligned}$$

which is a contradiction.

We define $\phi(x) = V(x, y_1(x) - y_0(x))$. Then, $\phi(x_1) = V(x_1, u_1 - y_0(x_1)) < z(x_1)$. We shall show that $\phi(x) \leq z(x)$, $0 < x \leq x_1$. If this is not true, then there exists an interval (x_2, x_3) where $x_0 < x_2 < x_3 < x_1$ such that

$$\phi(x) > z(x),\ x_2 < x < x_3 \ \text{ and } \ \phi(x_3) = z(x_3). \tag{3.26.1}$$

Since $V(x,y)$ is locally Lipschitz, it follows that

$$\begin{aligned} D^+[\phi(x) - z(x)] &= D^+\phi(x) - z'(x) \\ &\geq g(x, \phi(x)) - z'(x),\ x \in (x_2, x_3). \end{aligned}$$

Hence, the nondecreasing nature of $g(x,z)$ implies that

$$D^+[\phi(x) - z(x)] \geq g(x, z(x)) - z'(x) = 0.$$

Consequently, the function $\phi(x) - z(x)$ is nondecreasing in (x_2, x_3) and this leads to a contradiction to (3.26.1). Thus, it follows that $0 \leq \phi(x) \leq z(x)$, $0 <$

$x \le x_1$. This means that $V(0, y_1(0) - y_0(0)) = 0$ which implies that $y_1(0) = y_0$. Further, since

$$0 \le \lim_{x\to 0+} \frac{\phi(x)}{B(x)} \le \lim_{x\to 0+} \frac{z(x)}{B(x)} = 0$$

we find that $\lim_{x\to 0+} \frac{V(x, y_1(x) - y_0(x))}{B(x)} = 0$. ∎

Example 3.26.1. For the initial value problem $y' = \sqrt{y}$, $y(0) = 0$ the conditions of Theorem 3.26.1 are satisfied with $y_0(x) = 0$, $g(x,z) = \frac{1}{4}x^3$, $z(x) = \frac{1}{16}x^4$, $B(x) = x^3$ and $V(x,y) = y^2$. ∎

Example 3.26.2. Let $f(x,y)$ be as in Theorem 3.26.1 and let $y^{(i)}(0)$ exist and $y^{(i)}(0) = y_{0i}$, $i = 0, 1, \cdots, k$ $(k > 1)$. For this singular initial value problem we let $V(x,y) = \|y\|^2$, $g(x,z) = \frac{2(k+1)z}{x}$, $B(x) = x^{2k}$ and $z(x) = x^{2(k+1)}$. Thus, from Theorem 3.26.1 it follows that $y_1(0) = y_0$ and $\|y_1(x) - y_0(x)\|^2 = O(x^{2k})$ as $x \to 0$. This implies that if $y_0(x)$ is a solution of this problem then $y_1(x)$ is also a solution, which is different from $y_0(x)$. ∎

CHAPTER 4

HIGHER ORDER DIFFERENTIAL EQUATIONS

4.1 INTRODUCTION

Consider the initial value problem

$$y^{(n)} = f(x, y, y', \cdots, y^{(n-1)})$$

(4.1.1)

$$y^{(i-1)}(x_0) = y_{0i}, \ 1 \le i \le n.$$

If $f(x, z_1, z_2, \cdots, z_n)$ is continuous in a domain $D \subseteq R^{n+1}$, and $(x_0, y_{01}, y_{02}, \cdots, y_{0n}) \in D$ then by a **solution** of (4.1.1) in an interval J containing x_0, we mean a n times continuously differentiable function $y(x)$ in J satisfying :

(i) $y^{(i-1)}(x_0) = y_{0i}, \ 1 \le i \le n,$

(ii) for all $x \in J$, the points $(x, y(x), y'(x), \cdots, y^{(n-1)}(x)) \in D$; and

(iii) $y^{(n)}(x) = f(x, y(x), y'(x), \cdots, y^{(n-1)}(x)).$

If the function $f(x, z_1, z_2, \cdots, z_n)$ is only defined in the domain $D = J \times T \subseteq R^{n+1}$, where J is one of the intervals $[x_0, x_0+a], [x_0, x_0+a), (x_0, x_0+a], (x_0, x_0+$

a), and $(y_{01}, \cdots, y_{0n}) \in T \subseteq R^n$, then by a **classical solution** of (4.1.1) in the interval $\{x_0\} \bigcup J$, we mean a $(n-1)$ times continuously differentiable function $y(x)$ in $\{x_0\} \bigcup J$ satisfying the above conditions (i) and (ii), and

(iv) $y^{(n)}(x)$ exists in J^o; and

(v) for all $x \in J^o$, $y^{(n)}(x) = f(x, y(x), y'(x), \cdots, y^{(n-1)}(x))$.

We shall also consider the case when the function $f(x, z_1, z_2, \cdots, z_n)$ satisfies the Caratheodory conditions, i.e., $f(x, z_1, \cdots, z_n)$ is defined in D; measurable in x for each fixed $(z_1, \cdots, z_n)$; continuous in $(z_1, \cdots, z_n)$ for each fixed x, and there exists (Lebesgue) integrable function $M(x)$ in the interval $\bar{J}$ such that

$$|f(x, z_1, \cdots, z_n)| \leq M(x), \ (x, z_1, \cdots, z_n) \in D. \tag{4.1.2}$$

In such a case (or otherwise) by a **extended solution** of (4.1.1) in the interval $\bar{J}$, we mean an absolutely continuous function in $\bar{J}$ together with its $(n-1)$ derivatives satisfying the above conditions (i) and (ii), and

(vi) $y^{(n)}(x) = f(x, y(x), y'(x), \cdots, y^{(n-1)}(x))$ almost everywhere in $\bar{J}$, i.e., for all $x \in \bar{J}$ except on a set of measure zero.

Let $y_1 = y$, $y_2 = y'$, $\cdots$ $, y_n = y^{(n-1)}$. Then, the initial value problem (4.1.1) is equivalent to the first order system

$$y_i' = y_{i+1}, \ 1 \leq i \leq n-1$$

$$y_n' = f(x, y_1, \cdots, y_n)$$

$$y_i(x_0) = y_{0i}, \ 1 \leq i \leq n$$

which is a particular case of the problem we have considered in Chapter 3. Therefore, if we define $\bar{S}$ as a $(n+1)$ dimensional rectangle $|x - x_0| \leq a$, $\sum_{i=1}^{n} |y_i - y_{0i}| \leq b$ and analogously S, $\bar{S}_+$, S_+, $\bar{S}_-$, $\bar{\Delta}$, and Δ then all the results presented in Chapter 3 are equally valid for the problem (4.1.1) also. However, by considering the problem (4.1.1) as such it is possible to obtain some results which are more precise than the conclusions we can deduce from the results of Chapter 3.

4.2 LIPSCHITZ UNIQUENESS THEOREM

We begin with the following :

Definition 4.2.1. The function $f(x, z_1, z_2, \cdots, z_n)$ is said to satisfy the **uniform Lipschitz condition** in a domain D if

$$(4.2.1) \qquad |f(x, z_1, z_2, \cdots, z_n) - f(x, \bar{z}_1, \bar{z}_2, \cdots, \bar{z}_n)| \leq \sum_{k=1}^{n} L_k |z_k - \bar{z}_k|$$

for all $(x, z_1, z_2, \cdots, z_n)$, $(x, \bar{z}_1, \bar{z}_2, \cdots, \bar{z}_n)$ in D having the same x.

In establishing various uniqueness criteria for (1.1.1) Lemma 1.2.2 has played a fundamental role. This result for the problem (4.1.1) is extended in the following :

Lemma 4.2.1. Let $f(x, z_1, z_2, \cdots, z_n)$ be continuous in the domain D, then any solution of (4.1.1) is also a solution of the integral equation

$$(4.2.2) \quad y(x) = \sum_{i=1}^{n} \frac{(x - x_0)^{i-1}}{(i-1)!} y_{0i}$$

$$+ \frac{1}{(n-1)!} \int_{x_0}^{x} (x-t)^{n-1} f(t, y(t), y'(t), \cdots, y^{(n-1)}(t)) dt$$

and conversely.

Proof. If $y(x)$ is a solution of (4.1.1), then from Taylor's formula

$$y(x) = \sum_{i=1}^{n} \frac{(x - x_0)^{i-1}}{(i-1)!} y^{(i-1)}(x_0) \; + \; \frac{1}{(n-1)!} \int_{x_0}^{x} (x-t)^{n-1} y^{(n)}(t) dt$$

(4.2.2) follows immediately. Conversely, since

$$(4.2.3) \quad y^{(j-1)}(x) = \sum_{i=j}^{n} \frac{(x - x_0)^{i-j}}{(i-j)!} y_{0i} + \frac{1}{(n-j)!} \times$$

$$\int_{x_0}^{x} (x-t)^{n-j} f(t, y(t), y'(t), \cdots, y^{(n-1)}(t)) dt, \quad 1 \leq j \leq n$$

we find that $y^{(j-1)}(x_0) = y_{0j}$; $1 \leq j \leq n$ and $y^{(n)}(x) = f(x, y(x), y'(x), \cdots, y^{(n-1)}(x))$. ∎

Theorem 4.2.2 (Lipschitz Uniqueness Theorem). Let $f(x, z_1, z_2, \cdots, z_n)$ be continuous and satisfy the uniform Lipschitz condition (4.2.1) in $\bar{S}$. Then, the initial value problem (4.1.1) has at most one solution in $|x - x_0| \leq a$.

First Proof. Suppose $y(x)$ and $\bar{y}(x)$ are two solutions of (4.1.1) in $|x - x_0| \leq a$, then from (4.2.3) for the function $\phi(x) = y(x) - \bar{y}(x)$ it follows that

$$\phi^{(j-1)}(x) = \frac{1}{(n-j)!}\int_{x_0}^{x}(x-t)^{n-j}\left[f(t, y(t), y'(t), \cdots, y^{(n-1)}(t))\right.$$

$$\left.-f(t, \bar{y}(t), \bar{y}'(t), \cdots, \bar{y}^{(n-1)}(t))\right] dt, \ 1 \leq j \leq n$$

and hence

$$|\phi^{(j-1)}(x)| \leq \frac{1}{(n-j)!}\left|\int_{x_0}^{x}|x-t|^{n-j}\sum_{k=1}^{n}L_k|\phi^{(k-1)}(t)|dt\right|, \ 1 \leq j \leq n.$$

Now let $\alpha > 0$ be so small that $C = \sum_{k=1}^{n}\frac{L_k}{(n-k+1)!}\alpha^{n-k+1} < 1$. If $\|\phi\| = \max_{|x-x_0|\leq\alpha}\sum_{k=1}^{n}L_k|\phi^{(k-1)}(x)|$, then from the above inequality we get

$$\sum_{j=1}^{n}L_j|\phi^{(j-1)}(x)| \leq \|\phi\| \sum_{j=1}^{n}\frac{L_j}{(n-j)!}\left|\int_{x_0}^{x}|x-t|^{n-j}dt\right|,$$

which implies that

$$\|\phi\| \leq \|\phi\| \max_{|x-x_0|\leq\alpha}\sum_{j=1}^{n}\frac{L_j}{(n-j+1)!}|x-x_0|^{n-j+1},$$

i.e., $\|\phi\| \leq C\ \|\phi\|$. Since $C < 1$ it follows that $\phi(x) \equiv 0$, and hence $y(x) = \bar{y}(x)$, $x \in [x_0 - \alpha, x_0 + \alpha]$. The uniqueness of solutions of (4.1.1) to the whole interval $|x - x_0| \leq a$ is now straightforward. ∎

Second Proof. We shall consider the interval $[x_0, x_0 + a]$, whereas the proof in the interval $[x_0 - a, x_0]$ is similar. Suppose contrary, then there are two different solutions $y(x)$ and $\bar{y}(x)$ of (4.1.1). Let $\phi(x) = y(x) - \bar{y}(x)$. Then, the function $v(x) = \sum_{i=1}^{n}|\phi^{(i-1)}(x)|$ does not vanish for all x in $[x_0, x_0 + a]$. Since $v(x)$ is continuous and $v(x_0) = 0$, there exists a number ε with $x_0 \leq \varepsilon < x_0 + a$ such that $v(x) = 0$ for $x_0 \leq x \leq \varepsilon$ and $v(x)$ is not identically zero in an interval $\varepsilon \leq x \leq \varepsilon + l \leq x_0 + a$, where $l > 0$. Choose l so small that $l(1 + \max_{1\leq k\leq n}L_k) < 1$. Since the function $v(x)$ is continuous and not identically zero in the closed interval $\varepsilon \leq x \leq \varepsilon + l$, it has a positive maximum at some point $\hat{\varepsilon} > \varepsilon$ in this interval. However, since

$$\begin{aligned}
0 < v(\hat{\varepsilon}) &= \sum_{i=1}^{n} |\phi^{(i-1)}(\hat{\varepsilon}) - \phi^{(i-1)}(\varepsilon)| \\
&= \sum_{i=1}^{n} \delta_i(\phi^{(i-1)}(\hat{\varepsilon}) - \phi^{(i-1)}(\varepsilon)) \\
&= \sum_{i=1}^{n} \delta_i(\hat{\varepsilon} - \varepsilon)\phi^{(i)}(\varepsilon^*) \\
&\leq (\hat{\varepsilon} - \varepsilon)\left(\sum_{i=1}^{n-1} |\phi^{(i)}(\varepsilon^*)| + |\phi^{(n)}(\varepsilon^*)|\right) \\
&\leq (\hat{\varepsilon} - \varepsilon)\left(\sum_{i=1}^{n-1} |\phi^{(i)}(\varepsilon^*)| + \sum_{i=1}^{n} L_i|\phi^{(i-1)}(\varepsilon^*)|\right) \\
&= (\hat{\varepsilon} - \varepsilon)\left(L_1|\phi(\varepsilon^*)| + \sum_{i=1}^{n-1}(1 + L_{i+1})|\phi^{(i)}(\varepsilon^*)|\right) \\
&\leq (\hat{\varepsilon} - \varepsilon)\left[1 + \max_{1\leq k\leq n} L_k\right] v(\varepsilon^*) \\
&< v(x^*),
\end{aligned}$$

where $\varepsilon < \varepsilon^* < \hat{\varepsilon}$ and δ_i is 1 or -1, we find that $v(\hat{\varepsilon})$ is not the positive maximum of $v(x)$ in the interval $\varepsilon \leq x \leq \varepsilon + l$. This contradiction completes the proof. ∎

Third Proof. Once again suppose that $y(x)$ and $\bar{y}(x)$ are two solutions of (4.1.1) in $|x - x_0| \leq a$. Let $\phi(x) = y(x) - \bar{y}(x)$. Then, the function $v(x) = \sum_{i=1}^{n}(\phi^{(i-1)}(x))^2$ satisfies

$$\begin{aligned}
v'(x) &\leq 2\sum_{i=1}^{n-1} |\phi^{(i-1)}(x)|\ |\phi^{(i)}(x)| + 2|\phi^{(n-1)}(x)|\ |\phi^{(n)}(x)| \\
&\leq \sum_{i=1}^{n-1} \left((\phi^{(i-1)}(x))^2 + (\phi^{(i)}(x))^2\right) + 2|\phi^{(n-1)}(x)| \left(\sum_{i=1}^{n} L_i|\phi^{(i-1)}(x)|\right).
\end{aligned}$$

Therefore, it follows that

$$\begin{aligned} v'(x) &\le (\phi(x))^2 + 2\sum_{i=1}^{n-2}(\phi^{(i)}(x))^2 + (\phi^{(n-1)}(x))^2 + 2L_n(\phi^{(n-1)}(x))^2 \\ &\quad + \sum_{i=1}^{n-1} L_i \left((\phi^{(i-1)}(x))^2 + (\phi^{(n-1)}(x))^2\right) \\ &= (1+L_1)(\phi(x))^2 + \sum_{i=1}^{n-2}(2+L_{i+1})(\phi^{(i)}(x))^2 \\ &\quad + \left(1 + 2L_n + \sum_{i=1}^{n-1} L_i\right)(\phi^{(n-1)}(x))^2 \\ &\le 2K\, v(x), \end{aligned}$$

where $K = 1 + \sum_{i=1}^{n} L_i$. From the above inequality and $v(x_0) = 0$ we immediately obtain $v(x) \equiv 0$, i.e., $y(x) = \bar{y}(x)$ in $|x - x_0| \le a$. ∎

Finally, we note that the second proof of Theorem 4.2.2 does not require that the solutions $y(x)$ of (4.1.1) must be n times continuously differentiable. In fact, the existence of $y^{(n)}(x)$ is sufficient. Therefore, a result analogous to Theorem 4.2.2 for classical solutions also holds.

4.3 MERLI'S UNIQUENESS THEOREM

We shall present here an extension of Theorem 1.6.2 for the initial value problem (4.1.1). For this, we need the following :

Definition 4.3.1. The function $f(x, z_1, z_2, \cdots, z_n)$ is said to satisfy **Nagumo's condition** in a domain D if for all $(x, z_1, z_2, \cdots, z_n)$, $(x, \bar{z}_1, \bar{z}_2, \cdots, \bar{z}_n)$ in D with $x \ne x_0$

(4.3.1) $$|f(x, z_1, z_2, \cdots, z_n) - f(x, \bar{z}_1, \bar{z}_2, \cdots, \bar{z}_n)| \le \sum_{k=1}^{n} \frac{L_k}{|x - x_0|^{n-k+1}} |z_k - \bar{z}_k|,$$

where

(4.3.2) $$\sum_{k=1}^{n} \frac{L_k}{(n-k+1)!} \le 1.$$

Theorem 4.3.1 (Merli's Uniqueness Theorem). Let $f(x, z_1, z_2, \cdots, z_n)$ be continuous and satisfy Nagumo's condition (4.3.1) in $\bar{S}$. Then, the initial value problem (4.1.1) has at most one solution in $|x - x_0| \leq a$.

Proof. We shall consider the interval $[x_0, x_0 + a]$, whereas the proof in the interval $[x_0 - a, x_0]$ is similar. Suppose $y(x)$ and $\bar{y}(x)$ are two solutions of (4.1.1) in $[x_0, x_0 + a]$. Consider the function

$$v(x) = \begin{cases} \displaystyle\sum_{k=1}^{n} \frac{L_k}{(x - x_0)^{n-k+1}} |y^{(k-1)}(x) - \bar{y}^{(k-1)}(x)|, & x_0 < x \leq x_0 + a \\ 0, & x = x_0. \end{cases}$$

It will be first shown that this function is continuous at $x = x_0$. For this, on using the mean value theorem and the fact that $y^{(k-1)}(x_0) = \bar{y}^{(k-1)}(x_0)$, $1 \leq k \leq n$, we find

$$|y^{(k-1)}(x) - \bar{y}^{(k-1)}(x)| = \frac{(x - x_0)^{n-k+1}}{(n-k+1)!} |y^{(n)}(p_k) - \bar{y}^{(n)}(p_k)|, \ 1 \leq k \leq n$$

where $x_0 < p_k < x$. Thus, for $x > x_0$ we have

$$(4.3.3) \quad v(x) = \sum_{k=1}^{n} \frac{L_k}{(n-k+1)!} |f(p_k, y(p_k), y'(p_k), \cdots, y^{(n-1)}(p_k))$$
$$- f(p_k, \bar{y}(p_k), \bar{y}'(p_k), \cdots, \bar{y}^{(n-1)}(p_k))|.$$

The continuity of f at the point $(x_0, y_{01}, y_{02}, \cdots, y_{0n})$ now implies that $v(x) \to 0$ as $x \to x_0$.

Now we shall show that $v(x) = 0$ in $[x_0, x_0+a]$. If contrary, then there exists a number $x_m > x_0$ such that $v(x)$ attains its positive maximum in $[x_0, x_0 + a]$ at $x = x_m$. Furthermore, $v(x) < v(x_m)$, $x_0 < x < x_m$. However, in view of (4.3.3) with $x = x_m$ (and thus $x_0 < p_k < x_m$; $1 \leq k \leq n$) and Nagumo's condition (4.3.1) it follows that

$$0 < v(x_m) \leq \sum_{k=1}^{n} \frac{L_k}{(n-k+1)!} v(p_k).$$

But, then from the condition (4.3.2) and the intermediate value theorem we find that

$$0 < v(x_m) \leq v(x^*),$$

where $x_0 < \min_{1\le k\le n} p_k \le x^* \le \max_{1\le k\le n} p_k < x_m$. This contradicts the definition of x_m, and hence the proof is complete. ∎

Obviously, the above proof of Theorem 4.3.1 does not require that the solutions $y(x)$ of (4.1.1) must be n times continuously differentiable. In fact, the continuity of the function f at the point $(x_0, y_{01}, y_{02}, \cdots, y_{0n})$ and the existence of $y^{(n)}(x)$ are sufficient. Therefore, a result analogous to Theorem 4.3.1 for classical solutions also holds. We also note that in the above result condition (4.3.1) can easily be replaced by

$$(4.3.4)\quad |f(x,z_1,z_2,\cdots,z_n) - f(x,\bar{z}_1,\bar{z}_2,\cdots,\bar{z}_n)| \le \sum_{k=1}^{n} \frac{L_k(x)}{|x-x_0|^{n-k+1}}|z_k - \bar{z}_k|,$$

where the functions $L_k(x) \ge 0$ are continuous in $|x - x_0| \le a$, and

$$(4.3.5)\qquad \sum_{k=1}^{n} \frac{L_k(x)}{(n-k+1)!} \le 1.$$

4.4 KIKODZE'S UNIQUENESS THEOREM

Let the function $f(x, z_1, z_2, \cdots, z_n)$ satisfy the Carathèodory conditions in $\bar{S}_+$. We define S_M as the set of all absolutely continuous functions $y(x)$ in $[x_0, x_0 + a]$ including its derivatives up to $(n-1)$ - th order and satisfy the inequalities

$$(4.4.1)\quad |y^{(j-1)}(x) - y_{0j}| \le |h^{(j-1)}(x) - y_{0j}| + \frac{1}{(n-j)!}\int_{x_0}^{x} (x-t)^{n-j} M(t)dt, \ 1 \le j \le n$$

where the function $M(x)$ is the same as in (4.1.2), and

$$h(x) = \sum_{i=1}^{n} \frac{(x-x_0)^{i-1}}{(i-1)!} y_{0i}.$$

We also define

$$(4.4.2)\quad \sigma(x) = \frac{1}{(n-2)!}\int_{x_0}^{x} (x-t)^{n-2} \times \left(\sup_{y(x),\bar{y}(x)\in S_M} \int_{x_0}^{t} |f(s,y(s),\cdots,y^{(n-1)}(s)) - f(s,\bar{y}(s),\cdots,\bar{y}^{(n-1)}(s))| ds \right) dt.$$

Theorem 4.4.1 (Kikodze's Uniqueness Theorem). Suppose that the function $f(x, z_1, z_2, \cdots, z_n)$ satisfies the Carathèodory conditions in $\bar{S}_+$, and for all $(x, z_1, z_2, \cdots, z_n)$, $(x, \bar{z}_1, \bar{z}_2, \cdots, \bar{z}_n)$ in $\bar{S}_+$

$$(4.4.3)\quad |f(x, z_1, z_2, \cdots, z_n) - f(x, \bar{z}_1, \bar{z}_2, \cdots, \bar{z}_n)| \le h(x, |z_1 - \bar{z}_1|, |z_2 - \bar{z}_2|, \cdots, |z_n - \bar{z}_n|),$$

where the function $h(x, u_1, u_2, \cdots, u_n)$ satisfies the Carathèodory conditions for $x_0 \le x \le x_0 + a$, $u_i \ge 0$, $1 \le i \le n$; nondecreasing in u_i, $1 \le i \le n$ and $h(x, 0, 0, \cdots, 0) \equiv 0$. Further, the sequence $\{u_k(x)\}$ of extended solutions of the equation

$$(4.4.4)\qquad u^{(n)} = h(x, u, u', \cdots, u^{(n-1)})$$

tends to 0 as $k \to \infty$, and such that every extended solution $u_k(x)$ is defined in $[x_k, x_0 + a]$ and

$$(4.4.5)\qquad u_k^{(j-1)}(x_k) > \sigma^{(j-1)}(x_k),\ 1 \le j \le n$$

where $x_0 < x_k < x_0 + a$ and $x_k \to x_0$ as $k \to \infty$. Then, the initial value problem (4.1.1) has at most one extended solution in $[x_0, x_0 + a]$.

Proof. Suppose that $y(x)$ and $\bar{y}(x)$ are two extended solutions of (4.1.1) in $[x_0, x_0 + a]$. Then, from (4.1.2) and (4.2.3) it is clear that both $y(x)$ and $\bar{y}(x)$ satisfy the inequality (4.4.1). Hence, $y(x)$, $\bar{y}(x) \in S_M$. On the other hand, since

$$|y^{(j-1)}(x) - \bar{y}^{(j-1)}(x)| \le \frac{1}{(n-j-1)!}\int_{x_0}^{x}(x-t)^{n-j-1}|y^{(n-1)}(t) - \bar{y}^{(n-1)}(t)|dt$$

and

$$|y^{(n-1)}(x) - \bar{y}^{(n-1)}(x)| \le \int_{x_0}^{x}\Big|f(t, y(t), y'(t), \cdots, y^{(n-1)}(t)) - f(t, \bar{y}(t), \bar{y}'(t), \cdots, \bar{y}^{(n-1)}(t))\Big|\, dt$$

for each $1 \le j \le n$, we find that

$$|y^{(j-1)}(x) - \bar{y}^{(j-1)}(x)| \le \frac{1}{(n-j-1)!}\int_{x_0}^{x}(x-t)^{n-j-1}\times \left(\int_{x_0}^{t}|f(s, y(s), y'(s), \cdots, y^{(n-1)}(s)) - f(s, \bar{y}(s), \bar{y}'(s), \cdots, \bar{y}^{(n-1)}(s))|ds\right) dt.$$

Thus, from (4.4.2) we get

$$|y^{(j-1)}(x) - \bar{y}^{(j-1)}(x)| \le \sigma^{(j-1)}(x),\ 1 \le j \le n$$

for $x \in [x_0, x_0 + a]$.

Now in view of (4.4.5) for each k the following inequality holds

$$(4.4.6) \qquad |y^{(j-1)}(x_k) - \bar{y}^{(j-1)}(x_k)| < u_k^{(j-1)}(x_k),\ 1 \le j \le n.$$

Thus, from (4.4.3) and (4.4.6), for $x \in [x_k, x_0 + a]$ and each $1 \le j \le n$ it is clear that

$$|y^{(j-1)}(x) - \bar{y}^{(j-1)}(x)| < \sum_{i=j}^{n} \frac{(x - x_k)^{i-j}}{(i-j)!} u_k^{(i-j)}(x_k) + \frac{1}{(n-j)!} \times$$

$$\int_{x_k}^{x} (x-t)^{n-j}\, h(t, |y(t) - \bar{y}(t)|, |y'(t) - \bar{y}'(t)|, \cdots, |y^{(n-1)}(t) - \bar{y}^{(n-1)}(t)|)dt.$$

On the other hand from (4.4.4) for $x \in [x_k, x_0 + a]$ and each $1 \le j \le n$, we have

$$u_k^{(j-1)}(x) = \sum_{i=j}^{n} \frac{(x - x_k)^{i-j}}{(i-j)!} u_k^{(i-j)}(x_k) + \frac{1}{(n-j)!} \times$$

$$\int_{x_k}^{x} (x-t)^{n-j}\, h(t, u_k(t), u_k'(t), \cdots, u_k^{(n-1)}(t))dt.$$

Therefore, it is immediate that

$$(4.4.7) \quad |y^{(j-1)}(x) - \bar{y}^{(j-1)}(x)| - u_k^{(j-1)}(x) < \frac{1}{(n-j)!} \int_{x_k}^{x} (x-t)^{n-j} \times$$

$$\Big[h(t, |y(t) - \bar{y}(t)|, |y'(t) - \bar{y}'(t)|, \cdots, |y^{(n-1)}(t) - \bar{y}^{(n-1)}(t)|)$$

$$-h(t, u_k(t), u_k'(t), \cdots, u_k^{(n-1)}(t))\Big]dt$$

for $x \in [x_k, x_0 + a]$ and each $1 \le j \le n$.

We shall now show that for $x \in [x_k, x_0 + a]$ and each $1 \le j \le n$

$$(4.4.8) \qquad |y^{(j-1)}(x) - \bar{y}^{(j-1)}(x)| < u_k^{(j-1)}(x).$$

If contrary, then in view of (4.4.6) for some k and i there exists $x^* \in (x_k, x_0 + a)$ such that

$$|y^{(i-1)}(x^*) - \bar{y}^{(i-1)}(x^*)| = u_k^{(i-1)}(x^*)$$

and (4.4.8) holds in the interval $[x_k, x^*)$. However, since the function $h(x, u_1, u_2, \cdots, u_n)$ is nondecreasing in u_i, $1 \le i \le n$ from (4.4.7) we find that

$$|y^{(i-1)}(x^*) - \bar{y}^{(i-1)}(x^*)| < u_k^{(i-1)}(x^*).$$

This contradiction shows that (4.4.8) holds.

The conclusion that $y(x) = \bar{y}(x)$, $x \in [x_0, x_0 + a]$ now follows from (4.4.8) and the fact that the sequence $\{u_k(x)\}$ of extended solutions of the equation (4.4.4) tends to 0 as $k \to \infty$. ∎

Corollary 4.4.2. Suppose that the function $f(x, z_1, z_2, \cdots, z_n)$ satisfies the Carathèodory conditions in $\bar{S}_+$, and $f(x, z_1, z_2, \cdots, z_n)/\psi(x)$ is continuous at $(x_0, y_{01}, y_{02}, \cdots, y_{0n})$, where the function $\psi(x)$ is positive and integrable (Lebesgue) in (x_0, x_0+a). Further, suppose that for all $(x, z_1, z_2, \cdots, z_n)$, $(x, \bar{z}_1, \bar{z}_2, \cdots, \bar{z}_n)$ in $\bar{S}_+$

$$(4.4.9) \quad |f(x, z_1, z_2, \cdots, z_n) - f(x, \bar{z}_1, \bar{z}_2, \cdots, \bar{z}_n)| \le \psi(x) \sum_{k=1}^{n} \frac{L_k\, |z_k - \bar{z}_k|}{\int_{x_0}^{x} (x-t)^{n-k} \psi(t)dt},$$

where

$$(4.4.10) \qquad \sum_{k=1}^{n} \frac{L_k}{(n-k)!} \le 1.$$

Then, the initial value problem (4.1.1) has at most one extended solution in $[x_0, x_0 + a]$.

Proof. In the given case equation (4.4.4) has the form

$$(4.4.11) \qquad u^{(n)} = \psi(x) \sum_{k=1}^{n} \frac{L_k}{\int_{x_0}^{x} (x-t)^{n-k} \psi(t)dt} u^{(k-1)}.$$

Without loss of generality we can assume that in (4.4.10) the equality holds. Then, the function

$$u(x) = c \int_{x_0}^{x} (x-t)^{n-1} \psi(t)dt,$$

where c is an arbitrary constant, is a solution of (4.4.11). It is clear that the sequence $\{u_k(x)\}$ of solutions of the equation (4.4.11) defined by

$$(4.4.12) \qquad u_k(x) = \left(\frac{1}{k} + \sum_{i=1}^{n} \frac{\sigma^{(i-1)}(x_k)}{\int_{x_0}^{x_k} (x_k - t)^{n-i} \psi(t)dt}\right) \int_{x_0}^{x} (x-t)^{n-1} \psi(t)dt$$

satisfies the condition (4.4.5).

Thus, to complete the proof it remains to show that $\lim_{k\to\infty} u_k(x) \equiv 0$. But, in view of (4.4.2) and (4.4.12) it certainly holds provided

$$\lim_{x\to x_0} \frac{\sigma^{(n-1)}(x)}{\int_{x_0}^{x} \psi(t)dt} = 0. \tag{4.4.13}$$

From the continuity of the function $f_1(x, z_1, z_2, \cdots, z_n) = f(x, z_1, z_2, \cdots, z_n)/\psi(x)$ at the point $(x_0, y_{01}, y_{02}, \cdots, y_{0n})$ for any $\varepsilon > 0$ there exists a $\delta > 0$ such that $x - x_0 < \delta$, $|z_i - y_{0i}| < \delta$, $1 \le i \le n$ and

$$|f_1(x, z_1, z_2, \cdots, z_n) - f_1(x, z_{01}, z_{02}, \cdots, z_{0n})| < \frac{\varepsilon}{2}.$$

Further, since any function $y(x) \in S_M$ satisfies (4.4.1), there exists a $\eta > 0$ depending only on δ such that if $x - x_0 < \eta$ then

$$|y^{(j-1)}(x) - y_{0j}| < \delta,\ 1 \le j \le n.$$

Therefore, from the triangle inequality we get

$$\sup_{y(x),\bar{y}(x)\in S_M} \Big|f_1(x, y(x), y'(x), \cdots, y^{(n-1)}(x)) \\ - f_1(x, \bar{y}(x), \bar{y}'(x), \cdots, \bar{y}^{(n-1)}(x))\Big| < \varepsilon,\ x - x_0 < \eta. \tag{4.4.14}$$

Using (4.4.14) in (4.4.2), we get

$$\sigma^{(n-1)}(x) = \sup_{y(x),\bar{y}(x)\in S_M} \int_{x_0}^{x} \psi(t)|f_1(t, y(t), y'(t), \cdots, y^{(n-1)}(t)) \\ - f_1(t, \bar{y}(t), \bar{y}'(t), \cdots, \bar{y}^{(n-1)}(t))|dt \\ < \varepsilon \int_{x_0}^{x} \psi(t)dt,\ x - x_0 < \eta.$$

Hence, it follows that

$$\limsup_{x\to x_0} \frac{\sigma^{(n-1)}(x)}{\int_{x_0}^{x} \psi(t)dt} < \varepsilon.$$

Since $\varepsilon > 0$ is arbitrary, condition (4.4.13) follows. ∎

Corollary 4.4.3. Suppose that the function $f(x, z_1, z_2, \cdots, z_n)$ satisfies the Caratheòdory conditions in $\bar{S}_+$, and $(x - x_0)^{\alpha} f(x, z_1, z_2, \cdots, z_n)$ where $\alpha < 1$ is continuous at $(x_0, y_{01}, y_{02}, \cdots, y_{0n})$. Further, suppose that for all $(x, z_1, z_2, \cdots, z_n)$, $(x, \bar{z}_1, \bar{z}_2, \cdots, \bar{z}_n)$ in $\bar{S}_+$

$$|f(x, z_1, z_2, \cdots, z_n) - f(x, \bar{z}_1, \bar{z}_2, \cdots, \bar{z}_n)| \le \sum_{k=1}^{n} \frac{L_k \, |z_k - \bar{z}_k|}{(x - x_0)^{n-k+\alpha}}, \tag{4.4.15}$$

where

$$\sum_{k=1}^{n} \frac{L_k}{(1-\alpha)\cdots(n-k+1-\alpha)} \le 1. \tag{4.4.16}$$

Then, the initial value problem (4.1.1) has at most one extended solution in $[x_0, x_0 + a]$. ∎

We note that for $\alpha = 0$, Corollary 4.4.3 reduces to Theorem 4.3.1 for the extended solutions of (4.1.1). Further, the inequality (4.4.16) is the best possible. For this, we have the following :

Example 4.4.1. Consider the initial value problem

$$\begin{aligned} y' &= f(x, y) = x^{\beta} \sin \frac{Ly}{x^{1+\beta}} \\ y(0) &= 0 \end{aligned} \tag{4.4.17}$$

where $\beta > -1$ and $L > 1 + \beta$. It is clear that for any $\alpha > -\beta$ the function $x^{\alpha} f(x, y)$ is continuous at $(0, 0)$. However, the initial value problem (4.4.17) besides the trivial solution also has the solution $y(x) = cx^{1+\beta}$, where $c \neq 0$ is the solution of the equation $c(1 + \beta) = \sin Lc$. This is due to the fact that for all $\alpha > -\beta$ we have $L > 1 - \alpha$. ∎

4.5 WEND'S UNIQUENESS THEOREM

An extension of Theorem 1.19.1 for the initial value problem (4.1.1) where $y_{0i} \ge 0, \ 2 \le i \le n$ is the following :

Theorem 4.5.1 (Wend's Uniqueness Theorem). Let $f(x, z_1, z_2, \cdots, z_n)$ be continuous and nondecreasing in x and each z_i, $1 \le i \le n$ in S_+. Further, let

$f(x, z_1, z_2, \cdots, z_n)$ be nonnegative in $\bar{S}_+$, and $f(x, y_{01}, y_{02}, \cdots, y_{0n}) \neq 0$ for all $x \in (x_0, x_0 + a)$. Then, the initial value problem (4.1.1) where $y_{0i} \geq 0$ for $2 \leq i \leq n$ has at most one solution in $[x_0, x_0 + a]$.

Proof. We shall assume without loss of generality that $x_0 = 0$. Thus, from Lemma 4.2.1 for any solution $y(x)$ of (4.1.1) we have

$$(4.5.1) \qquad y^{(j-1)}(x) = \sum_{i=j}^{n} \frac{x^{i-j}}{(i-j)!} y_{0i} + \frac{1}{(n-j)!} \times$$
$$\int_0^x (x-t)^{n-j} f(t, y(t), y'(t), \cdots, y^{(n-1)}(t))dt, \quad 1 \leq j \leq n.$$

Hence, in view of the hypotheses $y(x)$ and its first $n-1$ derivatives are strictly increasing functions of x for $0 < x < a$. Therefore, for $\delta > 0$ sufficiently small there exists for each $1 \leq j \leq n$ a smallest η_j such that $0 < \eta_j < a$ and $y^{(j-1)}(\eta_j) = y_j + \delta$. Let $\eta = \max\{\eta_1, \cdots, \eta_n\}$ and define $y_\delta(x) = y(x + \eta)$, $0 \leq x \leq a - \eta$. Then, $y_\delta^{(j-1)}(0) = y^{(j-1)}(\eta) \geq y^{(j-1)}(\eta_j) = y_j + \delta$, $1 \leq j \leq n$. Using the hypothesis that $f(x, z_1, z_2, \cdots, z_n)$ is nondecreasing in x and the fact that $y_\delta^{(n)}(x) = y^{(n)}(x + \eta) = f(x + \eta, y_\delta(x), \cdots, y_\delta^{(n-1)}(x))$, it follows from (4.5.1) that

$$(4.5.2) \qquad y_\delta^{(j-1)}(x) - y^{(j-1)}(x) \geq \delta \sum_{i=j}^{n} \frac{x^{i-j}}{(i-j)!} + \frac{1}{(n-j)!} \times$$
$$\int_0^x (x-t)^{n-j} (f(t, y_\delta(t), \cdots, y_\delta^{(n-1)}(t)) - f(t, y(t), \cdots, y^{(n-1)}(t)))dt.$$

Since $y_\delta^{(j-1)}(0) > y^{(j-1)}(0)$, $y_\delta^{(j-1)}(x) > y^{(j-1)}(x)$ for each $1 \leq j \leq n$ and for nonnegative x sufficiently close to 0. Suppose $y_\delta^{(j-1)}(t) > y^{(j-1)}(t)$ for $0 \leq t < x$ for each $1 \leq j \leq n$, but for some j, $y_\delta^{(j-1)}(x) = y^{(j-1)}(x)$. Then, from (4.5.2) it follows that $f(t, y_\delta(t), \cdots, y_\delta^{(n-1)}(t)) - f(t, y(t), \cdots, y^{(n-1)}(t)) < 0$ for some t, $0 < t < x$, which is a contradiction since f is nondecreasing in the last n variables. Hence, $y_\delta(x) > y(x)$ for $0 \leq x \leq a - \eta$. Now let $r(x)$ be the maximal solution of (4.1.1) with $x_0 = 0$. Then, by the same argument, if $0 < \bar{\delta} < \delta$, we have

$$y(x) \leq r(x) \leq y_{\bar{\delta}}(x) < y_\delta(x), \quad 0 \leq x \leq a - \eta.$$

But, $y_\delta(x) \to y(x)$ as $\eta \to 0$ and $\delta \to 0$. So, $y(x) \equiv r(x)$, and hence (4.1.1) has at most one solution in $[x_0, x_0 + a]$. ∎

Example 4.5.1. The initial value problem

$$\text{(4.5.3)} \qquad \begin{aligned} y^{(n)} &= (y^{(n-1)})^{2/3} \\ y^{(i-1)}(0) &= 0, \ 1 \le i \le n \end{aligned}$$

has many solutions, e.g., $y(x) \equiv 0$ and $y(x) = \frac{2x^{n+2}}{9(n+2)!}$. Here, all the conditions of Theorem 4.5.1 are satisfied except that $f(x, 0, \cdots, 0) = 0$. ∎

Example 4.5.2. Consider the initial value problem

$$\text{(4.5.4)} \qquad \begin{aligned} y^{(n)} &= (y^{(n-1)})^{2/3} \exp(y^{(n-1)}) \\ y^{(i-1)}(0) &= 0, \ 1 \le i \le n \end{aligned}$$

for which $y(x) \equiv 0$ is a solution. Letting $z = y^{(n-1)}$ so that $z(x)$ is defined by the equation

$$x = \int_0^z t^{-2/3} e^{-t} dt.$$

As $z \to \infty$, $x \to r_0 = \int_0^\infty t^{-2/3} e^{-t} dt < \infty$. Thus, in the interval $[0, r_0)$ the function

$$y(x) = \frac{1}{(n-2)!} \int_0^x (x-t)^{n-2} z(t) dt$$

is also a solution of (4.5.4). ∎

Example 4.5.3. Consider the initial value problem

$$\text{(4.5.5)} \qquad \begin{aligned} y'' &= (y-x)^{2/3} \exp(y-x) \\ y(0) &= 0, \ y'(0) = 1 \end{aligned}$$

for which $y(x) = x$ is a solution. Letting $z = y - x$ so that $z'' = z^{2/3} e^z$, $z(0) = z'(0) = 0$ and hence another solution of (4.5.5) is defined by the equation

$$x = \int_0^{y-x} \frac{1}{\sqrt{2 \int_0^s t^{2/3} e^t dt}} ds.$$

Here also as in Example 4.5.2 as $y \to \infty$, $x \to r_0 < \infty$. The function $f(x, z_0, z_1)$ in (4.5.5) is decreasing in x for all $x \neq y$, but $f(x, 0, 1) > 0$ for all $x \neq 0$. ∎

4.6 REDHEFFER'S UNIQUENESS THEOREM

Consider the second order initial value problem

$$(4.6.1)\qquad \begin{aligned} &y'' = f(x, y, y') \\ &y(x_0) = y_{01},\ y'(x_0) = y_{02} \end{aligned}$$

where the function $f(x, z_1, z_2)$ is defined in $S_2 = \{(x, z_1, z_2) : x_0 < x < x_0 + a,\ |z_1 - y_{01}| + |z_2 - y_{02}| < b\}$ and for all (x, z_1, z_2) $(x, \bar{z}_1, \bar{z}_2) \in S_2$ it satisfies the generalized Lipschitz condition

$$(4.6.2)\qquad |f(x, z_1, z_2) - f(x, \bar{z}_1, \bar{z}_2)| \le L_0(x)|z_1 - \bar{z}_1| + L_1(x)|z_2 - \bar{z}_2|.$$

In (4.6.2) functions $L_0(x)$ and $L_1(x)$ are assumed to be defined and nonnegative in $(x_0, x_0 + a)$.

If $y(x)$ and $\bar{y}(x)$ are two classical solutions of (4.6.1) in $[x_0, x_0 + a)$ then the function $\phi(x) = y(x) - \bar{y}(x)$ satisfies

$$(4.6.3)\qquad \begin{aligned} &|\phi''(x)| \le L_0(x)|\phi(x)| + L_1(x)|\phi'(x)|,\ x_0 < x < x_0 + a \\ &\phi(x_0) = \phi'(x_0) = 0. \end{aligned}$$

The following result provides sufficient conditions so that $\phi(x) \equiv 0$, i.e., $y(x) = \bar{y}(x),\ x \in [x_0, x_0 + a)$.

Lemma 4.6.1. Let the function $\phi(x)$ satisfy (4.6.3). Further, let for each $x_1 \in [x_0, x_0 + a)$ there exist $x_2 > x_1$ and a function $\psi(x)$ such that

$$(4.6.4)\qquad \begin{aligned} &\psi''(x) > L_0(x)\psi(x) + L_1(x)\psi'(x),\ x_1 < x < x_2 \\ &\psi(x_1) = 0,\ \psi'(x_1) > 0. \end{aligned}$$

Then, $\phi(x) = 0,\ x_0 \le x < x_0 + a$.

Proof. Suppose the conclusion fails. Then, there exists the largest interval $[x_0, x_1]$ in which $\phi(x) = 0$ and for every $x_2 > x_1$, $|\phi(x)| > 0$ at some point of (x_1, x_2). Thus, for a sufficiently small $\delta > 0$ there exits a point $\bar{x} \in (x_1, x_2)$ such that $|\phi(\bar{x})| > \delta\psi(\bar{x})$. Since $\phi(x_1) = \phi'(x_1) = 0$ and $\psi(x_1) = 0,\ \psi'(x_1) > 0$ for $0 < x - x_1$ sufficiently small it follows that $|\phi(x)| < \delta\psi(x)$.

Let $\hat{x} > x_1$ be the first point where the graph of $y = |\phi(x)|$ crosses the graph of $y = \delta\psi(x)$. The above discussion shows that such a point $\hat{x} \in (x_1, x_2)$ exists. Changing the sign of $\phi(x)$, if necessary, we can suppose that $\phi(\hat{x}) = \delta\psi(\hat{x})$ rather than $\phi(\hat{x}) = -\delta\psi(\hat{x})$.

The function $\phi(x) - \delta\psi(x)$ is 0 at x_1 and $\hat{x}$ but is negative for $x_1 < x < x_1 + \varepsilon$, where $\varepsilon > 0$ is a small constant. Thus, it has a minimum at some point $x^* \in (x_1, \hat{x})$. Further, at this point

$$|\phi(x^*)| < \delta\psi(x^*),\ \phi'(x^*) = \delta\psi'(x^*),\ \phi''(x^*) \geq \delta\psi''(x^*)$$

and hence

$$|\phi''(x^*)| - L_1(x^*)|\phi'(x^*)| - L_0(x^*)|\phi(x^*)|$$
$$\geq \delta(\psi''(x^*) - L_1(x^*)\psi'(x^*) - L_0(x^*)\psi(x^*)) > 0.$$

But this contradicts (4.6.3) and therefore the result follows. ∎

If L_0 and L_1 are constant then we can take $\psi(x) = e^{\alpha x} - e^{\alpha x_1}$ where $\alpha^2 > L_0 + L_1\alpha$. Thus, Lemma 4.6.1 gives uniqueness of classical solutions of (4.6.1) if the function f satisfies the Lipschitz condition.

The importance of Lemma 4.6.1 lies in the fact that it holds even when neither $L_0(x)$ nor $L_1(x)$ is bounded, and even when $L_0(x)$ is not integrable. Furthermore, $\phi''(x)$ can fail to exist in a dense subset of $[x_0, x_0 + a)$.

We say that (4.6.3) holds mod E if $\phi'(x)$ exists for $x_0 \leq x < x_0 + a$, and (4.6.3) holds except for x in a countable subset of $[x_0, x_0 + a)$. Similarly, $L_0(x)$ and $L_1(x)$ are continuous mod E if for each the set of discontinuities in $[x_0, x_0 + a)$ is countable. The conclusion of Lemma 4.6.1 remains valid if the differential inequalities in (4.6.3) and (4.6.4) hold only mod E, because $\phi'(x) = \delta\psi'(x)$ at the minimum point, and the exceptional set can be avoided be the suitable choice of δ.

Theorem 4.6.2 (Redheffer's Uniqueness Theorem). Let (4.6.3) hold mod E, and let $L_0(x)$ and $L_1(x)$ be continuous mod E. Further, let for each $x_1 \in [x_0, x_0 + a)$ there exist $x_2 > x_1$ such that

(4.6.5) $$\int_{x_1}^{x_2} [(t - x_1)L_0(t) + L_1(t)]\, dt < \infty.$$

Then, $\phi(x) = 0,\ x \in [x_0, x_0 + a)$.

Proof. Setting $\psi'(x) = \exp(\mu(x))$ the result follows immediately. ∎

The two - sided inequality for $\phi''(x)$ in (4.6.3) allows replacement of $\phi(x)$ by $-\phi(x)$ in the proof of Lemma 4.6.1. Therefore, the conclusion of Theorem 4.6.2 remains valid even if (4.6.3) is replaced by

$$\text{(4.6.6)}\qquad \begin{aligned} &\phi''(x) \le L_0(x)\phi(x) + L_1(x)\phi'(x) \bmod \mathrm{E},\ \ \phi(x) \ge 0 \\ &\phi(x_0) = \phi'(x_0) = 0. \end{aligned}$$

Example 4.6.1. If for each $x_1 \in [x_0, x_0 + a)$ there exists a constant $\alpha < 1$ such that

$$\text{(4.6.7)}\qquad \limsup_{x \to x_1^+} \left[(x - x_1)^{1+\alpha} L_0(x) + (x - x_1)^{\alpha} L_1(x)\right] < \infty$$

then (4.6.5) holds. This is clear from the fact that $(x - x_1)^{-\alpha}$ is integrable in the right neighborhood of x_1. However, the function $\phi(x) = x(\ln x)^{-1},\ x_1 = 0$ suggests that the condition

$$\text{(4.6.8)}\qquad \limsup_{x \to x_1^+} \left[(x - x_1)^2 L_0(x) + (x - x_1) L_1(x)\right] = 0$$

is not sufficient for the uniqueness. ∎

Example 4.6.2. If the differential inequalities for $\phi(x)$ and $\psi(x)$ in (4.6.3) and (4.6.4) hold at the point x_1 then in Lemma 4.6.1 one can allow $\psi'(x_1) = 0$. The proof remains the same. The choice $\psi(x) = (x - x_1)^2 + (x - x_1)^3$ then shows that for the uniqueness the inequality

$$\text{(4.6.9)}\qquad \frac{(x - x_1)^2}{2} L_0(x) + (x - x_1) L_1(x) \le 1,\ \ x_1 < x < x_2$$

is sufficient. Obviously, (4.6.9) is weaker than (4.6.8) if (4.6.3) holds at $x = x_1$, and the same as Nagumo's condition. The choice $\phi(x) = (x - x_1)^2$ shows that (4.6.9) is essential, i.e., x_1 must not belong to E even if $\phi(x)$ is twice continuously differentiable. Finally, the function $\phi(x) = x^2(\ln x)^{-1}$ suggests that the condition (4.6.9) cannot be replaced by

$$\text{(4.6.10)}\qquad \limsup_{x \to x_1^+} \left[\frac{(x - x_1)^2}{2} L_0(x) + (x - x_1) L_1(x)\right] \le 1. \qquad ∎$$

4.7 HEIDEL'S UNIQUENESS THEOREM

Consider the differential equation

$$y'' + f(x)y^{\gamma} = 0, \tag{4.7.1}$$

where $f(x)$ is nonnegative and continuous in $[0, \infty)$ and $\gamma = p/q$ where p and q are odd positive integers. If $\gamma > 1$, then from the Lipschitz Uniqueness Theorem 4.2.2 for any set of initial conditions differential equation (4.7.1) has at most one solution. The uniqueness of the trivial solution of (4.7.1) for the case $0 < \gamma < 1$ is proved in the following :

Theorem 4.7.1 (Heidel's Uniqueness Theorem). Let $f(x)$ be positive and continuously differentiable in $[0, \infty)$ and $0 < \gamma < 1$. Then, the trivial solution of (4.7.1) is unique in $[0, \infty)$.

Proof. Suppose that $y(x)$ is a solution of (4.7.1) such that $(y(x_0))^2 + (y'(x_0))^2 > 0$ for some $x_0 \geq 0$. It suffices to show that $(y(x))^2 + (y'(x))^2 > 0$ for all $x \geq 0$. Let $x_1 > x_0$, and

$$\phi(x) = \frac{2f(x)}{\gamma+1}(y(x))^{\gamma+1} + (y'(x))^2.$$

Then, $\phi'(x) = \frac{2}{\gamma+1}f'(x)(y(x))^{\gamma+1}$, and therefore

$$\phi'(x) \geq -\frac{2}{\gamma+1}\frac{|f'(x)|}{f(x)}\phi(x).$$

On integrating the above inequality, we obtain

$$\phi(x) \geq \phi(x_0)\exp\left(-\int_{x_0}^{x_1} g(t)dt\right), \quad x_0 \leq x \leq x_1$$

where $g(x) = \frac{2|f'(x)|}{(\gamma+1)f(x)}$. However, since

$$\int_{x_0}^{x_1} \frac{|f'(t)|}{f(t)}dt < \infty$$

it follows that $\phi(x) > 0$ for $x_0 \leq x \leq x_1$. Therefore, $(y(x))^2 + (y'(x))^2 > 0$ for $x_0 \leq x \leq x_1$. But, x_1 is arbitrary, and hence, $(y(x))^2 + (y'(x))^2 > 0$ for $x \geq x_0$. A similar argument applies for $x_1 < x_0$. ∎

Using Theorem 3.20.1 it is possible to extend the above result. For this we transform the equation (4.7.1) into a system and generalize it to

$$
\begin{aligned}
y_1' &= p_1(x)g_2(y_2) \\
y_2' &= -p_2(x)g_1(y_1),
\end{aligned}
\tag{4.7.2}
$$

where p_i, $i = 1,2$ are defined in $[0,\infty)$ and g_i, $i = 1,2$ in R.

Theorem 4.7.2. Let p_i; $i = 1,2$ be positive and absolutely continuous in $[0,\infty)$. Further, let g_i; $i = 1,2$ be continuous in R and $yg_i(y) > 0$ for $y \neq 0$. Then, the trivial solution of (4.7.2) is unique in $[0,\infty)$.

Proof. Since we are looking the uniqueness of the trivial solution only, in Theorem 3.20.1 it suffices to define a Lyapunov function in $[0,\infty) \times R^2$ instead of in $[0,\infty) \times R^4$. For $x \in [0,\infty)$ and $y = (y_1, y_2) \in R^2$, we let

$$V(x,y) = p_1(x)\int_0^{y_2} g_2(t)dt + p_2(x)\int_0^{y_1} g_1(t)dt.$$

Then, $V(x,y) \geq 0$, with equality if and only if $y = 0$. Further, if $y(x)$ is any solution of (4.7.2) then in view of absolute continuity of $V(x, y(x))$ for almost all $x \geq 0$ it follows that

$$
\begin{aligned}
\frac{d}{dx}V(x,y(x)) &= p_1'(x)\int_0^{y_2(x)} g_2(t)dt + p_2'(x)\int_0^{y_1(x)} g_1(t)dt \\
&\leq \max\left[\frac{|p_1'(x)|}{p_1(x)}, \frac{|p_2'(x)|}{p_2(x)}\right] V(x,y(x)).
\end{aligned}
$$

Thus, the conditions of Theorem 3.20.1 with $B(x) \equiv 1$ are satisfied and the conclusion follows. ∎

Using continuation and nonoscillatory properties of the solutions of (4.7.1) it is possible to modify Theorem 4.7.1. We state such a result in the following :

Theorem 4.7.3. Let $f(x)$ be continuously differentiable in $[0,\infty)$ and positive except for a sequence of isolated points $\{x_i\}$, $i = 1, 2, \cdots$. Further, let for each i the function

$$\bar{f}(t) = \frac{f\left(x_i - \frac{1}{t}\right)}{|t|^{\gamma+3}}$$

satisfies $\bar{f}'(t) \leq 0$ for large $|t|$. Then, for $0 < \gamma < 1$ the trivial solution of (4.7.1) is unique in $[0, \infty)$. ∎

CHAPTER 5

DIFFERENTIAL EQUATIONS IN ABSTRACT SPACES

5.1 INTRODUCTION

Let Y be an abstract space (real or complex normed linear space, inner product space, real or complex Banach space; or real or complex Hilbert space). For $y \in Y$ in this chapter also we shall denote the norm as $|y|$, so that the definitions of $\bar{S}$, S, $\bar{S}_+$, S_+, $\bar{S}_-$, $\bar{\Delta}$, and Δ remain the same. We shall also use the Euclidean norm and as earlier to distinguish it from $|\cdot|$, we shall denote it as $\|\cdot\|$. For y, $\bar{y} \in Y$, $y \cdot \bar{y}$ represents the usual inner product.

Let $f(x, y)$ be a mapping from $R \times Y$ into Y. Further, let $f(x, y)$ be strongly continuous for $(x, y) \in J \times E \subseteq R \times Y$, and $(x_0, y_0) \in J \times E$. For the initial value problem (1.1.1) a function $y(x)$ is said to be an **abstract solution** in J if

(i) $y(x_0) = y_0$,

(ii) for each $x \in J$, the ordered pair $(x, y(x)) \in J \times E$,

(iii) $y(x)$ is strongly continuously differentiable for all $x \in J$; and

(iv) $y'(x) = f(x, y(x))$.

As in earlier chapters if the mapping $f(x, y)$ is only defined for $(x, y) \in J \times E$ and J is one of the intervals $[x_0, x_0 + a]$, $[x_0, x_0 + a)$, $(x_0, x_0 + a]$, $(x_0, x_0 + a)$, then by a **classical abstract solution** of (1.1.1) in the interval $\{x_0\} \bigcup J$, we mean a function $y(x)$ satisfying the above conditions (i) and (ii), and

(v) $y(x)$ is strongly continuous in $\{x_0\} \bigcup J$,

(vi) $y'(x)$ strongly exists in J^o; and

(vii) $y'(x) = f(x, y(x))$.

If the conditions (vi) and (vii) hold in $\{x_0\} \bigcup J$ instead of J^o then a classical abstract solution is called **restricted abstract classical solution.**

Now let Y be a real or complex Banach space. The mapping $y(x) \in Y$, $x \in J$ is said to be strongly absolutely continuous if and only if for each compact subinterval $[c, d]$ of J and for each $\varepsilon > 0$, there exists a $\delta > 0$ such that $\sum_{i=1}^{n} |y(d_i) - y(c_i)| < \varepsilon$ whenever $c \leq c_i \leq d_i \leq d$; $i = 1, 2, \cdots, n$ and $\sum_{i=1}^{n} (d_i - c_i) < \delta$. It is clear that if $y(x)$ is strongly absolutely continuous then the real valued function $|y(x)|$ is absolutely continuous. In particular, if $y(x)$ is strongly differentiable almost everywhere with strong derivative $y'(x)$, and $y(x) - y(c) = \int_c^x y'(t)dt$, where c, $x \in J$ and the integral is in the Bochner sense, then $y(x)$ is strongly absolutely continuous. The converse of this result is false in general, however if Y is reflexive then it can be proved to be true. As an another example, if $y(x)$ has a weakly continuous derivative everywhere, then $y(x) - y(c) = \int_c^x y'(t)dt$, and $y(x)$ is strongly absolutely continuous.

If the mapping $f(x, y)$ is defined for $(x, y) \in J \times E$ then for the initial value problem (1.1.1) a function $y(x)$ is said to be an **extended abstract solution** in $\bar{J}$ if the above conditions (i) and (ii) are satisfied, and

(viii) $y(x)$ is strongly absolutely continuous and has a strong derivative $y'(x)$ almost everywhere in $\bar{J}$; and

(ix) $y'(x) = f(x, y(x))$ almost everywhere in $\bar{J}$.

In Section 1.1 we have remarked that the continuity of $f(x,y)$ in $\bar{S}$ is sufficient for the existence of a solution of (1.1.1) in $|x - x_0| \leq h$. This classical result of Peano extends rather easily when (1.1.1) is finite dimensional. However, this result cannot be generalized to infinite dimensions even if Y is a Banach space or a Hilbert space. For this, counterexamples are known [44,177]. One of the main reason for this is that the closed unit ball in a Banach space is not necessarily compact. But, if Y is a Banach space and $f(x,y)$ is strongly continuous and satisfies the Lipschitz condition (1.2.1) in $\bar{S} \subseteq R \times Y$, and for all $(x,y) \in \bar{S}$, $|f(x,y)| \leq M$ then the existence and uniqueness of abstract solutions of (1.1.1) in $|x - x_0| \leq h$ is guaranteed. Another result which ensures the existence of abstract solutions of (1.1.1) when Y is a complex Hilbert space is due to Browder [27]. For this, let Y_w be the Hilbert space endowed with the weak topology. We say that $f(x,y)$ is weakly continuous from $[0,\infty) \times Y$ into Y if $f(x,y)$ is continuous from $[0,\infty) \times Y_w$ into Y_w.

Theorem 5.1.1 (Browder's Existence Theorem). Let Y be a Hilbert space and let $f(x,y)$ be weakly continuous from $[0,\infty) \times Y$ into Y. Then, for each $r > 0$ there exists $\delta(r) > 0$ such that for each $y_0 \in Y$ with $\|y_0\| < r$, there exists an abstract solution $y(x)$ of (1.1.1) with $x_0 = 0$ in $[0, \delta(r)]$. ∎

Since our main concern in this chapter is to provide uniqueness criteria of abstract solutions of (1.1.1), as in earlier chapters we shall separate the uniqueness conditions from the conditions sufficient to guarantee existence.

5.2 AZIZ - DIAZ'S UNIQUENESS THEOREM

An analog of Theorem 1.2.4 for abstract solutions is the following :

Theorem 5.2.1. Let Y be a Banach space and let $f(x,y)$ be strongly continuous and satisfy the Lipschitz condition (1.2.1) in $\bar{S} \subseteq R \times Y$. Then, the initial value problem (1.1.1) has at most one abstract solution in $|x - x_0| \leq a$.

Proof. First we note that as in the finite dimensional case the initial value problem (1.1.1) is equivalent to the integral equation (1.2.4), where the integral is being taken relative to the strong topology. Now let $y(x)$ and $\bar{y}(x)$ be two abstract solutions of (1.1.1) in $|x-x_0| \leq a$, then as in the first proof of Theorem 1.2.4 we have

$$\begin{aligned} |y(x) - \bar{y}(x)| &\leq \left| \int_{x_0}^{x} |f(t, y(t)) - f(t, \bar{y}(t))| dt \right| \\ &\leq L \left| \int_{x_0}^{x} |y(t) - \bar{y}(t)| dt \right|. \end{aligned}$$

Thus, from Lemma 1.2.3 it follows that $|y(x) - \bar{y}(x)| \equiv 0$, and hence $y(x) = \bar{y}(x)$ in $|x - x_0| \leq a$. ∎

In Section 3.2 we have remarked that the mean value theorem which plays a significant role in establishing various uniqueness criteria is no longer valid for the vector valued functions. Further, the modification made there to accomplish the second proof of Theorem 1.2.4 for n - solutions of (1.1.1) is not possible if Y is infinite dimensional. Therefore, to extend uniqueness results for the classical abstract solutions of (1.1.1) we need a suitable form of the mean value theorem. We state such an extension in the following :

Theorem 5.2.2 (Mean Value Theorem). Let Y be a normed linear space and for each $x \in [c, d]$, $y(x) \in Y$. Further, let

(i) $y(x)$ be strongly continuous in $[c, d]$; and

(ii) $y'(x)$ strongly exist in (c, d).

Then, there exists a number $\xi \in (c, d)$ such that

$$\left| \frac{y(d) - y(c)}{d - c} \right| \leq |y'(\xi)|. \quad ∎$$

Theorem 5.2.3 (Aziz - Diaz's Uniqueness Theorem). Let Y be a normed linear space and let $f(x, y)$ be defined and satisfy the Lipschitz condition (1.2.1) in $S_+ \subseteq R \times Y$. Then, the initial value problem (1.1.1) has at most one classical abstract solution in $[x_0, x_0 + a)$.

Proof. Let $y(x)$ and $\bar{y}(x)$ be two classical abstract solutions of (1.1.1) in $[x_0, x_0 + a)$. Suppose that the function $y(x) - \bar{y}(x)$ which is known to be strongly continuous in $[x_0, x_0+a)$, finitely strongly differentiable in (x_0, x_0+a) and vanishes at $x = x_0$ does not vanish for all x in $[x_0, x_0+a)$. Then, as in the second proof of Theorem 1.2.4 for the scalar function $|y(x) - \bar{y}(x)|$ we arrive at the numbers ε, $\hat{\varepsilon}$ and l. However, in view of Theorem 5.2.2 there exists a number $\varepsilon < \varepsilon^* < \hat{\varepsilon}$ such that

$$\begin{aligned} 0 < |y(\hat{\varepsilon}) - \bar{y}(\hat{\varepsilon})| &= |(y(\hat{\varepsilon}) - \bar{y}(\hat{\varepsilon})) - (y(\varepsilon) - \bar{y}(\varepsilon))| \\ &\le (\hat{\varepsilon} - \varepsilon)|y'(\varepsilon^*) - \bar{y}'(\varepsilon^*)| \\ &\le (\hat{\varepsilon} - \varepsilon)|f(\varepsilon^*, y(\varepsilon^*)) - f(\varepsilon^*, \bar{y}(\varepsilon^*))| \\ &\le lL|y(\varepsilon^*) - \bar{y}(\varepsilon^*)|. \end{aligned}$$

Thus, we arrive at the same contradiction. ∎

5.3 MEDEIROS' UNIQUENESS THEOREM

An analog of Osgood's Uniqueness Theorem 1.4.2 is the following :

Theorem 5.3.1 (Medeiros' Uniqueness Theorem). Let in addition to the weak continuity of $f(x,y)$ in Theorem 5.1.1 for all (x,y), $(x,\bar{y}) \in (0,\delta(r)] \times Y$ the following inequality hold

$$2Re[(f(x,y) - f(x,\bar{y})) \cdot (y - \bar{y})] \le g(\|y - \bar{y}\|^2), \tag{5.3.1}$$

where the function $g(z)$ is continuous and nondecreasing in the interval $[0, \infty)$, $g(0) = 0$, $g(z) > 0$ for $z > 0$ and satisfies the condition (1.4.1). Then, the initial value problem (1.1.1) with $x_0 = 0$ has at most one abstract solution in $[0, \delta(r)]$.

Proof. Let $y(x)$ and $\bar{y}(x)$ be two abstract solutions of (1.1.1) with $x_0 = 0$ in $[0, \delta(r)]$. Then, it follows that

$$y'(x) - \bar{y}'(x) = f(x, y(x)) - f(x, \bar{y}(x)). \tag{5.3.2}$$

Taking the inner product of both sides of (5.3.2) with $(y(x)-\bar{y}(x))$ and taking the real parts, we find

$$\begin{aligned}\frac{d}{dx}\|y(x)-\bar{y}(x)\|^2 &= \frac{d}{dx}[(y(x)-\bar{y}(x))\cdot(y(x)-\bar{y}(x))] \\ &= 2Re[(f(x,y(x))-f(x,\bar{y}(x)))\cdot(y(x)-\bar{y}(x))].\end{aligned}$$

Thus, in view of (5.3.1) it follows that

$$\phi'(x) \le g(\phi(x)),\ 0 < x \le \delta(r) \tag{5.3.3}$$

where $\phi(x) = \|y(x)-\bar{y}(x)\|^2$. However, since $\phi(0) = 0$ an integration of (5.3.3) from 0 to $x \le \delta(r)$ gives

$$0 \le \phi(x) \le \int_0^x g(\phi(t))dt.$$

The proof is now immediate from Lemma 1.4.1. ∎

In Theorem 5.3.1 the condition that the function $g(z)$ is nondecreasing is in fact superfluous. For this, instead of appealing to Lemma 1.4.1 to the above inequality we can follow as in Theorem 1.4.4 to get the same conclusion that $\phi(x) \equiv 0$. We also note that if $g(z)$ satisfies the conditions of Lemma 1.4.1, then so is $2g(z)$. Therefore, the factor 2 in the left side of (5.3.1) is unnecessary, i.e, the condition (5.3.1) can be replaced by

$$Re[(f(x,y)-f(x,\bar{y}))\cdot(y-\bar{y})] \le g(\|y-\bar{y}\|^2). \tag{5.3.4}$$

It is clear that (5.3.4) includes in particular the Lipschitz condition

$$Re[(f(x,y)-f(x,\bar{y}))\cdot(y-\bar{y})] \le \|y-\bar{y}\|^2. \tag{5.3.5}$$

An analog of Theorem 5.3.1 for classical abstract solutions of (1.1.1) is the following :

Theorem 5.3.2. Let Y be a inner product space, $y_0 \in Y$ and let $f(x,y)$ be defined and satisfy the condition (5.3.4) in $(0,a)\times Y$, where $g(z)$ is the same as in Theorem 5.3.1. Then, the initial value problem (1.1.1) with $x_0 = 0$ has at most one classical abstract solution in $[0,a)$. ∎

5.4 DIAZ - WEINACHT'S UNIQUENESS THEOREM

A result similar to Nagumo's Uniqueness Theorem 1.6.2 in the setting of Theorem 5.1.1 takes the following form.

Theorem 5.4.1. Let in addition to the weak continuity of $f(x,y)$ in Theorem 5.1.1 for all (x,y), $(x,\bar{y}) \in (0,\delta(r)] \times Y$ the following inequality hold

$$Re[(f(x,y) - f(x,\bar{y})) \cdot (y - \bar{y})] \leq \frac{1}{2x}\|y - \bar{y}\|^2. \tag{5.4.1}$$

Then, the initial value problem (1.1.1) with $x_0 = 0$ has at most one abstract solution in $[0, \delta(r)]$.

Proof. Using the same arguments as in Theorem 5.3.1 the condition (5.4.1) leads to

$$\phi'(x) \leq \frac{1}{x}\phi(x), \ 0 < x \leq \delta(r) \tag{5.4.2}$$

and hence

$$0 \leq \phi(x) \leq \int_0^x \frac{\phi(t)}{t}dt.$$

The proof is now immediate from Lemma 1.6.1. ∎

In the proof of the above theorem the factor 1/2 in the right side of (5.4.1) is essential. However, from Example 1.6.1 it appears to be disturbing. In fact, in our next result we shall show that the condition (5.4.1) can be replaced by the best possible condition

$$Re[(f(x,y) - f(x,\bar{y})) \cdot (y - \bar{y})] \leq \frac{1}{x}\|y - \bar{y}\|^2. \tag{5.4.3}$$

Theorem 5.4.2 (Diaz - Weinacht's Uniqueness Theorem). Let Y be a inner product space, $y_0 \in Y$ and let $f(x,y)$ be defined in $[0,a) \times Y$, and satisfy the condition (5.4.3) in $(0,a) \times Y$. Then, the initial value problem (1.1.1) with $x_0 = 0$ has at most one restricted classical abstract solution in $[0,a)$.

Proof. As usual suppose that $y(x)$ and $\bar{y}(x)$ are two restricted classical abstract solutions of (1.1.1) in $[0,a)$. Then, the real valued nonnegative function

$\|y(x) - \bar{y}(x)\|$ is continuous in $[0,a)$, and vanishes at $x = 0$. We shall show that this function has a finite derivative in $(0,a)$ given by

$$(5.4.4)\quad \frac{d}{dx}\|y(x)-\bar{y}(x)\| = \begin{cases} \dfrac{Re[(y'(x)-\bar{y}'(x))\cdot(y(x)-\bar{y}(x))]}{\|y(x)-\bar{y}(x)\|}, & y(x)\neq\bar{y}(x) \\ 0, & y(x)=\bar{y}(x). \end{cases}$$

For $x \in (0,a)$ the function $\|y(x) - \bar{y}(x)\|^2$ is differentiable is clear. In fact, as in Theorem 5.3.1 we have

$$\frac{d}{dx}\|y(x)-\bar{y}(x)\|^2 = 2Re[(y'(x)-\bar{y}'(x))\cdot(y(x)-\bar{y}(x))].$$

The differentiability of $\|y(x)-\bar{y}(x)\|$ follows from the differentiability of $\|y(x)-\bar{y}(x)\|^2$, but there are two cases to consider. If $x \in (0,a)$ and $y(x) \neq \bar{y}(x)$, then we have

$$\begin{aligned}\frac{d}{dx}\|y(x)-\bar{y}(x)\| &= \frac{d}{dx}[\|y(x)-\bar{y}(x)\|^2]^{1/2} \\ &= \frac{1}{2}\frac{1}{[\|y(x)-\bar{y}(x)\|^2]^{1/2}}\frac{d}{dx}\|y(x)-\bar{y}(x)\|^2,\end{aligned}$$

where the positive square root is taken. If $x \in (0,a)$ and $y(x) = \bar{y}(x)$, then the differential equation for $y(x)$ and $\bar{y}(x)$ yields that

$$y'(x) - \bar{y}'(x) = f(x,y(x)) - f(x,\bar{y}(x)) = 0,$$

and thus

$$\begin{aligned}\frac{d}{dx}\|y(x)-\bar{y}(x)\| &= \lim_{h\to 0}\frac{\|y(x+h)-\bar{y}(x+h)\| - \|y(x)-\bar{y}(x)\|}{h} \\ &= \lim_{h\to 0}\frac{\|y(x+h)-\bar{y}(x+h)\|}{h} \\ &= \lim_{h\to 0}\frac{[(y(x+h)-\bar{y}(x+h))\cdot(y(x+h)-\bar{y}(x+h))]^{1/2}}{h} \\ &= \lim_{h\to 0}\frac{|h|}{h}\left[\frac{y(x+h)-\bar{y}(x+h)}{h}\cdot\frac{y(x+h)-\bar{y}(x+h)}{h}\right]^{1/2} \\ &= [(y'(x)-\bar{y}'(x))\cdot(y'(x)-\bar{y}'(x))]^{1/2} = 0.\end{aligned}$$

This completes the proof of (5.4.4). Thus, if $x \in (0,a)$ and $y(x) \neq \bar{y}(x)$ then in view of (5.4.3) it follows that

$$\begin{aligned}\frac{d}{dx}\|y(x)-\bar{y}(x)\| &= \frac{Re[(f(x,y(x))-f(x,\bar{y}(x)))\cdot(y(x)-\bar{y}(x))]}{\|y(x)-\bar{y}(x)\|}\\ &\le \frac{1}{x}\|y(x)-\bar{y}(x)\|.\end{aligned}$$

However, from (5.4.4) this inequality also holds if $x \in (0,a)$ and $y(x) = \bar{y}(x)$. Therefore, for all $x \in (0,a)$ we have

$$\frac{d}{dx}\|y(x)-\bar{y}(x)\| \le \frac{1}{x}\|y(x)-\bar{y}(x)\|.$$

The above inequality is the same as

$$\frac{d}{dx}\left(\frac{1}{x}\|y(x)-\bar{y}(x)\|\right) \le 0. \tag{5.4.5}$$

Further, since

$$\begin{aligned}\lim_{x\to0^+}\frac{y(x)-\bar{y}(x)}{x} &= \lim_{x\to0^+}\frac{y(x)-y(0)}{x}-\frac{\bar{y}(x)-\bar{y}(0)}{x}\\ &= y'(0)-\bar{y}'(0) = f(0,y(0))-f(0,\bar{y}(0)) = 0\end{aligned}$$

it follows that

$$\lim_{x\to0^+}\left\|\frac{y(x)-\bar{y}(x)}{x}\right\| = 0. \tag{5.4.6}$$

Thus, the function $v(x)$ defined by

$$v(x) = \begin{cases}\dfrac{\|y(x)-\bar{y}(x)\|}{x}, & 0 < x < a\\[2ex] 0, & x = 0\end{cases}$$

is continuous in $[0,a)$ and nonincreasing in $(0,a)$. Since $v(0) = 0$ if follows that $v(x) \equiv 0$, i.e., $y(x) = \bar{y}(x),\ x \in [0,a)$. ∎

A variant of Theorem 5.4.2 is the following :

Theorem 5.4.3. Let Y be a inner product space, $y_0 \in Y$ and let $f(x,y)$ be defined and satisfy the condition (5.4.3) in $(0,a) \times Y$. Then, the initial value problem (1.1.1) with $x_0 = 0$ has at most one classical abstract solution $y(x)$ in $[0, a)$ which has a prescribed right hand strong derivative at $x = 0$, i.e.,

$$\lim_{x \to 0^+} \frac{y(x) - y(0)}{x} = y_1, \tag{5.4.7}$$

where $y_1 \in Y$ is fixed.

Proof. The proof of the inequality (5.4.5) remains the same. Further, in view of (5.4.7) we have

$$\begin{aligned}\lim_{x \to 0^+} \left\| \frac{y(x) - \bar{y}(x)}{x} \right\| &= \lim_{x \to 0^+} \left\| \frac{y(x) - y(0)}{x} - \frac{\bar{y}(x) - \bar{y}(0)}{x} \right\| \\ &= \|y_1 - y_1\| = 0.\end{aligned}$$

Thus, (5.4.6) also holds. ∎

Finally, we shall state two results analogous to Theorems 5.4.2 and 5.4.3 in normed linear spaces. The proof of both of these results is similar to that of the second proof of Theorem 1.6.2 (also see remark following Theorem 2.3.1) and uses the Mean Value Theorem 5.2.2.

Theorem 5.4.4. Let Y be a normed linear space and let $f(x, y)$ be defined and satisfy the Nagumo condition (1.6.1) in $S_+ \subseteq R \times Y$. Further, let the function $f(x, y)$ possesses a finite limit as (x, y) approaches (x_0, y_0), i.e., there exists $y_1 \in Y$ such that $\lim_{(x,y) \to (x_0, y_0)} f(x, y) = y_1$. Then, the initial value problem (1.1.1) has at most one classical abstract solution in $[x_0, x_0 + a)$. ∎

Theorem 5.4.5. Let Y be a normed linear space and let $f(x, y)$ be defined and satisfy the Nagumo condition (1.6.1) in $S_+ \subseteq R \times Y$. Then, the initial value problem (1.1.1) has at most one classical abstract solution in $[x_0, x_0 + a)$ which has a prescribed right hand derivative at $x = x_0$, i.e.,

$$\lim_{x \to x_0^+} \frac{y(x) - y(x_0)}{x - x_0} = y_1,$$

where $y_1 \in Y$ is fixed. ∎

5.5 PERRON'S UNIQUENESS THEOREM

Theorem 1.11.4 for abstract solutions of (1.1.1) can be stated as follows :

Theorem 5.5.1 (Perron's Uniqueness Theorem). Let Y be a Banach space and let $f(x,y)$ be strongly continuous in $\bar{S}_+ \subseteq R \times Y$. Further, let in addition to the hypothesis (i) of Theorem 1.11.4 for all (x,y), $(x,\bar{y}) \in \bar{S}_+$ Perron's condition (1.11.2) be satisfied. Then, the initial value problem (1.1.1) has at most one abstract solution in $[x_0, x_0 + a]$. ∎

In view of the results discussed in Section 5.3 we also note that if Y is a real Hilbert space then Perron's condition (1.11.2) in Theorem 5.5.1 can be replaced by the weaker condition

$$((f(x,y) - f(x,\bar{y})) \cdot (y - \bar{y})) \leq g(x, \|y - \bar{y}\|)\|y - \bar{y}\|. \tag{5.5.1}$$

5.6 OLECH'S UNIQUENESS THEOREM

Kamke's Uniqueness Theorem 3.8.1 for abstract solutions of (1.1.1) can be stated as follows :

Theorem 5.6.1 (Olech's Uniqueness Theorem). Let Y be a Banach space and let $f(x,y)$ be strongly continuous in $\bar{S}_+ \subseteq R \times Y$. Further, let in addition to the hypothesis (i) of Theorem 3.8.1 for all (x,y), $(x,\bar{y}) \in \bar{S}_+$ where $x \neq x_0$ the inequality (1.11.2) be satisfied. Then, the initial value problem (1.1.1) has at most one abstract solution in $[x_0, x_0 + a]$. ∎

5.7 GOLDSTEIN'S UNIQUENESS THEOREM

Here we shall prove a result which generalizes Theorems 5.4.1 and 5.4.2.

For this, let Y be a real or complex Banach space and let Y^* be the dual of Y, i.e., the space of all linear functionals on Y. The image of $y \in Y$ under $\varphi \in Y^*$ will be denoted by (y, φ). For each $y \in Y$ let $J(y)$ be the nonempty set of all $\varphi \in Y^*$ for which $(y, \varphi) = |y|^2 = |\varphi|^2$. A duality map of Y is a function $j(y)$ which maps Y into Y^* such that $j(y) \in J(y)$ for each $y \in Y$.

Lemma 5.7.1. Suppose that $y(x)$ has a weak derivative $y'(t) \in Y$, at $x = t$ and $|y(x)|$ is differentiable at $x = t$. Then,

$$|y(t)| \frac{d}{dx}|y(x)|\Big|_{x=t} = Re(y'(t), \varphi)$$

for each $\varphi \in J(y(t))$.

Proof. Since

$$Re(y(x), \varphi) \leq |y(x)|\ |\varphi| = |y(x)|\ |y(t)|$$

and

$$Re(y(t), \varphi) = |y(t)|^2,$$

we have

$$Re(y(x) - y(t), \varphi) \leq |y(t)|(|y(x)| - |y(t)|).$$

Dividing both sides by $x - t$ and letting $x \to t$ from above and from below, we obtain

$$Re(y'(t), \varphi) \lesseqgtr |y(t)| \frac{d}{dx}|y(x)|\Big|_{x=t}. \qquad \blacksquare$$

Theorem 5.7.2. (Goldstein's Uniqueness Theorem). Let Y be a real or complex Banach space, $y_0 \in Y$ and let $f(x, y)$ be defined in $[0, a) \times Y$. Further, let there is a duality map j of Y such that for all (x, y), $(x, \bar{y}) \in (0, a) \times Y$

$$Re(f(x, y) - f(x, \bar{y}), j(y - \bar{y})) \leq \frac{1}{x}\|y - \bar{y}\|^2. \tag{5.7.1}$$

Then, for the initial value problem (1.1.1) with $x_0 = 0$ there is at most one extended abstract solution $y(x)$ in $[0, a)$ satisfying $\lim_{x \to 0^+} \frac{y(x) - y(0)}{x} = y_1$, where $y_1 \in Y$ is prescribed.

Proof. Suppose that $y(x)$ and $\bar{y}(x)$ are two extended abstract solutions of (1.1.1) in $[0, a)$. Then, from Lemma 5.7.1 it follows that

$$\begin{aligned} |y(x)-\bar{y}(x)|\frac{d}{dx}|y(x)-\bar{y}(x)| &= Re(f(x,y(x))-f(x,\bar{y}(x)), j(y(x)-\bar{y}(x))) \\ &\leq \frac{1}{x}|y(x)-\bar{y}(x)|^2. \end{aligned}$$

We claim that for almost all $x \in [0, a)$ the following inequality holds

$$\frac{d}{dx}|y(x)-\bar{y}(x)| \leq \frac{1}{x}|y(x)-\bar{y}(x)|. \tag{5.7.2}$$

Indeed, it is immediate for almost all $x \in [0, a)$ for which $y(x) \neq \bar{y}(x)$, and if $y(x) = \bar{y}(x)$ then, in view of $y'(x) = \bar{y}'(x)$ it suffices to note that

$$\begin{aligned} \frac{d}{dx}|y(x)-\bar{y}(x)| &= \lim_{h\to 0}\frac{|y(x+h)-\bar{y}(x+h)|}{h} \\ &= \lim_{h\to 0}\left|\frac{y(x+h)-y(x)}{h} - \frac{\bar{y}(x+h)-\bar{y}(x)}{h}\right| = 0. \end{aligned}$$

As earlier, the inequality (5.7.2) for almost all $x \in [0, a)$ is the same as

$$\frac{d}{dx}\left(\frac{|y(x)-\bar{y}(x)|}{x}\right) \leq 0$$

and hence the function $|y(x)-\bar{y}(x)|/x$ is monotone nonincreasing in $[0, a)$. Further, since as in Theorem 5.4.3, $\lim_{x\to 0^+}|y(x)-\bar{y}(x)|/x = y_1 - y_1 = 0$ it follows that $y(x) = \bar{y}(x)$ for all $x \in [0, a)$. ∎

If Y is a real or complex Hilbert space then $(\cdot,\cdot)$ is the same as the inner product in Y and j is the identity operator. Thus, Theorem 5.7.2 generalizes Theorems 5.4.1 and 5.4.2.

5.8 ARRATE'S UNIQUENESS THEOREM

Here we shall prove an extension of Gard's Uniqueness Theorem 3.11.1 for classical abstract solutions of (1.1.1). For this, in addition to the notation used in Section 5.7, for an operator P mapping from Y into Y we will write its domain as $D(P) \subset Y$.

Definition 5.8.1. The operator P is said to be **dissipative** if for $y, \bar{y} \in D(P)$ there exists $\varphi \in J(y - \bar{y})$ such that

$$Re(P(y) - P(\bar{y}), \varphi) \leq 0.$$

If Y is a Hilbert space, then P is dissipative provided

$$Re(P(y) - P(\bar{y}), y - \bar{y}) \leq 0, \quad \text{for all} \quad y,\ \bar{y} \in D(P).$$

Theorem 5.8.1 (Arrate's Uniqueness Theorem). Let $\phi(x)$ be a continuous function defined in $[0, a)$, and differentiable in $(0, a)$ such that $\phi(0) = 0,\ \phi(x) > 0$ for $x > 0$. Further, let the function $f(x, y)$ be defined in $(0, a) \times Y$ and

(i) $f(x, \cdot) - \phi'(x)I/\phi(x)$ is dissipative for all $x \in (0, a)$; and

(ii) for any two classical abstract solutions $y(x)$ and $\bar{y}(x)$ of (1.1.1) with $x_0 = 0$ in $(0, a)$

$$|f(x, y(x)) - f(x, \bar{y}(x))| = o(\phi'(x)) \quad \text{as} \quad x \to 0^+.$$

Then, the initial value problem (1.1.1) has at most one classical abstract solution in $[0, a)$.

Proof. We note that for each pair $y,\ \bar{y} \in D(f(x, \cdot))$ the hypothesis (i) is equivalent to the existence of a $j \in J(y - \bar{y})$ such that

$$Re(f(x, y) - f(x, \bar{y}), j) \leq \frac{\phi'(x)}{\phi(x)} |y - \bar{y}|^2.$$

Let $y(x)$ and $\bar{y}(x)$ be two classical abstract solutions of (1.1.1). Then, as in Theorem 3.11.1 we define a function

$$v(x) = \begin{cases} \frac{1}{2}[|y(x) - \bar{y}(x)|/\phi(x)]^2, & x \in (0, a) \\ 0, & x = 0. \end{cases}$$

For each $\varphi \in Y^*$, l'Hospital's rule and the hypothesis (ii) yield that

$$\begin{aligned}\lim_{x\to 0^+} \varphi\left[\frac{y(x)-\bar{y}(x)}{\phi(x)}\right] &= \lim_{x\to 0^+} \frac{[\varphi(y(x)-\bar{y}(x))]'}{\phi'(x)} \\ &= \lim_{x\to 0^+} \varphi\left[\frac{f(x,y(x))-f(x,\bar{y}(x))}{\phi'(x)}\right] = 0\end{aligned}$$

and therefore $(y(x)-\bar{y}(x))/\phi(x) \to 0$ weakly as $x \to 0^+$.

As a consequence of the Banach - Steinhaus theorem [7], each weakly convergent sequence is bounded, and hence for any sequence $\{x_n\}$ with $x_n \to 0^+$ there exists $M > 0$ such that

$$\left|\frac{y(x_n)-\bar{y}(x_n)}{\phi(x_n)}\right| \le M, \quad \text{for} \quad n = 1, 2, \cdots. \tag{5.8.1}$$

In view of Lemma 5.7.1 for each $x \in (0,a)$ and each $\varphi \in J(y(x)-\bar{y}(x))$

$$|y(x)-\bar{y}(x)|\frac{d}{dx}|y(x)-\bar{y}(x)| = Re(y'(x)-\bar{y}'(x), \varphi).$$

Thus, on using l'Hospital's rule again, we obtain

$$\begin{aligned}\lim_{x\to 0^+} v(x) &= \frac{1}{2}\lim_{x\to 0^+} \frac{|y(x)-\bar{y}(x)|^2}{\phi^2(x)} \\ &= \frac{1}{2}\lim_{x\to 0^+} \frac{|y(x)-\bar{y}(x)|\frac{d}{dx}|y(x)-\bar{y}(x)|}{\phi(x)\phi'(x)} \\ &= \frac{1}{2}\lim_{x\to 0^+} Re\left(\frac{y'(x)-\bar{y}'(x)}{\phi'(x)}, \frac{\varphi}{\phi(x)}\right) \\ &= \frac{1}{2}\lim_{x\to 0^+} \left|\frac{y'(x)-\bar{y}'(x)}{\phi'(x)}\right|\left|\frac{y(x)-\bar{y}(x)}{\phi(x)}\right|.\end{aligned}$$

In the last term the first factor tends to zero by (ii) and upon comparing it with (5.8.1) we see that $\lim_{x\to 0^+} v(x) = 0$ and so $v(x)$ is continuous in $[0,a)$. Hence, for $x > 0$ the mean value theorem implies that there is a $t \in (0,x)$ such that

$$\begin{aligned}0 < v(x) &= xv'(t) \\ &= \frac{x}{\phi(t)}\left(Re(f(t,y(t)) - f(t,\bar{y}(t)), \varphi) - \frac{\phi'(t)}{\phi(t)}|y(t)-\bar{y}(t)|^2\right) \\ &\le 0.\end{aligned}$$

This is a contradiction and therefore $y(x) = \bar{y}(x),\ x \in [0, a)$. ∎

Corollary 5.8.2. In Theorem 5.8.1 the hypothesis (ii) can be replaced by

(ii)′ $f(x, y) = h(x) + o(\phi'(x))$ as $(x, y) \to (0^+, y_0)$. ∎

Corollary 5.8.3. Assume that for some positive integer n, $f(x, \cdot) - nI/x$ is dissipative for each $x \in (0, a)$. Then, given $y_i \in Y,\ 0 \leq i \leq n$ there is at most one classical abstract solution $y(x)$ of (1.1.1) in $[0, a)$ such that $y^{(i)}(0)$ exists and $y^{(i)}(0) = y_i,\ 0 \leq i \leq n$.

Proof. Let $\phi(x) = x^n$. Then, for the classical solutions $y(x)$ and $\bar{y}(x)$ of (1.1.1) Taylor's expansion reveals that

$$y'(x) = \sum_{k=1}^{n} \frac{x^{k-1} y^{(k)}(0)}{(k-1)!} + o(x^{n-1}),$$

and

$$\bar{y}'(x) = \sum_{k=1}^{n} \frac{x^{k-1} \bar{y}^{(k)}(0)}{(k-1)!} + o(x^{n-1}).$$

Therefore, in view of $y^{(k)}(0) = \bar{y}^{(k)}(0),\ 0 \leq k \leq n$ it follows that

$$|f(x, y(x)) - f(x, \bar{y}(x))| = o(x^{n-1}).$$

Now the Theorem 5.8.1 is applicable. ∎

5.9 RAMANKUTTY'S UNIQUENESS THEOREM

Here we shall prove an extension of Kamke's Uniqueness Theorem 3.8.1 for the limiting initial value problem

$$y' = f(x, y), \quad \lim_{x \to 0^-} y(x) = 0. \tag{5.9.1}$$

For this, let Y be a normed linear space. In addition to the notation used in Section 5.7, let $<,>$ be the semi - inner product in Y defined from $Y \times Y$ into R by $< y, \bar{y} > = \inf\{Re(y, j) : j \in J(\bar{y})\}$. A variant of Lemma 5.7.1 is the following :

Lemma 5.9.1. Suppose that for each $x \in [c,d]$, $y(x) \in Y$. Further, suppose that at $x \in (c,d)$, $y(x)$ has a weak derivative $y'(x) \in Y$. Then,

$$|y(x)|D^-|y(x)| \leq < y'(x), y(x) > .$$

Proof. Let $j \in J(y(x))$ and $t \in (c,x)$. Then,

$$(y(t),j) \leq |(y(t),j)| \leq |y(t)||j| = |y(t)||y(x)|$$

and

$$(y(x),j) = |y(x)|^2$$

so that

$$Re(y(t)-y(x),j) \leq |y(x)|(|y(t)|-|y(x)|).$$

Thus, it follows that

$$\limsup_{t\to x^-} Re\left(\frac{y(t)-y(x)}{t-x}, j\right) \geq |y(x)| \limsup_{t\to x^-} \frac{|y(t)|-|y(x)|}{t-x}$$

and hence

$$Re(y'(x),j) \geq |y(x)|D^-|y(x)|.$$

Since this is true for all $j \in J(y(x))$, the lemma is proved. ∎

Theorem 5.9.2 (Ramankutty's Uniqueness Theorem). Suppose that Y is a normed linear space and $f(x,y)$ is defined for $0 \leq x \leq a$, $y \in Y$. Further, suppose that

(i) $\mu(x) \geq 0$ is defined and nondecreasing for $(0,a]$ and $\mu(x) \to 0$ as $x \to 0^-$,

(ii) $g(x,z)$ is nonnegative and continuous for $0 < x \leq a$, $0 \leq z < \infty$, $g(x,0) = 0$ for all $x \in (0,a]$, and for every x_1, $0 < x_1 < a$, $z(x) \equiv 0$ is the only solution of (1.13.1) in $(0,x_1)$ satisfying $z(x) = o(\mu(x))$ as $x \to 0$; and

(iii) for each strongly continuous function $y = y(x) \in Y$, $0 \leq x \leq a$ with $y(0) = 0$, $f(x,y(x))$ as a function of x is strongly continuous in $[0,a]$, for all (x,y), $(x,\bar{y})$, $0 < x \leq a$, $y, \bar{y} \in Y$

(5.9.2) $$< f(x,y) - f(x,\bar{y}), y - \bar{y} > \leq \; g(x, |y - \bar{y}|)|y - \bar{y}|,$$

and there exists some $u \in Y$ such that

(5.9.3) $$f(x,y) - u = o(\mu'(x)) \quad \text{as} \quad (x,y) \to (0,0).$$

Then, the initial value problem (5.9.1) has at most one abstract solution in $(0,a]$.

Proof. Let $y(x)$ and $\bar{y}(x)$ be two solutions of (5.9.1) in $(0,a]$ and let $\phi(x) = |y(x) - \bar{y}(x)|$. Since $f(x,y(x))$ is continuous and $y(x) \to 0$ as $x \to 0^-$ the representation $y(x) = \int_0^x f(t,y(t))dt$ holds. A similar representation holds for the solution $\bar{y}(x)$. Thus, it follows that

(5.9.4) $$\phi(x) \leq \int_0^x |f(t,y(t)) - f(t,\bar{y}(t))|dt.$$

From the hypothesis (5.9.3) it is clear that for any $\varepsilon > 0$ there exists a $\delta > 0$ such that $|f(x,y(t)) - u| \leq \frac{\varepsilon}{2}\mu'(t)$ and $|f(x,\bar{y}(t)) - u| \leq \frac{\varepsilon}{2}\mu'(t)$ for almost all $t \in (0,\delta]$. Therefore, $|f(t,y(t)) - f(t,\bar{y}(t))| \leq \varepsilon\mu'(t)$ for almost all $t \in (0,\delta]$. Hence, from (5.9.4) we can conclude that $0 \leq \phi(x) \leq \int_0^x \varepsilon\mu(t)dt \leq \varepsilon\mu(x)$ for all $t \in (0,\delta]$. Thus, $\phi(x) = o(\mu(x))$ as $x \to 0$.

Now, applying Lemma 5.9.1 to the function $y(x) - \bar{y}(x)$ we obtain

$$\begin{aligned}
\phi(x)D^-\phi(x) &= < y'(x) - \bar{y}'(x), y(x) - \bar{y}(x) > \\
&= < f(x,y(x)) - f(x,\bar{y}(x)), y(x) - \bar{y}(x) > \\
&\leq \phi(x)g(x,\phi(x)).
\end{aligned}$$

Since $g(x,z) \geq 0$, this gives $D^-\phi(x) \leq g(x,\phi(x))$ at each $x \in (0,a]$ where $\phi(x) \neq 0$. If $\phi(x) = 0$ for some x, then at that point $g(x,\phi(x)) = g(x,0) = 0$, and

$$D^-\phi(x) = \limsup_{h \to 0^-} \frac{\phi(x+h) - \phi(x)}{h} = \limsup_{h \to 0^-} \frac{\phi(x+h)}{h} \leq 0.$$

Thus, $D^-\phi(x) \leq \phi(x,\phi(x))$ holds at all points $x \in (0,a]$.

Let $\sigma \in (0,a)$ be such that $\phi(\sigma) > 0$, then as in Lemma 1.15.2 it follows that $0 \leq \rho(x) \leq \phi(x)$ for all $x \in (0,\sigma]$. Since $\phi(x) = o(\mu(x))$ as $x \to 0$, we also have $\rho(x) = o(\mu(x))$ as $x \to 0$. But, then from the hypotheses $\rho(x) = 0$

for all $x \in (0, \sigma]$, and hence $\phi(\sigma) = \rho(\sigma) = 0$. This contradiction implies that $\phi(x) \equiv 0$, i.e., $y(x) = \bar{y}(x)$, $x \in (0, a]$. ∎

If the function $f(x, y)$ is continuous at $(0, 0)$ then we can choose $\mu(x) = x$. Then, the condition (5.9.3) is automatically satisfied by $u = f(0, 0)$.

5.10 MURAKAMI'S UNIQUENESS THEOREM

Let Y be a normed linear space, $D(x)$, $x \in [x_0, x_0 + a]$ a subset of Y, and $D = \{(x, y) : x \in [x_0, x_0 + a],\ y \in D(x)\}$. For classical abstract solutions of (1.1.1) a result analogous to Okamura's Uniqueness Theorem 3.15.1 is the following :

Theorem 5.10.1 (Murakami's Uniqueness Theorem). Suppose that $y_0 \in \overline{D(x_0)}$ and $f(x, y)$ is defined in the set D. Further, suppose that there exists a Lyapunov function $V(x, y, \bar{y}) \geq 0$ defined and continuous in $Q = \{(x, y, \bar{y}) : (x, y) \in D,\ (x, \bar{y}) \in D\}$ such that $V(x, y, \bar{y}) = 0$ if and only if $y = \bar{y}$, and in Q it has continuous (bounded additive Frèchet) derivatives and

$$\frac{\partial V}{\partial x} + \frac{\partial V}{\partial y} f(x, y) + \frac{\partial V}{\partial \bar{y}} f(x, \bar{y}) \leq 0.$$

Then, the initial value problem (1.1.1) has at most one classical abstract solution in $[x_0, x_0 + a]$.

Proof. Suppose that $y(x)$ and $\bar{y}(x)$ are two classical abstract solutions of (1.1.1) in $[x_0, x_0 + a]$. Then, for $x > x_0$ it follows that

$$\frac{dV(x, y(x), \bar{y}(x))}{dx} = \frac{\partial V}{\partial x} + \frac{\partial V}{\partial y} f(x, y(x)) + \frac{\partial V}{\partial \bar{y}} f(x, \bar{y}(x)) \leq 0.$$

Hence, it follows that

$$V(x, y(x), \bar{y}(x)) \leq V(x_0, y(x_0), \bar{y}(x_0)).$$

Since $V(x_0, y(x_0), \bar{y}(x_0)) = 0$, we find that $V(x, y(x), \bar{y}(x)) \equiv 0$, and hence $y(x) = \bar{y}(x)$, $x \in [x_0, x_0 + a]$. ∎

The importance of Theorem 5.10.1 is illustrated in the following :

Example 5.10.1. It is clear that for the scalar differential equation

$$y' = f(x,y) = \begin{cases} 1+\sqrt{y}, & y \geq 0 \\ 1, & y < 0 \end{cases}$$

condition (5.3.5) is not satisfied in R. However, the Lyapunov function

$$V(x,y,\bar{y}) = \begin{cases} (\sqrt{y}-\sqrt{\bar{y}}-\ln(1+\sqrt{y})+\ln(1+\sqrt{\bar{y}}))^2, & y \geq 0,\ \bar{y} \geq 0 \\ (\sqrt{y}-\ln(1+\sqrt{y})-\frac{1}{2}\bar{y})^2, & y \geq 0,\ \bar{y} < 0 \\ (\frac{1}{2}y-\sqrt{\bar{y}}+\ln(1+\sqrt{\bar{y}}))^2, & y < 0,\ \bar{y} \geq 0 \\ \frac{1}{4}(y-\bar{y})^2, & y < 0,\ \bar{y} < 0. \end{cases}$$

satisfies the conditions of Theorem 5.10.1.

Example 5.10.2. Consider the parabolic equation

$$y_x = y_{ss} + F(x,s,y) \tag{5.10.1}$$

in the region bounded by $x = x_0$, $x = x_0 + a$, $s = \lambda_1(x)$, and $s = \lambda_2(x)$, where $\lambda_1(x)$ and $\lambda_2(x)$ are differentiable in $[x_0, x_0+a]$ and $\lambda_1(x) < \lambda_2(x)$ for $x_0 < x \leq x_0 + a$. The initial and boundary conditions are : $y = g(s)$ on $x = x_0$, $y = h_1(x)$ $y = h_2(x)$ on $s = \lambda_1(x)$, $s = \lambda_2(x)$, respectively, where g, h_1, and h_2 are continuous and $g(\lambda_1(x_0)) = h_1(x_0)$, $g_2(\lambda_2(x_0)) = h_2(x_0)$. We shall show that if

$$F(x,s,\bar{y}) - F(x,s,y) \leq L(\bar{y}-y) \quad \text{for} \quad \bar{y} > y$$

then (5.10.1) has at most one solution. To prove this, let $Y = L_2(R)$. For each $x \in [x_0, x_0+a]$ define $D(x)$ to be the space of functions $y = y(s) \in Y$ which are continuous in $[\lambda_1(x), \lambda_2(x)]$, and twice differentiable in $(\lambda_1(x), \lambda_2(x))$, vanish outside $[\lambda_1(x), \lambda_2(x)]$, and take values $h_1(x)$ and $h_2(x)$ at $s = \lambda_1(x)$ and $s = \lambda_2(x)$, respectively. For $(x,y) \in [x_0, x_0+a] \times D(x)$ we define

$$f(x,y) = \frac{d^2y(s)}{ds^2} + F(x,s,y).$$

Then, (5.10.1) together with the initial and boundary conditions is equivalent to the initial value problem (1.1.1) with $y(x_0) = g$. For this problem consider the Lyapunov function

$$V(x,y,\bar{y}) = e^{-2Lx}\int_{\lambda_1(x)}^{\lambda_2(x)} |y(s)-\bar{y}(s)|^2 ds$$

for which we have

$$\frac{\partial V}{\partial x} + \frac{\partial V}{\partial y}f(x,y) + \frac{\partial V}{\partial \bar{y}}f(x,\bar{y})$$

$$= -2Le^{-2Lx}\int_{\lambda_1(x)}^{\lambda_2(x)} |y(s)-\bar{y}(s)|^2 ds$$

$$+e^{-2Lx}\left[|y(\lambda_2(x)) - \bar{y}(\lambda_2(x))|^2\,\lambda_2'(x) - |y(\lambda_1(x)) - \bar{y}(\lambda_1(x))|^2\lambda_1'(x)\right]$$

$$+e^{-2Lx}\int_{\lambda_1(x)}^{\lambda_2(x)} 2[y(s)-\bar{y}(s)]f(x,y)ds$$

$$-e^{-2Lx}\int_{\lambda_1(x)}^{\lambda_2(x)} 2[y(s)-\bar{y}(s)]f(x,\bar{y})ds$$

$$= J_1 + J_2 + J_3 + J_4.$$

Since $y(\lambda_i(x)) = \bar{y}(\lambda_i(x)) = h_i(x)$ for $i = 1,2$ it follows that $J_2 = 0$. Further, from the definition of $f(x,y)$ and an integration by parts, we have

$$J_3 + J_4$$

$$= 2e^{-2Lx}\int_{\lambda_1(x)}^{\lambda_2(x)} [y_{ss} + F(x,s,y) - \bar{y}_{ss} - F(x,s,\bar{y})]\,(y(s)-\bar{y}(s))ds$$

$$= 2e^{-2Lx}\int_{\lambda_1(x)}^{\lambda_2(x)} (f(x,y)-f(x,\bar{y}))(y(s)-\bar{y}(s))ds$$

$$-2e^{-2Lx}\int_{\lambda_1(x)}^{\lambda_2(x)} \left(\frac{d(y(s)-\bar{y}(s))}{ds}\right)^2 ds$$

$$\leq 2e^{-2Lx}\int_{\lambda_1(x)}^{\lambda_2(x)} (f(x,y)-f(x,\bar{y}))(y(s)-\bar{y}(s))ds.$$

Thus, it follows that

$$J_1 + J_2 + J_3 + J_4 \leq 0. \qquad \blacksquare$$

From Theorem 5.10.1 it is clear that for abstract solutions of (1.1.1) several other uniqueness criteria involving Lyapunov functions and similar to those discussed in Chapter 3 can be stated and proved rather easily.

5.11 LAKSHMIKANTHAM'S UNIQUENESS THEOREM

Suppose that Y is a Banach space and $\{A(x)\}$ is a one - parameter family of closed linear operators defined for each $x \in [0, \infty)$. We shall assume that the domain $D[A(x)] = D$ of $A(x)$ is independent of x. Assume also that, for each $x \in [0, \infty)$, the resolvent set of $A(x)$ includes all sufficiently small positive real numbers α and that the domain of the resolvent $R(\alpha, A(x)) = [I - \alpha A(x)]^{-1}$ is dense in Y. We observe that, since $A(x)$ is closed, $D[R(\alpha, A(x))] = Y$, where α is in the resolvent set.

Here we shall prove the uniqueness of abstract solutions of the problem involving the nonlinear evolution equation and the initial condition

$$y' = A(x)y + f(x, y), \quad y(0) = 0 \tag{5.11.1}$$

where $f(x, y)$ is strongly continuous in $[0, \infty) \times Y$.

Theorem 5.11.1 (Lakshmikantham's Uniqueness Theorem). Suppose that

(i) any two abstract solutions of (5.11.1) satisfy (2.8.1), where $B(x)$ is continuous, positive in $(0, \infty)$ and $B(0^+) = 0$,

(ii) the function $g(x, z)$ is continuous and nonnegative for $0 \le x < \infty$, $0 \le z < \infty$ and the only solution $z(x)$ of (1.13.1) in $0 \le x < \infty$ such that $\lim_{x \to 0^+} \frac{z(x)}{B(x)} = 0$ is the trivial solution,

(iii) for each $x \in [0, \infty)$ and every $y \in Y$, $\lim_{h \to 0^+} R(h, A(x))y = y$; and

(iv) for each $x \in (0, \infty)$, each y, $\bar{y} \in Y$, and for all sufficiently small $h > 0$, depending on x and y

$$(5.11.2)\qquad |R(h,A(x))y - R(h,A(x))\bar{y} + h[f(x,y) - f(x,\bar{y})]|$$
$$\leq |y-\bar{y}| + hg(x, |y-\bar{y}|).$$

Then, the initial value problem (5.11.1) has at most one abstract solution in $[0,\infty)$.

Proof. Suppose that there are two abstract solutions of (5.11.1) in $[0,\infty)$. Let $\phi(x) = |y(x) - \bar{y}(x)|$. For small $h > 0$, we have

$$(5.11.3)\qquad \phi(x+h)$$
$$\leq |(y(x+h) - \bar{y}(x+h)) - (R(h,A(x))y(x) - R(h,A(x))\bar{y}(x))$$
$$-h[f(x,y(x)) - f(x,\bar{y}(x))]|$$
$$+|(R(h,A(x))y(x) - R(h,A(x))\bar{y}(x)) + h[f(x,y(x)) - f(x,\bar{y}(x))]|.$$

Since, for each $x \in [0,\infty)$, $A(x)$ is closed and $D[R(h,A(x))]$ is dense in Y, it follows that $D[R(h,A(x))] = Y$ and $R(h,A(x))(I - hA(x))y = y$ for every $y \in D$. Hence,

$$(5.11.4)\qquad (R(h,A(x))y(x) - R(h,A(x))\bar{y}(x))$$
$$= (y(x) - \bar{y}(x)) + hA(x)(y(x) - \bar{y}(x))$$
$$+h[(R(h,A(x))A(x)y(x) - R(x,A(x))A(x)\bar{y}(x)) - A(x)(y(x) - \bar{y}(x))].$$

Combining (5.11.2) - (5.11.4) and using the hypotheses, we obtain

$$\limsup_{h\to 0^+} \frac{\phi(x+h) - \phi(x)}{h} \leq g(x, \phi(x)).$$

Rest of the proof is similar to that of Theorem 2.8.3. ∎

CHAPTER 6

COMPLEX DIFFERENTIAL EQUATIONS

6.1 INTRODUCTION

Let C^n be the complex Euclidean n space. For $y \in C^n$ we shall denote the norm $\|y\| = \sum_{i=1}^{n} |y_i|$. The initial value problem involving complex differential system and the initial condition can be written as

$$\frac{dy}{dz} = f(z, y), \; y(0) = y_0 \tag{6.1.1}$$

where $y_0 \in C^n$ and $f(z,y)$ is an analytic complex valued vector function defined in the domain

$$S_c = \{(z, y) \in C^{n+1} : 0 \leq |z| < a, \; \|y - y_0\| < b\}.$$

A complex valued vector function $y(z)$ is said to be a **solution** of (6.1.1) in $0 \leq |z| < a$ if the following conditions are satisfied :

(i) $y(0) = y_0$,

(ii) for all z such that $0 \leq |z| < a$, the points $(z, y(z)) \in S_c$,

(iii) $y(z)$ is regular - analytic in $0 \leq |z| < a$; and

(iv) $\frac{dy}{dz} = f(z, y(z))$.

In the present chapter we shall provide only one criterion for the uniqueness of solutions of (6.1.1). This result is very general and can be used as a model to obtain analogs of several results obtained earlier .

6.2 DEO - LAKSHMIKANTHAM'S UNIQUENESS THEOREM

An analog of Theorem 2.8.3 for the initial value problem (6.1.1) is the following :

Theorem 6.2.1 (Deo - Lakshmikantham's Uniqueness Theorem). Suppose that $\arg z =$ constant, and $|z| = x$, and

(i) any two solutions $y_1(z)$ and $y_2(z)$ of (6.1.1) satisfy

$$\lim_{x \to 0^+} \frac{\|y_1(z) - y_2(z)\|}{B(x)} = 0, \tag{6.2.1}$$

where $B(x)$ is continuous, positive in $(0, a)$ and $B(0^+) = 0$,

(ii) the function $g(x, u)$ is continuous and nonnegative for $0 < x < a$, $u \geq 0$ and $g(x, 0) \equiv 0$, and the only solution $u(x)$ of $u' = g(x, u)$ such that $\lim_{x \to 0^+} \frac{u(x)}{B(x)} = 0$ is the trivial solution; and

(iii) the function $f(z, y)$ is an analytic complex valued vector function defined in the domain S_c and for all (z, y_1), (z, y_2) in S_c where $z \neq 0$ the inequality

$$\|f(z, y_1) - f(z, y_2)\| \leq g(|z|, \|y_1 - y_2\|) \tag{6.2.2}$$

is satisfied.

Then, the initial value problem (6.1.1) has at most one solution in $0 \leq |z| < a$.

Proof. Let $y_1(z)$ and $y_2(z)$ be two solutions of (6.1.1) in $0 \le |z| < a$. Further, let $\arg z$ = constant, and $z = xe^{i\theta}$. We define, $\phi(x) = \|y_1(z) - y_2(z)\| = \left\|y_1(xe^{i\theta}) - y_2(xe^{i\theta})\right\|$. For small $h > 0$, we obtain

$$\phi(x+h) - \phi(x) \le \left\|y_1((x+h)e^{i\theta}) - y_1(xe^{i\theta}) - y_2((x+h)e^{i\theta}) + y_2(xe^{i\theta})\right\|,$$

also, we have

$$\left\|\frac{dy_1(xe^{i\theta})}{dx} - \frac{dy_2(xe^{i\theta})}{dx}\right\| = \left\|\frac{dy_1(z)e^{i\theta}}{dx} - \frac{dy_2(z)e^{i\theta}}{dx}\right\| = \left\|\frac{dy_1(z)}{dx} - \frac{dy_2(z)}{dx}\right\|$$

and hence, it follows that

$$\phi'_+(x) \le \left\|\frac{dy_1(z)}{dx} - \frac{dy_2(z)}{dx}\right\|,$$

where as earlier $\phi'_+(x)$ is the right derivative of $\phi(x)$ with respect to x. Therefore, in view of condition (6.2.2) we find that

$$\begin{aligned}\phi'_+(x) &\le \|f(z, y_1(z)) - f(z, y_2(z))\| \\ &\le g(|z|, \|y_1(z) - y_2(z)\|) \\ &= g(x, \phi(x)), \ 0 < x < a.\end{aligned}$$

The rest of the proof is similar to that of Theorem 2.8.3. ∎

Example 6.2.1. Consider the initial value problem

$$(6.2.3) \qquad \frac{dy}{dz} = \begin{cases} \dfrac{2y}{z}, & 0 \le |z| < a \\ 0, & z = 0 \end{cases}$$

$$y(0) = 0.$$

This problem has a unique solution $y(z) \equiv 0$ in the sense defined above, whereas, for each complex c, $y(z) = cz^2$ represents a classical solution.

In the latter case the conditions of Theorem 6.2.1 are not satisfied. For this, we note that for the function $B(x) = x^{2-\epsilon}$, where $\varepsilon > 0$ condition (6.2.1) is satisfied. However, the differential equation $u' = g(x, u)$ takes the form $u' = 2u/x$ for which, for each nonnegative c_1, $u(x) = c_1x^2$ is the nontrivial solution which satisfies the requirement that $\lim_{x\to 0+} \frac{u(x)}{B(x)} = 0$. ∎

CHAPTER 7

FUNCTIONAL DIFFERENTIAL EQUATIONS

7.1 INTRODUCTION

For a given number $\tau > 0$, let $\hat{C_n} = C[[-\tau, 0], R^n]$, $(\hat{C_1} = \hat{C})$ denote the space of continuous functions with domain $[-\tau, 0]$ and range in R^n, $(R^1 = R)$. For any element $\varphi \in \hat{C_n}$, $(\hat{C})$ we define the norm $\|\varphi\|_0 = \max_{-\tau \le t \le 0} |\varphi(t)|$, $(|\varphi|_0)$. For any continuous function $y(s)$ defined for $-\tau \le s < \eta$ $(\eta > 0)$ and for a fixed x, $0 \le x < \eta$, the symbol y_x denotes a translation of the restriction of $y(s)$ to the interval $[x - \tau, x]$, i.e., y_x is an element of $\hat{C_n}$ defined by $y_x(t) = y(x+t)$, $-\tau \le t \le 0$. In other words, the graph of y_x is the graph of $y(s)$ in $[x - \tau, x]$ shifted to the interval $[-\tau, 0]$. Let $\mu > 0$ be a given constant, and let $C_\mu = \{\varphi \in \hat{C_n} : \|\varphi\|_0 < \mu\}$. With this notation we can write a n - dimensional functional differential system in the form

$$y'(x) = f(x, y_x). \tag{7.1.1}$$

Let $x_0 \ge 0$ be fixed and let $\varphi_0 \in C_\mu$ at $x = x_0$ be the initial function. A function $y(x, x_0, \varphi_0)$ is said to be a **solution** of (7.1.1) in $[x_0, x_0 + a]$ provided

(i) $y_{x_0}(x_0, \varphi_0) = \varphi_0$,

(ii) $y(x, x_0, \varphi_0)$ is defined and continuous in $[x_0 - \tau, x_0 + a]$,

(iii) $y_x(x_0, \varphi_0) \in C_\mu$ for $x_0 \le x \le x_0 + a$,

(iv) for all $x \in [x_0, x_0 + a)$, $y'(x, x_0, \varphi_0)$ exists; and

(v) $y'(x, x_0, \varphi_0) = f(x, y_x(x_0, \varphi_0))$.

We shall also consider the n - dimensional functional differential system of the form

$$y'(x) = f(x, y(x), y_x) \tag{7.1.2}$$

for which a solution is defined similarly.

In this chapter we shall provide a few uniqueness criteria for the solutions of the functional differential systems (7.1.1) and (7.1.2). The obtained theorems are analogs of some of the results proved in earlier chapters. Several other similar theorems can be stated and proved rather easily.

7.2 LAKSHMIKANTHAM'S UNIQUENESS THEOREM

Here we shall prove an analog of Perron's Uniqueness Theorem 1.11.4 for the functional differential system (7.1.1). For this, we need the following :

Lemma 7.2.1 [89]. Let the scalar function $\phi(x) \ge 0$ be continuous in $[x_0 - \tau, x_0 + a]$, and for every $\bar{x} \in (x_0, x_0 + a]$ for which $|\phi_{\bar{x}}|_0 = \phi(\bar{x})$ the inequality

$$D_-\phi(\bar{x}) \le g(\bar{x}, |\phi_{\bar{x}}|_0) = g(\bar{x}, \phi(\bar{x}))$$

holds, where the function $g(x, z) \ge 0$ is continuous for $x_0 \le x \le x_0 + a$, $z \ge 0$. Further, let the maximal solution $r(x)$ of the scalar initial value problem $z' = g(x, z)$, $z(x_0) = z_0 \ge 0$ exist in $[x_0, x_0 + a]$. Then, $\phi(x) \le r(x)$, $x_0 \le x \le x_0 + a$ provided $|\phi_{x_0}|_0 \le z_0$. ∎

Theorem 7.2.2 (Lakshmikantham's Uniqueness Theorem). Assume that

(i) condition (i) of Theorem 1.11.4 holds; and

(ii) the function $f(x,\varphi)$ is continuous in

$$\bar{S}_{\tau+} : x_0 \le x \le x_0 + a, \ \|\varphi - \varphi_0(0)\|_0 \le b,$$

and for all (x,α), $(x,\beta) \in \bar{S}_{\tau+}$ and α, $\beta \in \bar{C}$, where

$$\bar{C} = \{\alpha, \ \beta \in C_\mu : \|\alpha - \beta\|_0 = |\alpha(0) - \beta(0)|\}$$

the following inequality holds

$$|f(x,\alpha) - f(x,\beta)| \le g(x, |\alpha(0) - \beta(0)|). \tag{7.2.1}$$

Then, the functional differential system (7.1.1) has at most one solution $y(x)$ in $[x_0, x_0 + a]$ such that $y_{x_0} = \varphi_0$.

Proof. Suppose that $y(x, x_0, \varphi_0)$ and $\bar{y}(x, x_0, \varphi_0)$ are two solutions of (7.1.1) with the same initial function φ_0 at $x = x_0$. Consider the function $\phi(x) = |y(x, x_0, \varphi_0) - \bar{y}(x, x_0, \varphi_0)|$, so that for $x \ge x_0$, $\phi_x = |y_x(x_0, \varphi_0) - \bar{y}_x(x_0, \varphi_0)|$. Then, for $x > x_0$, we have

$$\begin{aligned} D_-\phi(x) \ &\le |y'(x, x_0, \varphi_0) - \bar{y}'(x, x_0, \varphi_0)| \\ &= |f(x, y_x(x_0, \varphi_0)) - f(x, \bar{y}_x(x_0, \varphi_0))|. \end{aligned}$$

Suppose that, for some $\bar{x}$, $x_0 < \bar{x} < x_0 + a$, $\|\phi_{\bar{x}}\|_0 = \phi(\bar{x})$. Such a point definetely exists if $\phi(x) \not\equiv 0$. Setting $\alpha = y_{\bar{x}}(x_0, \varphi_0)$ and $\beta = \bar{y}_{\bar{x}}(x_0, \varphi_0)$, we see that α, $\beta \in \bar{C}$ and $\phi(\bar{x}) = |\alpha(0) - \beta(0)|$. Then, from the above inequality and (7.2.1), we obtain

$$D_-\phi(\bar{x}) \le g(\bar{x}, \phi(\bar{x})).$$

Hence, the conditions of Lemma 7.2.1 are satisfied and we conclude that for any x_1 such that $x_0 < x_1 < x_0 + a$, $\phi(x) \le r(x)$, $x_0 \le x < x_1$, where $r(x)$ is the maximal solution of (1.11.1). By assumption, however, we have $r(x) \equiv 0$, and hence $y(x, x_0, \varphi_0) = \bar{y}(x, x_0, \varphi_0)$, $x \in [x_0, x_1)$. ∎

Corollary 7.2.3. The function $g(x,z) = Lz$, $L > 0$ is admissible in Theorem 7.2.2. ∎

7.3 LAKSHMIKANTHAM - LEELA'S UNIQUENESS THEOREM

An analog of Kamke's Uniqueness Theorem 1.15.1 is the following :

Theorem 7.3.1 (Lakshmikantham - Leela's Uniqueness Theorem). Assume that

(i) condition (i) of Theorem 1.15.1 holds; and

(ii) condition (ii) of Theorem 7.2.2 holds except that the inequality (7.2.1) is satisfied for all (x, α), $(x, \beta) \in \bar{S}_{\tau+}$, α, $\beta \in \bar{C}$ with $x \neq x_0$.

Then, the functional differential system (7.1.1) has at most one solution $y(x)$ in $[x_0, x_0 + a]$ such that $y_{x_0} = \varphi_0$.

Proof. The proof is similar to that of Theorems 1.15.1 and 7.2.2. ∎

Corollary 7.3.2. The function $g(x, z) = g(z)$ where $g(z)$ is continuous in the interval $[0, \infty)$ and $g(0) = 0$, $g(z) > 0$ for $z > 0$ and satisfies (1.4.1) is admissible in Theorem 7.3.1. ∎

7.4 LAKSHMIKANTHAM - SHENDGE'S UNIQUENESS THEOREM

To obtain a result similar to Theorem 7.2.2 for the functional differential system (7.1.2), we need to define the set $\hat{C}_+$ of all nonnegative functions belonging to $\hat{C}$. The result analogous to Lemma 7.2.1 is the following :

Lemma 7.4.1 [91]. Let the scalar function $\phi(x) \geq 0$ be continuous in $[x_0 - \tau, x_0 + a]$, and for $x \in (x_0, x_0 + a]$ satisfy the inequality

$$D_-\phi(x) \leq g(x, \phi(x), \phi_x),$$

where the function $g(x, z, w)$ is continuous for $x_0 \leq x \leq x_0 + a$, $z \geq 0, w \in \hat{C}_+$, and nondecreasing in w for each fixed (x, z). Further, let for the given initial function $\varphi_0 \in \hat{C}_+$ at $x = x_0$ the maximal solution $r(x)$ of the scalar functional differential equation

$$z'(x) = g(x, z(x), z_x) \tag{7.4.1}$$

exist in $[x_0, x_0+a]$. Then, $\phi(x) \le r(x)$, $x_0 \le x \le x_0+a$ provided $\phi_{x_0} \le \varphi_0$. ∎

Theorem 7.4.2 (Lakshmikantham - Shendge's Uniqueness Theorem). Assume that

(i) the function $g(x, z, w)$ is as in Lemma 7.4.1 and the maximal solution of (7.4.1) with the initial function $\varphi_0 = 0$ at $x = x_0$ is identically zero; and

(ii) the function $f(x, y, \varphi)$ is continuous for $x_0 \le x \le x_0 + a$, $y \in R^n$, $\varphi \in \hat{C}_n$ and for all $x_0 \le x \le x_0+a$, $y, \bar{y} \in R^n$, $\alpha, \beta \in \hat{C}_n$ the following inequality holds

$$|f(x, y, \alpha) - f(x, \bar{y}, \beta)| \le g(x, |y - \bar{y}|, |\alpha - \beta|). \tag{7.4.2}$$

Then, the functional differential system (7.1.1) has at most one solution $y(x)$ in $[x_0, x_0 + a]$ such that $y_{x_0} = \varphi_0$. ∎

7.5 FURUMOCHI'S UNIQUENESS THEOREM

Results of Section 3.15 will be extended here for the functional differential system (7.1.1). For this, here in R^n we shall consider the Euclidean norm $\|\cdot\|$. Futher, in addition to the notation used earlier we define the set $C_\mu(L)$ of all functions $\varphi \in C_\mu$ such that $\|\varphi(t_1) - \varphi(t_2)\| \le L|t_1 - t_2|$ for all $t_1,\ t_2 \in [-\tau, 0]$.

Throughout, we shall assume that in (7.1.1) the function $f(x, \varphi)$ is continuous and bounded by $M > 0$ in $[0, a] \times C_\mu$. Further, either of the sets C_μ and $C_\mu(L)$ where $M \le L < \infty$ will be represented by $\tilde{S}$.

Let $P_1(x_1, \varphi_1)$ and $P_2(x_2, \varphi_2)$ with $x_1 < x_2$ be two points in $[0, a] \times \tilde{S}$. We shall define the function $v(P_1, P_2)$ as follows : Consider the division $\Delta_\nu : x_1 =$

$\xi_0 \le \xi_1 \le \cdots \le \xi_\nu = x_2$ and that for each ξ_i, $0 \le i \le \nu - 1$, there exists a ψ_i in $\tilde{S}$ for which a solution $y(x, \xi_i, \psi_i)$ of (7.1.1) through (ξ_i, ψ_i) is continuable to ξ_{i+1}. For each ψ_i, $0 \le i \le \nu - 1$ let ζ_{i+1} be an element in $\tilde{S}$ such that $\zeta_{i+1} = y_{\xi_{i+1}}(\xi_i, \psi_i)$, where $y_{\xi_{i+1}}(\xi_i, \psi_i)$ denotes the restriction of $y(x, \xi_i, \psi_i)$ to the interval $[\xi_{i+1} - \tau, \xi_{i+1}]$. We set

$$K = \sum_{i=0}^{\nu} \|\psi_i - \zeta_i\|_0,$$

where $\zeta_0 = \varphi_1$ and $\psi_\nu = \varphi_2$. For the given two points P_1 and P_2 we can choose $\nu - 1$ points of division $\xi_1, \cdots, \xi_{\nu-1}$ and the points (ξ_i, ψ_i), (ξ_i, ζ_i) in $[0, a] \times \tilde{S}$ in an infinite different ways. We define $v(P_1, P_2)$ by the infimum of all possible sums K.

If $x_1 = x_2$, we set $v(P_1, P_2) = \|\varphi_1 - \varphi_2\|_0$ and if $x_1 > x_2$ then $v(P_1, P_2) = v(P_2, P_1)$.

For the functional differential system (7.1.1) now we shall state four results which correspond to Lemmas 3.15.4 - 3.15.7.

Lemma 7.5.1. Let $P_1(x_1, \varphi_1)$ and $P_2(x_2, \varphi_2)$ with $x_1 < x_2$ be two points in $[0, a] \times \tilde{S}$. In order that P_1 and P_2 are on one solution of (7.1.1) it is necessary and sufficient that $v(P_1, P_2) = 0$. ∎

Lemma 7.5.2. For two points $P_1(x_1, \varphi_1)$ and $P_2(x_2, \varphi_2)$ with $x_1 < x_2$ in $[0, a] \times \tilde{S}$, we have

$$|v(P_1, P_2) - \|T_\alpha \varphi_1 - \varphi_2\|_0\,| \le M(x_2 - x_1),$$

where $\alpha = \min(\tau, x_2 - x_1)$ and T_α is a mapping from $\hat{C}_n$ into $\hat{C}_n$ defined by

$$(T_\alpha \varphi)(t) = \begin{cases} \varphi(t + \alpha), & -\tau \le t < -\alpha \\ \varphi(0), & -\alpha \le t \le 0. \end{cases}$$

Particularly, in case $\tilde{S} = C_\mu(L)$, we have

$$|v(P_1, P_2) - \|\varphi_1 - \varphi_2\|_0\,| \le (L + M)(x_2 - x_1). \quad ∎$$

Lemma 7.5.3. For the points $P_i(x_i, \varphi_i)$, $i = 1, 2, 3$ in $[0, a] \times \tilde{S}$, where $x_1 \le x_2 \le x_3$ the inequality (3.15.12) holds. ∎

Lemma 7.5.4. Let $\tilde{S} = C_\mu(L)$ where $M \le L < \infty$, and let $P_i(x_i, \varphi_i)$, $i = 1, 2, 3$ be the points in $[0, a] \times \tilde{S}$, where $x_1 \le x_2 \le x_3$. Then, we have

$$v(P_1, P_3) \ge v(P_1, P_2) - \|\varphi_3 - \varphi_2\|_0 - (L + M)(x_3 - x_2)$$

and

$$v(P_1, P_3) \ge v(P_2, P_3) - \|\varphi_2 - \varphi_1\|_0 - (L + M)(x_2 - x_1). \quad \blacksquare$$

From the above lemmas it immediately follows that if $\tilde{S} = C_\mu(L)$ where $M \le L < \infty$, then for the points $P_i(x_i, \varphi_i)$, $i = 1, 2, 3, 4$ in $[0, a] \times \tilde{S}$ the following inequality holds

$$(7.5.1) \quad |v(P_1, P_2) - v(P_3, P_4)|$$
$$\le \|\varphi_1 - \varphi_3\|_0 + \|\varphi_2 - \varphi_4\|_0 + (L + M)(|x_1 - x_3| + |x_2 - x_4|).$$

Theorem 7.5.5 (Furumochi's Uniqueness Theorem). Suppose that the function $f(x, \varphi)$ is continuous and bounded by $M > 0$ in $[0, a] \times C_\mu$ and $f(x, 0) \equiv 0$. Then, for the functional differential system (7.1.1) with the initial function $\varphi_0 = 0$ at $x = 0$, $y(x, 0, 0) \equiv 0$ is the only solution in $[0, a]$ if and only if there exists a Lyapunov function $V(x, \varphi)$ which is defined in $[0, a] \times C_\mu$ and satisfies the following conditions

(i) $V(x, 0) \equiv 0$,

(ii) $V(x, \varphi) > 0$ for $\|\varphi\|_0 \ne 0$; and

(iii) for any solution $y(x, t, \varphi)$, $(t, \varphi) \in [0, a] \times C_\mu$, $x \ge t$ of (7.1.1) the function $V(x, y_x(t, \varphi))$ is nonincreasing in x.

Proof. Suppose that the functional differential system (7.1.1) with the initial function $\varphi_0 = 0$ at $x = 0$ has a solution $y(x, 0, 0)$ which is different from zero. Then, there are points x_1, x_2, $0 \le x_1 < x_2 \le a$, such that $y(x_1, 0, 0) = 0$ and that $\|y_{x_2}(0, 0)\|_0 > 0$. However, since by (ii) $V(x_2, y_{x_2}(0, 0)) > 0$, by (iii) $V(x_2, y_{x_2}(0, 0)) \le V(x_1, y_{x_1}(0, 0))$, and by (i) $V(x_1, y_{x_1}(0, 0)) = 0$ we arrive at a contradiction.

To show the necessary part, we let $P_1 = (0,0)$ and $P_2 = (x,\varphi)$ in $[0,a]\times C_\mu$ and set $V(x,\varphi) = v(P_1,P_2)$. By Lemma 7.5.1, we have $V(x,0) \equiv 0$ since $y(x,0,0) \equiv 0$ is a solution of (7.1.1) with the initial function $\varphi_0 = 0$ at $x = 0$. For any point (x,φ) in $[0,a]\times C_\mu$ such that $\|\varphi\|_0 \neq 0$, there is no solution of (7.1.1) passing through $(0,0)$ and (x,φ), and hence from Lemma 7.5.1 we also have $V(x,\varphi) > 0$. Moreover, from Lemmas 7.5.1 and 7.5.3, it follows that for any solution $y(x,t,\varphi)$, $(t,\varphi) \in [0,a]\times C_\mu$, $x \geq t$ of (7.1.1), $V(x, y_x(t,\varphi))$ is a nonincreasing function of x. ∎

In the above theorem it is clear that for (x,φ) and $(x,\bar{\varphi})$ in $[0,a]\times C_\mu$, we have

$$|V(x,\varphi) - V(x,\bar{\varphi})| \leq \|\varphi - \bar{\varphi}\|_0.$$

Theorem 7.5.6. Suppose that the function $f(x,\varphi)$ is continuous and bounded by $M > 0$ in $[0,a]\times C_\mu$ and $f(x,0) \equiv 0$. Further, suppose that $\tilde{S} = C_\mu(L)$, where $M \leq L < \infty$. Then, for the functional differential system (7.1.1) with the initial function $\varphi_0 = 0$ at $x = 0$, $y(x,0,0) \equiv 0$ is the only solution in $[0,a]$ if and only if there exists a continuous Lyapunov function $V(x,\varphi)$ which is defined in $[0,a]\times\tilde{S}$ and satisfies the following conditions

(i) $V(x,0) \equiv 0$ and $V(x,\varphi) > 0$ for $\|\varphi\|_0 \neq 0$,

(ii) for (x,φ), $(\bar{x},\bar{\varphi})$ in $[0,a]\times\tilde{S}$

$$|V(x,\varphi) - V(\bar{x},\bar{\varphi})| \leq \|\varphi - \bar{\varphi}\|_0 + (L+M)|x - \bar{x}|$$

(iii) $D^+V_f(x,\varphi) \leq 0$, where

$$D^+V_f(x,\varphi) = \limsup_{h\to 0+} \frac{1}{h}[V(x+h, y_{x+h}(x,\varphi)) - V(x,\varphi)].$$

Proof. The proof is similar to that of Theorem 7.5.5.

We note that in Theorem 7.5.5 the Lyapunov function $V(x,\varphi)$ is defined in $[0,a]\times C_\mu$, but $V(x,\varphi)$ is not necessarily continuous. However, the segment of a solution of (7.1.1) starting from a point in $[0,a]\times C_\mu(L)$ belongs to $C_\mu(L)$ because $L \geq M$. Therefore, it is sufficient that $V(x,\varphi)$ is defined in $[0,a]\times$

$C_\mu(L)$. But, in this case as mentioned in Theorem 7.5.2, we have a continuous Lyapunov function $V(x,\varphi)$ which satisfies locally Lipschitz condition (3.15.1) with respect to φ. Thus, in Theorem 7.5.5 we can replace the condition that $V(x,\varphi_x)$ is nonincreasing with respect to x by $D^+V_f(x,\varphi) \le 0$.

Finally, we shall prove a generalization of the above theorems.

Theorem 7.5.7. Suppose that the function $f(x,\varphi)$ is continuous and bounded by $M > 0$ in $[0,a]\times C_\mu$. Further, suppose that $\tilde{S} = C_\mu(L)$, where $M \le L < \infty$. Then, for the functional differential system (7.1.1) every solution starting from a point in $[0,a]\times\tilde{S}$ is unique to the right if and only if there exists a continuous Lyapunov function $V(x,\varphi,\bar{\varphi})$ which is defined in $[0,a]\times\tilde{S}\times\tilde{S}$ and satisfies the following conditions

(i) $V(x,\varphi,\bar{\varphi}) = 0$ if $\varphi = \bar{\varphi}$, and $V(x,\varphi,\bar{\varphi}) > 0$ for $\varphi \ne \bar{\varphi}$,

(ii) for $(x_1,\varphi_1,\bar{\varphi}_1)$, $(x_2,\varphi_2,\bar{\varphi}_2)$ in $[0,a]\times\tilde{S}\times\tilde{S}$

$$|V(x_1,\varphi_1,\bar{\varphi}_1) - V(x_2,\varphi_2,\bar{\varphi}_2)|$$
$$\le \|\varphi_1-\varphi_2\|_0 + \|\bar{\varphi}_1-\bar{\varphi}_2\|_0 + 2(L+M)|x_1-x_2|$$

(iii) $D^+V_f(x,\varphi,\bar{\varphi}) \le 0$, where

$$D^+V_f(x,\varphi,\bar{\varphi}) = \limsup_{h\to 0+}\frac{1}{h}[V(x+h, y_{x+h}(x,\varphi), \bar{y}_{x+h}(x,\bar{\varphi})) - V(x,\varphi,\bar{\varphi})]$$

and $(y(x,t,\varphi), \bar{y}(x,t,\bar{\varphi}))$ is a solution of the associated system

$$\begin{aligned} y'(x) &= f(x,y_x) \\ \bar{y}'(x) &= f(x,\bar{y}_x). \end{aligned} \tag{7.5.2}$$

Proof. As in Theorem 7.5.5 the sufficient part is immediate, and therefore we shall prove only the necessary part. The system (7.5.2) associated with (7.1.1) is a functional differential system in $[0,a]\times C_\mu\times C_\mu$. For two points $P_i(x_i,\varphi_i,\bar{\varphi}_i)$, $i=1,2$ in $[0,a]\times\tilde{S}\times\tilde{S}$ we define a function $v(P_1,P_2)$. Then, $v(P_1,P_2)$ has the properties described in Lemmas 7.5.1 - 7.5.4. Let $Q(x,\varphi,\bar{\varphi})$ be a point in $[0,a]\times\tilde{S}\times\tilde{S}$. We define a function $v^*(Q) = v^*(x,\varphi,\bar{\varphi})$ by

$$v^*(Q) = \inf_{P\in\pi(x)} v(P,Q),$$

where $\pi(x) = \{(t, \xi, \zeta) : 0 \le t \le x,\ \xi = \zeta \in \tilde{S}\}$. Then, $v^*(Q)$ is defined in $[0, a] \times \tilde{S} \times \tilde{S}$. Since $\pi(x)$ is compact and $v(P_1, P_2)$ is continuous by (7.5.1), there exists a point $\bar{Q}$ in $\pi(x)$ such that $v^*(Q) = v(\bar{Q}, Q)$. Let $V(x, \varphi, \bar{\varphi}) = v^*(Q)$.

If $\varphi = \bar{\varphi}$, then $P(x, \varphi, \varphi)$ belongs to $\pi(x)$. Since $V(x, \varphi, \varphi) = v^*(P) \le v(P, P) = 0$, we have $V(x, \varphi, \varphi) = 0$. Now suppose that there exists a point $Q(s, \varphi_0, \bar{\varphi}_0)$ in $[0, a] \times \tilde{S} \times \tilde{S}$ such that $V(s, \varphi_0, \bar{\varphi}_0) = 0$ and $\varphi_0 \ne \bar{\varphi}_0$. Then, there exists a point $P(x, \varphi, \varphi)$ in $\pi(s)$ such that $v(P, Q) = 0$. Hence, in view of Lemma 7.5.1 both the points P and Q should be on one solution of (7.5.2). On the other hand, since the solution $(y(x, t, \varphi), \bar{y}(x, t, \varphi))$, $t \le x \le s$ of (7.5.2) through $P(t, \varphi, \varphi)$ is unique to the right, $(x, y_x(t, \varphi), \bar{y}_x(t, \varphi))$ must be in $\pi(s)$ for all $x \in [t, s]$. Therefore, we have $\varphi_0 = y_s(t, \varphi) = \bar{y}_s(t, \varphi) = \bar{\varphi}_0$. This contradiction completes the proof of (i).

Let $Q_i(x_i, \varphi_i, \bar{\varphi}_i)$, $i = 1, 2$ be two points in $[0, a] \times \tilde{S} \times \tilde{S}$ such that $v^*(Q_1) \le v^*(Q_2)$. If $x_1 \le x_2$, then corresponding to $v^*(Q_1)$ there exists a point R in $\pi(x_1)$ such that $v^*(Q_1) = v(R, Q_1)$. Since R belongs to $\pi(x_2)$, we have $v^*(Q_2) \le v(R, Q_2)$. Further, from (7.5.1) it follows that

$$\begin{aligned} 0 &\le V(x_2, \varphi_2, \bar{\varphi}_2) - V(x_1, \varphi_1, \bar{\varphi}_1) \\ &\le v^*(Q_2) - v^*(Q_1) \\ &\le v(R, Q_2) - v(R, Q_1) \\ &\le \|\varphi_1 - \varphi_2\|_0 + \|\bar{\varphi}_1 - \bar{\varphi}_2\|_0 + 2(L + M)(x_2 - x_1). \end{aligned} \tag{7.5.3}$$

In case $x_1 > x_2$, there exists a point $R(x, \varphi, \varphi)$ in $\pi(x_1)$ such that $v^*(Q_1) = v(R, Q_1)$. If R belongs to $\pi(x_2)$, then the proof of (7.5.3) remains similar as for the case $x_1 \le x_2$. If R does not belong to $\pi(x_2)$, i.e., $x_1 > x_2$, then we consider a point $\bar{R}(x_2, \varphi, \varphi)$ in $\pi(x_2)$. Now in view of Lemma 7.5.2, it follows that

$$\begin{aligned} 0 &\le V(x_2, \varphi_2, \bar{\varphi}_2) - V(x_1, \varphi_1, \bar{\varphi}_1) \\ &\le v^*(Q_2) - v^*(Q_1) \\ &\le v(\bar{R}, Q_2) - v(R, Q_1) \end{aligned}$$

$$\le |||(\varphi_2 - \varphi, \bar\varphi_2 - \varphi)||| - v(R, Q_1)$$

$$\le |||(\varphi_2 - \varphi, \bar\varphi_2 - \varphi)||| - |||(\varphi - \varphi_1, \varphi - \bar\varphi_1)||| + 2(L+M)(x_1 - x_2)$$

$$\le |||(\varphi_2 - \varphi_1, \bar\varphi_2 - \bar\varphi_1)||| + 2(L+M)(x_1 - x_2)$$

$$\le \|\varphi_1 - \varphi_2\|_0 + \|\bar\varphi_1 - \bar\varphi_2\|_0 + 2(L+M)(x_1 - x_2),$$

where

$$|||(\varphi, \bar\varphi)||| = |||(\varphi^1, \cdots, \varphi^n, \bar\varphi^1, \cdots, \bar\varphi^n)|||$$

$$= \max_{-\tau \le t \le 0} \sum_{i=1}^{n} [(\varphi^i(t))^2 + (\bar\varphi^i(t))^2]^{1/2}.$$

Hence, in case $V(x_1, \varphi_1, \bar\varphi_1) \le V(x_2, \varphi_2, \bar\varphi_2)$, we have (7.5.3). The other case can be proved similarly. This completes the proof of (ii).

Finally, let $(y(x,t,\varphi), \bar y(x,t,\bar\varphi))$, $t \le x \le \delta$, $t < \delta$ be any solution of (7.5.2) through $(x, \varphi, \bar\varphi)$ in $[0,a] \times \tilde S \times \tilde S$. Let $Q_1(x_1, y_{x_1}(t,\varphi), \bar y_{x_1}(t, \bar\varphi))$ and $Q_2(x_2, y_{x_2}(t,\varphi), \bar y_{x_2}(t,\bar\varphi))$, where $t \le x_1 < x_2 \le \delta$ be two points. Since there exists a point R in $\pi(x_1)$ such that $v^*(Q) = v(R, Q_1)$ and since R belongs to $\pi(x_2)$, we have $v^*(Q_2) \le v(R, Q_2)$. Moreover, from Lemmas 7.5.1 and 7.5.3, it follows that $v^*(Q_2) \le v(R,Q_2) \le v(R,Q_1) + v(Q_1,Q_2) \le v(R,Q_1) + 0 = v^*(Q_1)$, and hence $V(x,\varphi,\bar\varphi)$ is nonincreasing with respect to x along any solution of (7.5.2) starting from a point in $[0,a] \times \tilde S \times \tilde S$. By this, and (ii) it is clear that $D^+V_f(x, \varphi, \bar\varphi) \le 0$. ∎

Example 7.5.1. If there exists a $L > 0$ such that $\|f(x,\varphi) - f(x,\bar\varphi)\| \le L\|\varphi - \bar\varphi\|_0$, $\varphi, \bar\varphi \in C_\mu$, then $V(x,\varphi,\bar\varphi) = e^{-Lx}\|\varphi - \bar\varphi\|_0$ can be used as a Lyapunov function. Indeed, it is clear that this $V(x,\varphi,\bar\varphi)$ satisfies (i) and (ii). Let $y(x,0,\varphi)$ and $\bar y(x,0,\bar\varphi)$ be two solutions of (7.1.1). Then, we have

$$D^+V_f(x, y_x, \bar y_x)$$

$$= \limsup_{h\to 0^+} \frac{1}{h}[e^{-L(x+h)}\|y_{x+h} - \bar y_{x+h}\|_0 - e^{-Lx}\|y_x - \bar y_x\|_0]$$

$$= \limsup_{h\to 0^+} \frac{1}{h}[(e^{-L(x+h)} - e^{-Lx})\|y_x - \bar y_x\|_0$$

$$+e^{-L(x+h)}(\|y_{x+h} - \bar{y}_{x+h}\|_0 - \|y_x - \bar{y}_x\|_0)]$$

$$\leq -e^{-Lx}L\|y_x - \bar{y}_x\|_0 + e^{-Lx}\limsup_{h\to 0+} \frac{1}{h}\int_x^{x+h} \|f(t, y_t) - f(t, \bar{y}_t)\|dt$$

$$\leq e^{-Lx}[-L\|y_x - \bar{y}_x\|_0 + L\|y_x - \bar{y}_x\|_0] = 0.$$

Hence, this function $V(x, \varphi, \bar{\varphi})$ also satisfies (iii). ∎

CHAPTER 8

IMPULSIVE DIFFERENTIAL EQUATIONS

8.1 INTRODUCTION

Let $T \subset R^n$ be an open set and $U = [x_0, \infty) \times T$, $x_0 \geq 0$. Let us suppose that for each $k = 1, 2, \cdots$, $\tau_k(y) > 0$, $y \in T$ be continuous and map into R^n, $\tau_k(y) < \tau_{k+1}(y)$ and $\lim_{k\to\infty} \tau_k(y) = \infty$ for all $y \in T$. It is clear that $x = \tau_k(y)$ are the surfaces.

The initial value problem for impulsive differential system is given by

$$y' = f(x, y), \quad x \neq \tau_k(y)$$

(8.1.1) $$\Delta y = I_k(y), \quad x = \tau_k(y)$$

$$y(x_0^+) = y_0$$

where the function $f(x, y)$ is defined in U, $I_k(y)$ maps T into R^n, and for $x = \tau_k(y(x))$, $\Delta y(x) = y(x^+) - y(x)$.

The term impulsive comes from the fact that at the moments where $x = \tau_k(y(x))$ the system (8.1.1) is subject to short term perturbations whose duration is negligible in comparison with the duration of the process.

By a **solution** of (8.1.1) in the interval $[x_0, x_0 + a)$ we mean a function $y(x)$ satisfying :

(i) $y(x_0^+) = y_0$,

(ii) for all $x \in [x_0, x_0 + a)$, the points $(x, y(x)) \in U$,

(iii) $y(x)$ is continuously differentiable and satisfies $y'(x) = f(x, y(x))$ for all $x \in [x_0, x_0 + a)$ and $x \neq \tau_k(y(x))$; and

(iv) if $x \in [x_0, x_0 + a)$ and $x = \tau_k(y(x))$, then $y(x^+) = y(x) + I_k(y(x))$, and at such x's we assume that $y(x)$ is left continuous and $t \neq \tau_j(y(t))$ for any j, $x < t < \delta$, for some $\delta > 0$.

In (8.1.1) instead of the usual initial condition $y(x_0) = y_0$, we have assumed the limiting condition $y(x_0^+) = y_0$. This is justified because the point (x_0, y_0) may be such that $x_0 = \tau_k(y_0)$ for some k. Thus, whenever $x_0 \neq \tau_k(y_0)$ for any k we shall understand the initial condition $y(x_0^+) = y_0$ in the usual sense, i.e., $y(x_0) = y_0$.

It is clear that unlike for (1.1.1) the initial value problem (8.1.1) may not have any solution even if the function $f(x, y)$ is continuous. Further, a solution of (8.1.1) may be a continuously differentiable function, a piecewise continuously differentiable function having finite number of discontinuities of the first kind, or a piecewise continuously differentiable function having a countable number of discontinuities of the first kind.

In this chapter we shall also consider the initial value problem

$$(8.1.2) \qquad \begin{aligned} Dy &= f(x, y) + g(x, y)Du \\ y(x_0) &= y_0 \end{aligned}$$

where the functions $f(x, y)$ and $g(x, y)$ are defined in U, the scalar function $u(x)$ is right - continuous and of bounded variation in $[x_0, \infty)$, and Dy, Du denote the distributional derivatives. Since $u(x)$ is a function of bounded variation, Du can be identified with a Stieltjes measure which has the effect of instantaneous changes of the state of the system (8.1.2) at the points of discontinuity of $u(x)$. Because of this reason the system (8.1.2) is of impulsive type, however, in literature it is better known as a 'measure differential system'.

By a **solution** of (8.1.2) in the interval $[x_0, x_0 + a)$ we mean a function $y(x)$ satisfying :

(i) $y(x_0) = y_0$,

(ii) for all $x \in [x_0, x_0 + a)$, the points $(x, y(x)) \in U$,

(iii) for all $x \in [x_0, x_0 + a)$ the function $y(x)$ is right - continuous and of bounded variation; and

(iv) for each $x \in (x_0, x_0 + a)$ the distributional derivative of $y(x)$ exists and satisfies $Dy(x) = f(x, y(x)) + g(x, y(x))Du(x)$.

8.2 HU - LAKSHMIKANTHAM'S UNIQUENESS THEOREM

Here we shall obtain an analog of Perron's Uniqueness Theorem 1.11.4 for the initial value problem (8.1.1). For this, as in earlier chapters, in R^n we shall denote the norm as $|\cdot|$ so that the definition of $\bar{S}_+$ remains the same.

Theorem 8.2.1 (Hu - Lakshmikantham's Uniqueness Theorem). Assume that

(i) the function $g(x, z)$ is continuous and nonnegative in $x_0 \le x \le x_0 + a$, $0 \le z \le 2b$ and for every x^*, $x_0 \le x^* < x_0 + a$, $z(x) = 0$, $x \in [x^*, x_0 + a]$ is the only solution of the initial value problem

$$z' = g(x, z), \ z(x^*) = 0 \tag{8.2.1}$$

(ii) condition (ii) of Theorem 1.11.4 holds.

Then, the initial value problem (8.1.1) has at most one solution in $[x_0, x_0 + a]$.

Proof. Suppose that $y(x)$ and $\bar{y}(x)$ are two solutions of (8.1.1) in $[x_0, x_0 + a]$. Then, for the function $\phi(x) = |y(x) - \bar{y}(x)|$ the set $A = \{x \ge x_0 : \phi(x) > 0\}$

is nonempty. Let $x^* = \inf A$ so that $\phi(x) = 0$ for $x_0 \le x < x^*$ and $\phi(x^*) = 0$ since both $y(x)$ and $\bar{y}(x)$ are left continuous. If $x^* \ne \tau_k(y(x^*))$ for all $k \ge 1$, then we also have $x^* \ne \tau_k(\bar{y}(x^*))$. If, on the other hand $x^* = \tau_k(y(x^*))$ for some $k \ge 1$, then we also have $x^* = \tau_k(\bar{y}(x^*))$. Thus, in both the cases we find that $\phi(x^*) = 0$ and $\phi(x^{*+}) = 0$. Also, in view of (1.11.2) it follows that

$$D_-\phi(x) \le g(x, \phi(x)), \; x \in [x_0, x_0 + a).$$

Hence, as earlier $\phi(x) \le r(x)$, $x^* \le x < x_0 + \delta$ $(\delta > 0)$, where $r(x)$ is the maximal solution of (8.2.1). Since $r(x) = 0$ in $[x^*, x^* + \delta)$ by our hypotheses, we conclude that $\phi(x) = 0$, $x \in [x^*, x^* + \delta)$, which contradicts the definition of x^*, and hence $y(x) = \bar{y}(x)$, $x \in [x_0, x_0 + a]$. ∎

Corollary 8.2.2. If for the differential system $y' = f(x, y)$ every solution starting from a point in $\bar{S}_+$ is unique to the right then the initial value problem (8.1.1) has at most one solution in $[x_0, x_0 + a]$. ∎

8.3 PANDIT'S UNIQUENESS THEOREM

A nontrivial analog of Lemma 1.2.2 for the initial value problem (8.1.2) is stated in the following :

Theorem 8.3.1 [37]. For the initial value problem (8.1.2), $y(x)$ is a solution in $[x_0, x_0 + a)$ if and only if it satisfies the integral equation

$$y(x) = y_0 + \int_{x_0}^{x} f(t, y(t))dt + \int_{x_0^+}^{x} g(t, y(t))du(t). \qquad (8.3.1)$$ ∎

The fact that $u(x)$ may have discontinuities creates difficulties in establishing uniqueness criterion for the solutions of (8.1.2). In fact, the following simple examples show a diverse behavior even when the functions $f(x, y)$ and $g(x, y)$ are linear.

Example 8.3.1. Consider the initial value problem

$$Dy = 1 + DH, \; y(0) = 0 \qquad (8.3.2)$$

where $H(x)$ is the Heaviside function defined by $H(x) = \begin{cases} 0, & x < 0 \\ 1, & x \geq 0 \end{cases}$. The problem (8.3.2) has a unique solution $y(x) = \begin{cases} x-1, & x < 0 \\ x, & x \geq 0 \end{cases}$.

Example 8.3.2. Consider the initial value problem

$$(8.3.3) \qquad Dy = \frac{2}{x+1} y Du, \ x \in [0,2], \ y(0) = 0$$

where $u(x) = \begin{cases} x, & 0 \leq x < 1 \\ x+1, & 1 \leq x \leq 2 \end{cases}$. The problem (8.3.3) has an infinite number of solutions $y(x) = \begin{cases} 0, & 0 \leq x < 1 \\ c(x+1)^2/4, & 1 \leq x \leq 2 \end{cases}$, where c is an arbitrary constant.

Definition 8.3.1. A scalar function $u(x)$ defined in $[x_0, \infty)$ is said to be of type Υ if the following hold

(i) $u(x)$ is right - continuous and of bounded variation in every compact subinterval of $[x_0, \infty)$,

(ii) the discontinuities $x_1 < x_2 < \cdots$ are isolated and are such that $x_1 > x_0$ and $x_k \to \infty$ as $k \to \infty$; and

(iii) $u(x)$ is differentiable in each subinterval $[x_k, x_{k+1})$ of $[0, \infty)$, where the derivative at x_k is the right hand derivative.

Lemma 8.3.2. Assume that

(i) the functions $\phi(x)$ and $g(x)$ are nonnegative, defined and du integrable in $[x_0, \infty)$; and

(ii) the function $u(x)$ is of type Υ, $u'(x)$ is nonnegative in $[x_0, \infty) - \{x_k\}$, the jumps $a_k = u(x_k) - u(x_k^-)$ are such that

$$(8.3.4) \qquad |a_k| g(x_k) < 1, \ k = 1, 2, \cdots$$

and

$$(8.3.5) \qquad \sum_{k=1}^{\infty} |a_k| g(x_k) < \infty.$$

Then, for any positive constant c the inequality

$$\phi(x) \le c + \int_{x_0}^{x} g(t)\phi(t)du(t), \ x \in [x_0, \infty) \tag{8.3.6}$$

implies that

$$\phi(x) \le cP^{-1} \exp\left(\int_{x_0}^{x} g(t)u'(t)dt\right), \ x \in [x_0, \infty) \tag{8.3.7}$$

where

$$P = \prod_{k=1}^{\infty}(1 - |a_k|g(x_k)). \tag{8.3.8}$$

Proof. Let the right side of the inequality (8.3.6) be $\psi(x)$. Since the function $u(x)$ is differentiable in $[x_0, x_1)$, in view of the Remark 1.12.1 it follows that

$$\psi(x) \le c\exp\left(\int_{x_0}^{x} g(t)u'(t)dt\right), \ x \in [x_0, x_1).$$

At $x = x_1$, we have for sufficiently small $h > 0$ that

$$\psi(x_1) = \psi(x_1 - h) + \int_{x_1-h}^{x_1} g(t)\phi(t)du(t)$$

$$\le c\exp\left(\int_{x_0}^{x_1-h} g(t)u'(t)dt\right) + \int_{x_1-h}^{x_1} g(t)\psi(t)du(t). \tag{8.3.9}$$

We shall prove that

$$\lim_{h\to 0^+}\left|\int_{x_1-h}^{x_1} g(t)\psi(t)du(t)\right| = |a_1|g(x_1)\psi(x_1). \tag{8.3.10}$$

For this, we consider the set function λ defined by

$$\lambda(A) = \left|\int_A g(t)\psi(t)du(t)\right|.$$

Let $\{h_k\}$ be a decreasing sequence of positive real numbers tending to zero and let $A_k = [x_1 - h_k, x_1]$. Then, $A_1 \supset A_2 \supset \cdots$ and $\bigcap_{k=1}^{\infty} A_k = \{x_1\}$. Therefore, $\lambda(A_k) \to \lambda(\{x_1\}) = |a_1|g(x_1)\psi(x_1)$, and hence (8.3.10) holds. Taking limit as $h \to 0_+$ in (8.3.9), we obtain

$$\psi(x_1) \le c\exp\left(\int_{x_0}^{x_1} g(t)u'(t)dt\right) + |a_1|g(x_1)\psi(x_1),$$

which in view of (8.3.4) yields

$$\psi(x_1) \le cP_1^{-1}\exp\left(\int_{x_0}^{x_1} g(t)u'(t)dt\right), \tag{8.3.11}$$

where

$$P_j = \prod_{k=1}^{j}(1-|a_k|g(x_k)).$$

For $x \in [x_1, x_2)$, we note that

$$\begin{aligned}\psi(x) &= \psi(x_1) + \int_{x_1}^{x} g(t)\phi(t)du(t)\\ &\le \psi(x_1) + \int_{x_1}^{x} g(t)\psi(t)du(t).\end{aligned}$$

Thus, following as above and using (8.3.11), this gives

$$\begin{aligned}\psi(x) &\le \psi(x_1)\exp\left(\int_{x_1}^{x} g(t)u'(t)dt\right)\\ &\le cP_1^{-1}\exp\left(\int_{x_0}^{x} g(t)u'(t)dt\right), \; x \in [x_0, x_2).\end{aligned}$$

Now by induction it is easy to establish that

$$\psi(x) \le cP_{k-1}^{-1}\exp\left(\int_{x_0}^{x} g(t)u'(t)dt\right), \; x \in [x_0, x_k). \tag{8.3.12}$$

From (8.3.4) and (8.3.5) we observe that $P_k \ge P_{k+1}$ for each k, and $\lim_{k\to\infty} P_k = P$. Therefore, for any $x \in [x_0, \infty)$, (8.3.12) implies that

$$\phi(x) \le \psi(x) \le cP^{-1}\exp\left(\int_{x_0}^{x} g(t)u'(t)dt\right). \quad \blacksquare$$

Corollary 8.3.3. If $f(x)$ is a nonnegative function defined and integrable in $[x_0, \infty)$, then under the hypotheses of Lemma 8.3.2 the inequality

$$\phi(x) \le c + \int_{x_0}^{x} f(t)\phi(t)dt + \int_{x_0}^{x} g(t)\phi(t)du(t), \; x \in [x_0, \infty)$$

implies that

$$\phi(x) \le cP^{-1}\exp\left(\int_{x_0}^{x}[f(t) + g(t)u'(t)]dt\right). \quad \blacksquare$$

We note that if the function $u(x)$ is of type Υ, so is its total variation function, say, $v_u(x)$. Further, $v_u(x)$ is nondecreasing in (x_{k-1}, x_k) for each k, so that $v_u'(x) \geq 0$ in $[x_0, \infty) - \{x_k\}$. Let $b_k = v_u(x_k) - v_u(x_k^-)$, $k = 1, 2, \cdots$.

Theorem 8.3.4 (Pandit's Uniqueness Theorem). Assume that

(i) $u(x)$ is of type Υ,

(ii) there exist nonnegative functions $f(x)$ and $g(x)$ in $[x_0, \infty)$ such that for all $x \in [x_0, \infty)$, $|y|$, $|\bar{y}| < \infty$

$$|f(x,y) - f(x,\bar{y})| \leq f(x)|y - \bar{y}|$$

and

$$|g(x,y) - g(x,\bar{y})| \leq g(x)|y - \bar{y}|$$

(iii) b_k satisfies

$$|b_k| g(x_k) < 1, \; k = 1, 2, \cdots, \tag{8.3.13}$$

$$\sum_{k=1}^{\infty} |b_k| g(x_k) < \infty,$$

and

$$Q = \prod_{k=1}^{\infty} (1 - |b_k| g(x_k)),$$

(iv) $f(x)$ and $g(x)$ satisfy the conditions of Corollary 8.3.3 and $f(x)$, $g(x) \times v_u'(x)$ are integrable in $[x_0, \infty)$.

Then, the initial value problem (8.1.2) has at most one solution in $[x_0, \infty)$.

Proof. Let $y(x)$ and $\bar{y}(x)$ be two solutions of (8.1.2) in $[x_0, \infty)$. Then, in view of Theorem 8.3.1 and the hypotheses it follows that

$$\phi(x) < \varepsilon + \int_{x_0}^{x} f(t)\phi(t)dt + \int_{x_0}^{x} g(t)\phi(t)dv_u(t), \; x \in [x_0, \infty)$$

where $\phi(x) = |y(x) - \bar{y}(x)|$ and $\varepsilon > 0$ is arbitrary. Thus, an application of Corollary 8.3.3 gives

$$\phi(x) \leq \varepsilon Q^{-1} \exp\left(\int_{x_0}^{x} [f(t) + g(t)v_u'(t)]dt\right), \; x \in [x_0, \infty).$$

The conclusion now follows from the hypothesis (iv) and the fact that $\varepsilon > 0$ is arbitrary. ∎

For the initial value problem (8.3.3) the conditions of Theorem 8.3.4 are not satisfied. In fact, at the point of discontinuity $x = 1$, we have $b_1 = -1$ so that $|b_1|g(1) = 1$, which contradicts (8.3.13).

Example 8.3.3. Consider the initial value problem

$$Dy = e^{-x}y^3 + \frac{1}{2}e^{-2x}(x^3+1)^{-1}y^2Du, \; x \in [1,\infty), \; |y| \leq 1 \tag{8.3.14}$$

where

$$u(x) = e^x + \sum_{k=2}^{\infty}\left(\sum_{i=1}^{k-1} e^i\right)\chi_{[k-1,k)}(x).$$

It is clear that for $k = 2, 3, \cdots$

$$v_u(x) = e^x - \sum_{k=2}^{\infty} e^{k-1}\chi_{[k-1,k)}(x), \; x_k = k, \; b_k = e^{k-1} - e^k.$$

Thus, if we choose the functions $f(x) = 3e^{-x}$ and $g(x) = e^{-2x}(x^3+1)^{-1}$ then all the conditions of Theorem 8.3.4 are satisfied. ∎

CHAPTER 9

DIFFERENTIAL EQUATIONS WITH HYSTERESIS

9.1 INTRODUCTION

In this chapter we shall investigate uniqueness criteria for the initial value problem

$$\begin{aligned} & y'(x) = f(x, y(x), u(x)) \\ (9.1.1) \qquad & u(x) = T(x_0, y_0, u_0)y(x) \\ & y(x_0) = y_0, \ u(x_0) = u_0 \end{aligned}$$

where $y(x)$ and $u(x)$ are continuous scalar functions in $[x_0, x_0 + a]$. For fixed x_0, y_0 and u_0 the operator $T(x_0, y_0, u_0)$ is single - valued on $C[x_0, x_0 + a]$, and maps each $y(x) \in C[x_0, x_0 + a]$ such that $y(x_0) = y_0$ into a function $u(x) \in C[x_0, x_0 + a]$ for which $u(x_0) = u_0$. Further, $T(x_0, y_0, u_0)$ in $C[x_0, x_0 + a]$ satisfies the Lipschitz condition

$$(9.1.2) \qquad |T(x_0, y_0, u_0)y(x) - T(x_0, y_0, u_0)\bar{y}(x)| \le L \max_{x_0 \le t \le x} |y(t) - \bar{y}(t)|.$$

Operators describing hysteresis transformations have the above properties [82].

If the function $f(x,y,u)$ is continuous in

$$\bar{S}_{+h} : x_0 \le x \le x_0 + a,\ |y - y_0| \le b,\ |u - u_0| \le c$$

then by a **solution** of (9.1.1) in the interval $[x_0, x_0 + a]$ we mean a pair $\{y(x), u(x)\}$ of functions satisfying :

(i) $y(x_0) = y_0,\ u(x_0) = u_0,$

(ii) for all $x \in [x_0, x_0 + a]$, the points $(x, y(x), u(x)) \in \bar{S}_{+h}$,

(iii) $y(x)$ is continuously differentiable and $u(x)$ is continuous in $[x_0, x_0 + a]$; and

(iv) $y'(x) = f(x, y(x), u(x)),\ u(x) = T(x_0, y_0, u_0)y(x).$

Further, if the function $f(x,y,u)$ is only defined in

$$S_{+h} : x_0 < x < x_0 + a,\ |y - y_0| < b,\ |u - u_0| < c$$

then by a **classical solution** of (9.1.1) in the interval $[x_0, x_0 + a)$ we mean a pair $\{y(x), u(x)\}$ of functions satisfying (i), and

(v) for all $x \in (x_0, x_0 + a)$, the points $(x, y(x), u(x)) \in S_{+h}$,

(vi) $y(x)$ is continuous in $[x_0, x_0 + a)$ and differentiable in $(x_0, x_0 + a)$, and $u(x)$ is continuous in $[x_0, x_0 + a)$; and

(vii) for all $x \in (x_0, x_0 + a)$, $y'(x) = f(x, y(x), u(x)),\ u(x) = T(x_0, y_0, u_0)y(x).$

9.2 BORZDYKO'S UNIQUENESS THEOREM 1

We begin by proving the following :

Lemma 9.2.1. If $\phi(x)$ is absolutely continuous in $[x_0, x_0 + a]$ then $\psi(x) = \max_{x_0 \le t \le x} |\phi(t)|$ is nondecreasing and absolutely continuous. Further, if $\phi(x)$ and $\psi(x)$ have derivatives $\phi'(x)$ and $\psi'(x)$ at a point $x \in [x_0, x_0 + a]$ then

$$0 \le \psi'(x) \le |\phi'(x)|. \tag{9.2.1}$$

Proof. From the absolute continuity of $\phi(x)$ in $[x_0, x_0 + a]$ it follows that the function $\psi(x) = \max_{x_0 \le t \le x} |\phi(t)|$ is also absolutely continuous in $[x_0, x_0 + a]$ and nondecreasing. Now since $\phi(x)$ and $\psi(x)$ are absolutely continuous in $[x_0, x_0 + a]$, they are differentiable almost everywhere. Suppose $x \in (x_0, x_0 + a]$ is a point where $\phi'(x)$ and $\psi'(x)$ exist. At this point the first part of the inequality (9.2.1) is obvious and to prove the second part we need to consider the following two cases :

(1) There is $x_1 \in [x_0, x)$ at which $|\phi(x_1)| = \max_{x_0 \le t \le x} |\phi(t)|$. Then, $\psi(t) = \psi(x_1)$, $x_0 \le t \le x$ and so $\psi'(x^-) = 0$. Hence, $\psi'(x) = 0$ and (9.2.1) holds.

(2) $|\phi(x)| > |\phi(t)|$, $x_0 \le t < x$. In this case there exists a sequence $\{x_n\}$ of distinct points such that $x_n < x$, $n = 1, 2, \cdots$, $x_n \to x$ as $n \to \infty$ and $|\phi(x_n)| > |\phi(t)|$, $x_0 \le t < x_n$. Thus, it follows that $\psi(x) = |\phi(x)| > 0$ and $\psi(x_n) = |\phi(x_n)|$. If $\phi(x) > 0$ then we can assume that $\phi(x_n) > 0$, and obtain

$$\frac{\psi(x_n) - \psi(x)}{x_n - x} = \frac{\phi(x_n) - \phi(x)}{x_n - x}.$$

Since $\phi'(x)$ and $\psi'(x)$ exist by assumption, as $n \to \infty$ we find that $\psi'(x) = \phi'(x)$. If $\phi(x) < 0$ then, similarly, it follows that $\psi'(x) = -\phi'(x)$. Finally, when $x = x_0$ the proof is similar, except that now $\phi'(x)$ and $\psi'(x)$ are the right derivatives. ∎

A generalization of Theorem 1.4.6 is the following :

Theorem 9.2.2 (Borzdyko's Uniqueness Theorem 1). Assume that the operator $T(x_0, y_0, u_0)$ in $C[x_0, x_0+a]$ satisfies the Lipschitz condition (9.1.2), and the function $f(x, y, u)$ is continuous and for all (x, y, u), $(x, \bar{y}, \bar{u}) \in \bar{S}_{+h}$ satisfies

$$|f(x, y, u) - f(x, \bar{y}, \bar{u})| \le g(|y - \bar{y}| + |u - \bar{u}|), \tag{9.2.2}$$

where $g(z) > 0$ is continuous for $0 < z \le \alpha \le x_0 + a$, nondecreasing for $0 < z < \delta \le \alpha$, $g(0) = 0$, and satisfies the condition (1.4.1). Then, the initial value problem (9.1.1) has at most one solution in a small right interval containing x_0.

Proof. Let the problem (9.1.1) have two solutions $\{y(x), u(x)\}$ and $\{\bar{y}(x), \bar{u}(x)\}$ in $[x_0, x_0 + a]$. Then, $y(x) \not\equiv \bar{y}(x)$. Let $\phi(x) = y(x) - \bar{y}(x)$ and $\psi(x) =$

$\max_{x_0 \le t \le x} |\phi(t)|$. Clearly, there exists a $x_1 \in [x_0, x_0 + a)$ such that $\psi(x_1) = 0$, and $\psi(x) > 0$ for $x_1 < x \le x_0 + a$. Now the conditions (9.1.2), (9.2.2) and the properties of the function $g(z)$ imply that

$$|\phi'(x)| \le g((L+1)\psi(x)) \tag{9.2.3}$$

in some interval $[x_1, x_1 + \delta_1]$, $\delta_1 > 0$. Since the continuously differentiable functions $y(x)$ and $\bar{y}(x)$ are absolutely continuous, in view of Lemma 9.2.1 it follows that the function $\psi(x)$ is absolutely continuous, nondecreasing and has a finite derivative almost everywhere such that $0 \le \psi'(x) \le |\phi'(x)|$. Thus, from (9.2.3), we find that

$$0 \le \psi'(x) \le g((L+1)\psi(x)) \tag{9.2.4}$$

almost everywhere in $[x_1, x_1 + \delta_1]$. Now let $\{t_n\}$ be a decreasing sequence such that $x_1 < t_n < x_1 + \delta_1$ and $t_n \to x_1$ as $n \to \infty$. Then, (9.2.4) implies that

$$0 \le \int_{t_n}^{x_1+\delta_1} \frac{\psi'(t)dt}{g((L+1)\psi(t))} \le \int_{t_n}^{x_1+\delta_1} dt = x_1 + \delta_1 - t_n. \tag{9.2.5}$$

However, from the properties of the function $\psi(x)$ and the hypotheses on the function $g(z)$, we have

$$\int_{t_n}^{x_1+\delta_1} \frac{\psi'(t)dt}{g((L+1)\psi(t))} = \frac{1}{L+1} \int_{(L+1)\psi(t_n)}^{(L+1)\psi(x_1+\delta_1)} \frac{dz}{g(z)}$$

and hence, in view of (1.4.1), as $n \to \infty$ in (9.2.5) we arrive at a contradiction.

∎

9.3 BORZDYKO'S UNIQUENESS THEOREM 2

An extension of Theorem 2.9.1 is the following :

Theorem 9.3.1 (Borzdyko's Uniqueness Theorem 2). Suppose that the operator $T(x_0, y_0, u_0)$ in $C[x_0, x_0 + a]$ satisfies the Lipschitz condition (9.1.2), and it is possible to find two functions f_1 and f_2 such that

$$f(x, y, u) = f_1(x, y) f_2(x, y, u), \tag{9.3.1}$$

where $f_1(x,y)$ is defined and continuous in $\bar{S}_+$, and $f_1(x,y) > 0$, $\frac{\partial f_1(x,y)}{\partial x}$ exists, nonnegative, and dominated by an (Lebesgue) integrable function of y. The function $f_2(x,y,u)$ is defined in $\bar{S}_{+h}$ and for all (x,y,u), $(x,\bar{y},\bar{u}) \in \bar{S}_{+h}$ satisfies the condition

$$|f_2(x,y,u) - f_2(x,\bar{y},\bar{u})| \le \lambda(x)[|y-\bar{y}| + |u-\bar{u}|], \tag{9.3.2}$$

where $\lambda(x)$ is an (Lebesgue) integrable function in $[x_0, x_0+a]$. Then, the initial value problem (9.1.1) has at most one solution in $[x_0, x_0+a]$.

Proof. As usual let the problem (9.1.1) have two solutions $\{y(x), u(x)\}$ and $\{\bar{y}(x), \bar{u}(x)\}$ in $[x_0, x_0+a]$. Then, $y(x) \not\equiv \bar{y}(x)$. Consider the function

$$H(x) = \frac{1}{2}\left[\int_{y(x)}^{\bar{y}(x)} \frac{dt}{f_1(x,t)}\right]^2. \tag{9.3.3}$$

By the hypotheses on the function f_1 there exists an (Lebesgue) integrable function $g(y)$ in $|y - y_0| \le b$ such that

$$0 \le \frac{\partial f_1(x,y)}{\partial x} \le g(y) \quad \text{for all} \quad (x,y) \in \bar{S}_+.$$

Thus, the function $H(x)$ is differentiable, and as in Theorem 3.10.1 we find that

$$H'(x) \le \lambda(x)[|y(x) - \bar{y}(x)| + |u(x) - \bar{u}(x)|]\left|\int_{y(x)}^{\bar{y}(x)} \frac{dt}{f_1(x,t)}\right|. \tag{9.3.4}$$

Let M and m be numbers such that $M \ge f_1(x,y) \ge m > 0$, for $x_0 \le x \le x_0 + a$, $|y - y_0| \le b$. Then, from (9.1.2) and (9.3.4) it follows that

$$H'(x) \le \frac{L+1}{m}\lambda(x)\left[\max_{x_0 \le t \le x} |y(t) - \bar{y}(t)|\right]^2. \tag{9.3.5}$$

Further, from (9.3.3) we have

$$\left[\max_{x_0 \le t \le x} |y(t) - \bar{y}(t)|\right]^2 \le 2M^2 \max_{x_0 \le t \le x} H(t). \tag{9.3.6}$$

Hence, on combining (9.3.5) and (9.3.6), we get

$$H'(x) \le 2M^2 \frac{L+1}{m}\lambda(x) \max_{x_0 \le t \le x} H(t).$$

Now let $\psi(x) = \max_{x_0 \le t \le x} H(t)$. Then, from the above inequality and Lemma 9.2.1, we obtain

$$(9.3.7) \qquad -\psi'(x) + 2M^2 \frac{L+1}{m} \lambda(x)\psi(x) \ge 0$$

almost everywhere in $[x_0, x_0 + a]$. Let

$$(9.3.8) \qquad 0 \ge \varphi(x) = -\psi(x) \exp\left(-\frac{2M^2(L+1)}{m} \int_{x_0}^{x} \lambda(t)dt \right).$$

Then, from (9.3.7) we find that $\varphi'(x) \ge 0$ almost everywhere in $[x_0, x_0 + a]$, and it is easy to see that $D^+\varphi(x) > -\infty$ everywhere in $[x_0, x_0 + a]$. Therefore, $0 \le \int_{x_0}^{x} \varphi'(t)dt \le \varphi(x) - \varphi(x_0)$. However, $\varphi(x_0) = -\psi(x_0) = -H(x_0) = 0$, and so $\varphi(x) \ge 0$. But, this contradicts (9.3.8) and hence $\psi(x) \equiv 0$, i.e., $y(x) = \bar{y}(x)$, $x \in [x_0, x_0 + a]$. ∎

9.4 BORZDYKO'S UNIQUENESS THEOREM 3

We need the following :

Lemma 9.4.1. Assume that

(i) $\phi(x)$ is continuously differentiable in some neighborhood of x_0, and $\phi(x) > 0$, $\phi'(x) > 0$ for $x > x_0$,

(ii) $\psi(x)$ is continuous in some neighborhood of x_0, and

(a) $\psi(x) \ge 0$ for $x \ge x_0$ and $\int_{x_0}^{x} \psi(t)dt > 0$ for $x > x_0$; and

(b) $\psi(x) = o((\phi^k(x))')$ for $x \to x_0^+$, where $k > 0$ is a constant.

Then, there exists a sequence $\{x_n\}$ such that $x_n \to x_0^+$ as $n \to \infty$ and $\Phi(x_n) > k$, where

$$\Phi(x) = \frac{\phi(x)\psi(x)}{\phi'(x) \int_{x_0}^{x} \psi(t)dt}.$$

Proof. Let $\varphi(x) = \psi(x)/(\phi^k(x))'$. In view of the hypothesis (b) it is clear that $\varphi(x) \to 0$ as $x \to x_0^+$. The condition therefore has a meaning when either

$(\phi^k(x_0))' \neq 0$, or $(\phi^k(x))' \to \infty$ as $x \to x_0^+$. Let $\varphi(x_0) = 0$, so that the function $\varphi(x)$ is continuous and nonnegative for $x \geq x_0$.

Let $\{x_n\}$ be a sequence such that $x_n \to x_0^+$ as $n \to \infty$, and let $\bar{x}_n$ be one of the points such that $0 < \bar{x}_n \leq x_n$ and $\varphi(\bar{x}_n) = \max_{x_0 \leq t \leq x_n} \varphi(t)$. Since $\varphi(x)$ is continuous such a point $\bar{x}_n$ exists.

Properties of $\phi(x)$ and $\psi(x)$ imply that $\varphi(\bar{x}_n) > 0$. Thus,

$$(9.4.1) \qquad \int_{x_0}^{\bar{x}_n} \psi(t)dt = \int_{x_0}^{\bar{x}_n} \varphi(t)(\phi^k(t))'dt$$

$$< \int_{x_0}^{\bar{x}_n} \varphi(\bar{x}_n)(\phi^k(t))'dt = \varphi(\bar{x}_n)\phi^k(\bar{x}_n).$$

We rewrite $\bar{x}_n$ as x_n, $n = 1, 2, \cdots$. Thus, from (9.4.1) it follows that

$$\Phi(x_n) > \frac{\phi(x_n)(\phi^k(x_n))'\varphi(x_n)}{\phi'(x_n)\phi^k(x_n)\varphi(x_n)} = k. \qquad \blacksquare$$

An analog of Theorem 2.11.1 is the following :

Theorem 9.4.2 (Borzdyko's Uniqueness Theorem 3). Assume that the operator $T(x_0, y_0, u_0)$ in $C[x_0, x_0 + a]$ satisfies the Lipschitz condition (9.1.2), and the function $f(x, y, u)$ is continuous in $\bar{S}_{+h}$ and

(i) for all (x, y, u), $(x, \bar{y}, \bar{u}) \in \bar{S}_{+h}$ with $x \neq x_0$ it satisfies

$$(9.4.2) \qquad |f(x, y, u) - f(x, \bar{y}, \bar{u})| \leq l \frac{\phi'(x)}{\phi(x)} \sup(|y - \bar{y}|, |u - \bar{u}|),$$

where $l > 0$ is a constant and the function $\phi(x)$ is continuously differentiable in $[x_0, x_0 + a]$, $\phi(x) > 0$ and $\phi'(x) > 0$ for $x > x_0$; and

(ii) if $\{y(x), u(x)\}$ and $\{\bar{y}(x), \bar{u}(x)\}$ are solutions of (9.1.1), then

$$|f(x, y(x), u(x)) - f(x, \bar{y}(x), \bar{u}(x))| = o((\phi^k(x))')$$

as $x \to x_0^+$. Here $k = l\mu$ and $\mu = \max\{1, L\}$.

Then, the initial value problem (9.1.1) has at most one solution in $[x_0, x_0 + a]$.

Proof. Let $\{y(x), u(x)\}$ and $\{\bar{y}(x), \bar{u}(x)\}$ be two solutions of (9.1.1) in $[x_0, x_0 + a]$. Then, there are two cases to consider.

1. These two solutions are distinct on each interval $[x_0, x_0 + h]$ for sufficiently small $h > 0$. Then, $\max_{x_0 \le t \le x} |y(t) - \bar{y}(t)| > 0$ for $x > x_0$. If we introduce the function $\psi(x) = |f(x, y(x), u(x)) - f(x, \bar{y}(x), \bar{u}(x))|$, then it is clear that $\psi(x) \ge 0$ for $x \ge x_0$. Further, since

$$|y(t) - \bar{y}(t)| \le \int_{x_0}^{t} |f(s, y(s), u(s)) - f(s, \bar{y}(s), \bar{u}(s))| ds \tag{9.4.3}$$

we find that

$$\int_{x_0}^{x} \psi(s) ds = \max_{x_0 \le t \le x} \int_{x_0}^{t} \psi(s) ds \ > \ 0, \quad x > x_0.$$

Thus, this function $\psi(x)$ satisfies the hypothesis (ii) of Lemma 9.4.1.

From the hypothesis (i) of Theorem 9.4.1, (9.1.2) and (9.4.3), we find that for $x > x_0$

$$\psi(x) \le k \frac{\phi'(x)}{\phi(x)} \int_{x_0}^{x} \psi(t) dt. \tag{9.4.4}$$

Therefore, from the definition of $\Phi(x)$ it follows that $\Phi(x) \le k$ for $x > x_0$. But, this contradicts the conclusion of Lemma 9.4.1. Hence, this case is not possible.

2. There is $h_0 > 0$ for which $y(x) = \bar{y}(x)$ when $x \in [x_0, x_0 + h_0]$, but for some sequence $\{x_n\}$ such that $x_n \to x_0 + h_0^+$ as $n \to \infty$, we have $y(x_n) \ne \bar{y}(x_n)$. Then, $\psi(x) = 0$ for $x \in [x_0, x_0 + h_0]$ and

$$\int_{x_0}^{x} \psi(t) dt = \int_{x_0 + h_0}^{x} \psi(t) dt \ > \ 0, \quad x > x_0 + h_0.$$

Now as in case 1 for $x > x_0 + h_0$ it easily follows that

$$\Phi(x) = \frac{\phi(x)\psi(x)}{\phi'(x) \int_{x_0 + h_0}^{x} \psi(t) dt} \le k. \tag{9.4.5}$$

However, since $\psi(x)$ is continuous, $\psi(x_0 + h_0) = 0$ and $(\phi^k(x_0 + h_0))' \ne 0$ it follows that $\psi(x) = o((\phi^k(x))')$ as $x \to x_0 + h_0^+$. Therefore, Lemma 9.4.1 is applicable and we conclude that (9.4.5) must be violated by some sequence $\{\bar{x}_n\}$, $\bar{x}_n \to x_0 + h_0^+$ as $n \to \infty$. Thus, this case is also impossible. ∎

The following particular cases of the function $\phi(x)$ in Theorem 9.4.2 provide various uniqueness criteria for the initial value problem (9.1.1).

(i) **Lipschitz Uniqueness**

$$\phi(x_0) \neq 0 \quad \text{and} \quad \phi'(x_0) \neq 0.$$

The hypothesis (ii) of Theorem 9.4.2 is a consequence of the continuity of the function $f(x, y, u)$ in $\bar{S}_{+h}$.

(ii) **Nagumo's Uniqueness**

$$\phi(x) = x - x_0.$$

Once again for $l = 1$ and $L = 1$ the hypothesis (ii) of Theorem 9.4.2 is a consequence of the continuity of the function $f(x, y, u)$ in $\bar{S}_{+h}$.

(iii) **Roger's Uniqueness**

$$\phi(x) = \exp\left(-\frac{1}{(\alpha - 1)(x - x_0)^{\alpha - 1}}\right), \quad \alpha > 1, \ x > x_0.$$

Clearly, in this case $\phi'(x)/\phi(x) = 1/(x - x_0)^\alpha$.

9.5 BORZDYKO'S UNIQUENESS THEOREM 4

The proof of the following result is similar to that of Theorem 3.20.1.

Theorem 9.5.1 (Borzdyko's Uniqueness Theorem 4). Assume that the function $f(x, y, z)$ is defined in S_{+h}, and

(i) hypothesis (i) of Theorem 3.20.1 hold; and

(ii) there exists a Lyapunov function $V(x, y, \bar{y}, u, \bar{u})$ defined and continuous in $(x_0, x_0 + a) \times R^4$ such that $V(x, y, \bar{y}, u, \bar{u}) = 0$ if and only if $y = \bar{y}$, $u = \bar{u}$; for any two solutions $\{y(x), u(x)\}$, $\{\bar{y}(x), \bar{u}(x)\}$ of (9.1.1) in $[x_0, x_1)$, $\lim_{x \to x_0^+} \frac{V(x, y(x), \bar{y}(x), u(x), \bar{u}(x))}{B(x)} = 0$, and $D_+ V(x, y(x), \bar{y}(x), u(x), \bar{u}(x)) \leq g(x, V(x, y(x), \bar{y}(x), u(x), \bar{u}(x)))$ whenever $x_0 < x < x_1$ and $V(x, y(x), \bar{y}(x), u(x), \bar{u}(x)) > 0$.

Then, the initial value problem (9.1.1) has at most one classical solution in $[x_0, x_1)$. ∎

We note that in the above result the Lipschitz condition (9.1.2) is not required.

We conclude this chapter with the remark that analogs of several other uniqueness criteria discussed earlier for the problem (1.1.1) as well as extension of almost all the results to finite systems with hysteresis of the type (9.1.1) is now straightforward.

CHAPTER 10

GENERALIZED DIFFERENTIAL EQUATIONS

10.1 INTRODUCTION

The concept of generalized differential equations was introduced in the year 1957 by Kurzweil [83]. Although, the motivation of this concept has come in the study of continuous dependence on parameters for differential equations, it has proved to be very useful in topological dynamics, Volterra integral equations, and impulsive as well as measure differential equations. Generalized differential equations are based on Kurzweil - Henstock integral (Generalized Perron integral).

Let $U(t,x)$ be a real valued function defined in $[c,d] \times [c,d]$. A partition P of $[c,d]$ is a finite collection $P = \{x_0, \cdots, x_n; t_1 \cdots, t_n\}$, where $c = x_0 < x_1 < \cdots < x_n = d$, and $t_i \in [x_{i-1}, x_i]$ for each $1 \leq i \leq n$. A gauge in $[c,d]$ is a positive function $\delta(t)$. Given a gauge δ, the partition P is called δ - fine if $[x_{i-1}, x_i] \subset [t_i - \delta(t_i), t_i + \delta(t_i)]$. For a given gauge δ in $[c,d]$ there always exists a δ - fine partition P of $[c,d]$. With the partition P we define the integral sum

$$S(U,P) = \sum_{i=1}^{n} [U(t_i, x_i) - U(t_i, x_{i-1})] \,.$$

Definition 10.1.1. The real valued function $U(t,x)$ defined in $[c,d] \times [c,d]$ is said to be **Kurzweil - Henstock integrable** to I if for every $\varepsilon > 0$ there is a gauge $\delta(t)$ such that for any δ - fine partition P

$$|S(U,P) - I| < \varepsilon.$$

We write $I = \int_c^d DU(t,x)$, and if, I exists then define $\int_d^c DU(t,x) = -\int_c^d DU(t,x)$, and set $\int_c^d DU(t,x) = 0$ if $c = d$. It is clear that Kurzweil - Henstock integral is nonlinear. If we set $U(t,x) = f(t)x$ where $f(t)$ is a real valued function defined in $[c,d]$ then for $t \in [c,d]$, x_1, $x_2 \in [c,d]$ we have $U(t,x_2) - U(t,x_1) = f(t)(x_2 - x_1)$. Therefore, in this case

$$\sum_{i=1}^{n} [U(t_i,x_i) - U(t_i,x_{i-1})] = \sum_{i=1}^{n} f(t_i)(x_i - x_{i-1}).$$

Hence, what we have defined above is the linear Kurzweil - Henstock integral (Generalized Riemann integral). For the geometric interpretation, if we suppose that $f(t) \geq 0$ then this linear integral is in contrast with the classical Riemann integral, here we choose t_i first and according to the behavior of $f(t)$ around t_i decide how close x_{i-1} and x_i should be to t_i so that $f(t_i)(x_i - x_{i-1})$ is a good approximation of the area under the graph of $f(t)$ and between $x = x_{i-1}$, $x = x_i$.

The relation between the Kurzweil - Henstock and the Lebesgue integrals is the following : If $f(t)$ is Lebesgue integrable in $[c,d]$ then it is Kurzweil - Henstock integrable there; conversely, if $f(t)$ is Kurzweil - Henstock integrable in $[c,d]$ and positive then it is Lebesgue integrable there.

More generally, if $U(t,x) = f(t)h(x)$ where the function $h(x)$ is also defined in $[c,d]$ then

$$\sum_{i=1}^{n} [U(t_i,x_i) - U(t_i,x_{i-1})] = \sum_{i=1}^{n} f(t_i)(h(x_i) - h(x_{i-1})).$$

Thus, in this case the Kurzweil - Henstock integral reduces to the generalized Riemann - Stieltjes integral. This integral we will write as $\int_c^d f(t)dh(t)$ instead of the symbol $\int_c^d D[f(t)h(x)]$.

Let the set of all Kurzweil - Henstock integrable functions $U(t,x)$ be $KH[c,d]$. The following properties of the Kurzweil - Henstock integral are fundamental.

P1. If $U(t,x)$, $V(t,x) \in KH[c,d]$ and α, β are constants then $\alpha U(t,x) + \beta V(t,x) \in KH[c,d]$ and

$$\int_c^d D[\alpha U(t,x) + \beta V(t,x)] = \alpha \int_c^d DU(t,x) + \beta \int_c^d DV(t,x).$$

P2. If $U(t,x) \in KH[c,d]$ then $U(t,x) \in KH[c_1,d_1]$ for every interval $[c_1,d_1] \subset [c,d]$.

P3. If $\alpha \in (c,d)$ and $U(t,x) \in KH[c,\alpha]$ as well as $U(t,x) \in KH[\alpha,d]$, then $U(t,x) \in KH[c,d]$ and

$$\int_c^d DU(t,x) = \int_c^\alpha DU(t,x) + \int_\alpha^d DU(t,x).$$

P4. The function $U(t,x) \in KH[c,d]$ if and only if for every $\varepsilon > 0$ there exists a gauge $\delta(t)$ such that

$$|S(U,P_1) - S(U,P_2)| < \varepsilon$$

for all δ - fine partitions P_1, P_2 of $[c,d]$.

P5. Let $U(t,x)$, $V(t,x) \in$ KH$[c,d]$, and $\delta(t)$ be a gauge in $[c,d]$ such that

$$|x-t|\ |U(t,x) - U(t,t)| \le (x-t)(V(t,x) - V(t,t)) \tag{10.1.1}$$

for every $x \in [t-\delta(t), t+\delta(t)] \bigcap [c,d]$ then the inequality

$$\left|\int_c^d DU(t,x)\right| \le \int_c^d DV(t,x) \tag{10.1.2}$$

holds.

In particular, if $V(t,x) = f(t)h(x)$, then (10.1.1) and (10.1.2) reduce to

$$|x-t|\ |U(t,x) - U(t,t)| \le (x-t)f(t)(h(x) - h(t))$$

and

$$\left|\int_c^d DU(t,x)\right| \le \int_c^d f(t)dh(t),$$

respectively.

P6. Let $U(t,x) \in \mathrm{KH}[c,d]$, $\alpha \in [c,d]$. Then,

$$\lim_{s\to\alpha, s\in[c,d]} \left[\int_c^s DU(t,x) - U(\alpha,s) + U(\alpha,\alpha)\right] = \int_c^\alpha DU(t,x).$$

P7. Assume that $\xi(z)$ is continuous and increasing (decreasing) function in $[c,d]$ and one of the integrals $\int_{\xi(c)}^{\xi(d)} DU(t,x)$, $\int_c^d DU(\xi(\sigma),\xi(\eta))$ exists. Then, the other integral exists and

$$\int_{\xi(c)}^{\xi(d)} DU(t,x) = \int_c^d DU(\xi(\sigma),\xi(\eta)).$$

P8. If $f(t)$ and $h(x)$ are functions of bounded variation in $[c,d]$ then the integral $\int_c^d f(t)dh(t)$ exists.

P9. Let $f(t)$ be defined in $[c,d]$ and $|f(t)| \le c$, and let $h(x)$ be defined and of bounded variation in $[c,d]$. Further, let the integral $\int_c^d f(t)dh(t)$ exists then

$$\left|\int_c^d f(t)dh(t)\right| \le c \operatorname{var}_c^d h.$$

Let the function $F(x,y)$ be defined in a open set $S \subset R^2$. A function $y(x)$ is said to be a solution of the generalized differential equation

$$\frac{dy}{dt} = DF(x,y) \tag{10.1.3}$$

in an interval J if for all $x \in J$, $(x, y(x)) \in S$ and if for every x_1, $x_2 \in J$ the identity

$$y(x_2) = y(x_1) + \int_{x_1}^{x_2} DF(x, y(t)) \tag{10.1.4}$$

holds. Here the integral in the right side means that $U(t,x) = F(x,y(t))$ as in Definition 10.1.1.

The notation (10.1.3) for the generalized differential equation is purely symbolical. The letter D indicates that (10.1.3) is a generalized differential equation which is defined only via its solutions. Even the symbol $\frac{dy}{dt}$ on the left side of (10.1.3) does not mean that the solution has a derivative. For example, let $f(x)$ be a continuous, but nowhere differentiable function in $[0,1]$. We set $F(x,y) = f(x)$, to have

$$\int_{x_1}^{x_2} DF(x,y(t)) = \int_{x_1}^{x_2} Df(x) = f(x_2) - f(x_1),$$

i.e., the function $f(x)$ is a solution of the generalized differential equation $\frac{dy}{dt} = Df(x)$ though it has no derivative at any point in $[0,1]$.

A relation between generalized and ordinary differential equations is the following : Let $B \subset R$ be an open set and let the function $f(x,y)$ satisfy the Caratheòdory conditions in $[c,d] \times B$. Suppose that for $y \in B$, x, $x_0 \in [c,d]$ the Lebesgue integral $F(x,y) = \int_{x_0}^{x} f(t,y)dt$ exists. Then, the equation (10.1.3) is the generalized differential equation associated with the ordinary differential equation $y' = f(x,y)$.

Theorem 10.1.1 [83]. Let $f(x,y)$ be as above. Suppose that for every compact subset $A \subseteq B$, there is a locally Lebesgue integrable function $M_A(x)$ such that $|f(x,y)| \le M_A(x)$ for all $y \in A$ and all $x \in [c,d]$. Then, $y(x)$ is a extended solution of the ordinary differential equation $y' = f(x,y)$ if and only if it is a solution of the associated generalized differential equation (10.1.3). ∎

Let $H = (c,d) \times B$. Assume that the function $\psi(x)$ is nondecreasing in $[c,d]$ and left continuous in $(c,d]$ and that the function $g(z)$ is continuous, increasing in $[0,\infty)$ and $g(0) = 0$. The function $F(x,y)$ defined in H belongs to the class $\bar{F}(H,\phi,g)$ if for all (x_1,y), (x_2,y), $(x_1,\bar{y})$, $(x_2,\bar{y}) \in H$ the following inequalities hold

$$|F(x_2,y) - F(x_1,y)| \le |\psi(x_2) - \psi(x_1)| \tag{10.1.5}$$

and

$$|F(x_2,y) - F(x_1,y) - F(x_2,\bar{y}) + F(x_1,\bar{y})| \le g(|y-\bar{y}|)|\psi(x_2) - \psi(x_1)|. \tag{10.1.6}$$

10.2 SCHWABIK'S UNIQUENESS THEOREM

We begin with the following :

Lemma 10.2.1. Assume that

(i) $\psi(x)$ is a nonnegative, nondecreasing and left continuous function in $[c,d]$,

(ii) $g(z)$ is a nonnegative, nondecreasing and continuous function in $[0,\infty)$; and

(iii) $G(z)$ is defined in $[0,\infty)$, and $G'(z) = g(z)$, $z \in [0,\infty)$.

Then, the integral $\int_c^d g(\psi(t))d\psi(t)$ exists and

$$\int_c^d g(\psi(t))d\psi(t) \le G(\psi(d)) - G(\psi(c)). \tag{10.2.1}$$

Proof. Since $g(\psi(t))$ is nondecreasing, it is of bounded variation. Thus, the existence of the integral $\int_c^d g(\psi(t))d\psi(t)$ follows from P8.

Let $\varepsilon > 0$ be given. By the definition of the derivative, for every $z \in [0,\infty)$ there exists $\Delta(z) > 0$ such that if $0 \le |\eta| < \Delta(z)$ then

$$|G(z+\eta) - G(z) - g(z)\eta| \le \varepsilon|\eta|. \tag{10.2.2}$$

Since $\lim_{x\to t+}\psi(x) = \psi(t+)$ for $t \in [c,d)$, there exists a $\delta^r(t) > 0$, $\delta^r(d) = 1$ such that for $x \in (t, t+\delta^r(t)]\bigcap[c,d]$ we have $0 \le \psi(x) - \psi(t+) \le \Delta(\psi(t+))$, and (10.2.2) with $z = \psi(t+)$, $\eta = \psi(x) - \psi(t+)$ implies that

$$g(\psi(t+))(\psi(x) - \psi(t+)) \le G(\psi(x)) - G(\psi(t+)) + \varepsilon(\psi(x) - \psi(t+)).$$

Further, in view of $g(\psi(t)) \le g(z)$, $z \in [\psi(t), \psi(t+)]$, we have

$$g(\psi(t))(\psi(t+) - \psi(t)) - [G(\psi(t+)) - G(\psi(t))] = \int_{\psi(t)}^{\psi(t+)} [g(\psi(t)) - g(z)]dz \le 0.$$

Thus, for $x \in (t, t+\delta^r(t)]\bigcap[c,d]$ it follows that

$$\begin{aligned} |g(\psi(t))(\psi(x) - \psi(t))| &= g(\psi(t))(\psi(x) - \psi(t+)) + g(\psi(t))(\psi(t+) - \psi(t)) \\ &\le g(\psi(t+))(\psi(x) - \psi(t+)) + G(\psi(t+)) - G(\psi(t)) \\ &\le G(\psi(x)) - G(\psi(t)) + \varepsilon(\psi(x) - \psi(t)). \end{aligned}$$

Now by (10.2.2) and the left continuity of the function $\psi(x)$ for $x \in (c,d]$ we find that for $t \in (c,d]$ there is $\delta^l(t) > 0$, $\delta^l(c) = 1$ such that for $x \in [t-\delta^l(t), t]\bigcap[c,d]$ the inequality

$$|g(\psi(t))(\psi(x) - \psi(t))| \le G(\psi(t)) - G(\psi(x)) + \varepsilon(\psi(t) - \psi(x))$$

holds. Let $\delta(t) = \min(\delta^l(t), \delta^r(t)) > 0,\ t \in [c,d]$ then for $t \in [c,d],\ x \in [t-\delta(t), t+\delta(t)]\bigcap[c,d]$, we get

$$|x-t|\ |g(\psi(t))\psi(x) - g(\psi(t))\psi(t)| \le (x-t)[G(\psi(x)) + \varepsilon\psi(x) - G(\psi(t)) - \varepsilon\psi(t)]$$

for which an application of P5 gives

$$\int_c^d g(\psi(t))d\psi(t) \le \int_c^d d[G(\psi(t)) + \varepsilon\psi(t)] = G(\psi(d)) - G(\psi(c)) + \varepsilon(\psi(d) - \psi(c)).$$

Since $\varepsilon > 0$ can be arbitrarily small, the inequality (10.2.1) follows. ∎

Corollary 10.2.2. Let $\psi(x)$ be a nonnegative, nondecreasing and left continuous function in $[c,d]$, and $\alpha \ge 0$. Then,

$$\int_c^d \psi^\alpha(t)d\psi(t) \le \frac{1}{\alpha+1}\left[\psi^{\alpha+1}(d) - \psi^{\alpha+1}(b)\right]. \quad ∎$$

The following result extends Corollary 1.12.2 in the present setting.

Lemma 10.2.3. Assume that

(i) $\phi(x)$ is a nonnegative and bounded function in $[c,d]$,

(ii) $\psi(x)$ is a nonnegative, nondecreasing and left continuous function in $[c,d]$; and

(iii) $g(z)$ is a nondecreasing and continuous function in $[0,\infty)$, such that $g(0) = 0,\ g(z) > 0$ for $z > 0$.

For $\phi > 0$ and $\phi_0 > 0$ the function $G(\phi)$ defined in (1.12.8) maps $(0,\infty)$ into R, increasing, $G(\phi_0) = 0$ and $\lim_{\phi\to 0+} G(\phi) = \alpha \ge -\infty$ and $\lim_{\phi\to\infty} G(\phi) = \beta \le \infty$.

Then, for an arbitrary positive constant k the inequality

$$\phi(x) \le k + \int_c^x g(\phi(t))d\psi(t),\ c \le x \le d \tag{10.2.3}$$

implies that

$$\phi(x) \le G^{-1}(G(k) + \psi(x) - \psi(c)),\ c \le x \le d \tag{10.2.4}$$

provided $G(k)+\psi(d)-\psi(c)<\beta$, where G^{-1} is the inverse function of G and maps (α,β) into R.

Proof. If for some $l>0$, $G(l)+\psi(d)-\psi(c)<\beta$ then for every $t\in[c,d]$ we have $\alpha<G(l)+\psi(t)-\psi(c)<\beta$. Thus, $G(l)+\psi(t)-\psi(c)$, $t\in[c,d]$ belongs to the domain of G^{-1}. For $t\in[c,d]$ we define $w_l(t)=G^{-1}(G(l)+\psi(t)-\psi(c))$. Further, we set

$$\varphi(z)=g(G^{-1}(G(l)+z)),\ z\in[0,\beta-G(l)]. \tag{10.2.5}$$

If $G^{-1}(G(l)+z)>0$ then the derivative G' of the function G exists and

$$G'(G^{-1}(G(l)+z))=\frac{1}{g(G^{-1}(G(l)+z))}\neq 0.$$

Therefore, it follows that

$$\begin{aligned}\frac{d}{dz}[G^{-1}(G(l)+z)]&=\frac{1}{G'(G^{-1}(G(l)+z))}\\&=g(G^{-1}(G(l)+z))=\varphi(z),\ z\in[0,\beta-G(l)].\end{aligned} \tag{10.2.6}$$

If $x\in[c,d]$ then on using (10.2.5), we get

$$\begin{aligned}\int_c^x g(w_l(t))d\psi(t)&=\int_c^x g(G^{-1}(G(l)+\psi(t)-\psi(c)))d\psi(t)\\&=\int_c^x \varphi(\psi(t)-\psi(c))d(\psi(t)-\psi(c)).\end{aligned}$$

Using Lemma 10.2.1 and (10.2.6) in the above equality, we obtain

$$\int_c^x g(w_l(t))d\psi(t)\le G^{-1}(G(l)+\psi(x)-\psi(c))-G^{-1}(G(l))=w_l(x)-l,$$

i.e., for $x\in[c,d]$ we have $l+\int_c^x g(w_l(t))d\psi(t)\le w_l(x)$. For a sufficiently small $\varepsilon>0$ we set $l=k+\varepsilon$ so that $k+\varepsilon+\int_c^x g(w_{k+\varepsilon}(t))d\psi(t)\le w_{k+\varepsilon}(x)$. This is possible for $\varepsilon<\varepsilon_0$ provided $G(k+\varepsilon_0)+\psi(d)-\psi(c)<\beta$. This inequality together with (10.2.3) implies that for all $x\in[c,d]$

$$\begin{aligned}\phi(x)-w_{k+\varepsilon}(x)&\le k+\int_c^x g(\phi(t))d\psi(t)-k-\varepsilon-\int_c^x g(w_{k+\varepsilon}(t))d\psi(t)\\&=-\varepsilon+\int_c^x[g(\phi(t))-g(w_{k+\varepsilon}(t))]d\psi(t).\end{aligned} \tag{10.2.7}$$

Hence, $\phi(c) - w_{k+\varepsilon}(c) \le -\varepsilon$ and also $g(\phi(c)) - g(w_{k+\varepsilon}(c)) \le 0$ because g is nondecreasing. Since ϕ and $w_{k+\varepsilon}$ are bounded there is a constant $C > 0$ such that $|g(\phi(t)) - g(w_{k+\varepsilon}(t))| < C,\ t \in [c,d]$. Thus, in view of P6 and P9 we have

$$\begin{aligned}\phi(x) - w_{k+\varepsilon}(x) &\le -\varepsilon + [g(\phi(c)) + g(w_{k+\varepsilon}(c))](\psi(c+) - \psi(c)) \\ &\quad + \lim_{\delta\to 0+}\int_{c+\delta}^{x}[g(\phi(t)) - g(w_{k+\varepsilon}(t))]d\psi(t) \\ &\le -\varepsilon + C\lim_{\delta\to 0+}[\psi(x) - \psi(c+\delta)] \\ &= -\varepsilon + C[\psi(x) - \psi(c+)].\end{aligned}$$

Since $\lim_{x\to c+}\psi(x) = \psi(c)$, there is an $\eta > 0$ such that for $x \in (c, c+\eta)$ the inequality $\psi(x) - \psi(c+) < \varepsilon/(2C+1)$ holds. Thus, for $x \in (c, c+\eta)$ we get

$$\phi(x) - w_{k+\varepsilon}(x) < -\varepsilon + \frac{C\varepsilon}{2C+1} < -\frac{\varepsilon}{2} < 0.$$

Let $T = sup\{t \in [c,d] : \phi(x) - w_{k+\varepsilon}(x) < 0 \text{ for } x \in [c,t]\}$. It is clear that $T > c$ and for $x \in [c,T)$ we have $\phi(x) - w_{k+\varepsilon}(x) \le 0$. Thus, from the fact that g is nondecreasing it follows that $g(\phi(x)) - g(w_{k+\varepsilon}(x)) \le 0,\ x \in [c,T)$. Therefore, by (10.2.7), P6, and the fact that $\psi(T) - \psi(T-) = \psi(T) - \lim_{t\to T-}\psi(t) = 0$, we have

$$\begin{aligned}\phi(T) - w_{k+\varepsilon}(T) &\le -\varepsilon + \lim_{\delta\to 0+}\int_{c}^{T-\delta}[g(\phi(t)) - g(w_{k+\varepsilon}(t))]d\psi(t) \\ &\quad + [g(\phi(T)) - g(w_{k+\varepsilon}(T))](\psi(T) - \psi(T)) \le -\varepsilon < 0\end{aligned}$$

and

$$\lim_{\delta\to 0+}\int_{c}^{T-\delta}[g(\phi(t)) - g(w_{k+\varepsilon}(t))]d\psi(t) \le 0.$$

If $T < d$ then this procedure can be repeated for $x > T$ by the inequality

$$\phi(x) - w_{k+\varepsilon}(x) \le -\varepsilon + \int_{T}^{x}[g(\phi(t)) - g(w_{k+\varepsilon}(t))]d\psi(t),$$

thus obtaining $\phi(x) - w_{k+\varepsilon}(x) < 0$ where $x \in [T, T+\eta]$ with $\eta > 0$. Hence, $T = d$ and $\phi(x) < w_{k+\varepsilon}(x) = G^{-1}(G(k+\varepsilon) + \psi(x) - \psi(c))$ for $x \in [c,d]$. Finally, since $\varepsilon > 0$ is arbitrarily small and G is continuous, the required inequality (10.2.4) follows. ∎

Corollary 10.2.4. Assume that the function $\psi(x)$ is nonnegative, nondecreasing and right continuous in $[c,d]$. Then, for an arbitrary positive constant k the inequality

$$\phi(x) \le k + \int_x^d g(\phi(t))d\psi(t), \ c \le x \le d \tag{10.2.8}$$

implies that

$$\phi(x) \le G^{-1}(G(k) + \psi(d) - \psi(x)), \ c \le x \le d \tag{10.2.9}$$

provided $G(k) + \psi(d) - \psi(c) < \beta$.

Proof. Let $\bar{\psi}(s) = \psi(d) - \psi(-s), \ s \in [-d,-c]$. Evidently, $\bar{\psi}$ maps $[-d,-c]$ into $[0,\infty)$, left continuous and nondecreasing. Thus, in view of P7 we have

$$\int_x^d g(\phi(t))d\psi(t) = \int_{-x}^{-d} g(\phi(-t))d\psi(-t)$$

$$= -\int_{-d}^{-x} g(\phi(-t))d\psi(-t) = \int_{-d}^{-x} g(\phi(-t))d\bar{\psi}(t)$$

and hence the inequality (10.2.8) can be written as

$$\phi(-x) \le k + \int_{-d}^{-x} g(\phi(-t))d\bar{\psi}(t), \ c \le x \le d$$

which is the same as

$$\bar{\phi}(x) \le k + \int_{-d}^{x} g(\bar{\phi}(t))d\bar{\psi}(t), \ -d \le x \le -c$$

where $\bar{\phi}(t) = \phi(-t), \ t \in [-d,-c]$. Consequently, Lemma 10.2.3 can be used to obtain the inequality

$$\bar{\phi}(x) \le G^{-1}(G(k) + \bar{\psi}(x) - \bar{\psi}(-d)) = G^{-1}(G(k) + \psi(d) - \psi(-x)), \ -d \le x \le -c.$$

Hence, for $x \in [c,d]$ we find the required inequality $\phi(x) \le G^{-1}(G(k) + \psi(d) - \psi(x))$. ∎

Corollary 10.2.5. Let $\phi, \ \psi, \ k$ be as in Lemma 10.2.3 and $L \ge 0$ be a constant. Then, for $x \in [c,d]$ the inequality

$$\phi(x) \le k + L\int_c^x \phi(t)d\psi(t)$$

implies that

$$\phi(x) \le k\exp(L(\psi(x)-\psi(c))). \qquad \blacksquare$$

Corollary 10.2.6. Let ϕ, ψ, k be as in Corollary 10.2.4 and $L \ge 0$ be a constant. Then, for $x \in [c,d]$ the inequality

$$\phi(x) \le k + L\int_x^d \phi(t)d\psi(t)$$

implies that

$$\phi(x) \le k\exp(L(\psi(d)-\psi(x))). \qquad \blacksquare$$

An extension of Györi's Uniqueness Theorem 1.12.3 is the following :

Theorem 10.2.7 (Schwabik's Uniqueness Theorem). Assume that $F(x,y) \in \bar{F}(H,\psi,g)$, and

(i) for $(x_0,y_0) \in H_F = \{(x,y) : (x, y + F(x+,y) - F(x,y)) \in H\}$ the generalized differential equation (10.1.3) with $y(x_0) = y_0$ has a solution $y(x)$ in the interval $[x_0, x_0+a] \subset (c,d)$; and

(ii) for every $k > 0$ there exists a $\Delta > 0$, $\Delta < k$ such that

$$\int_\Delta^k \frac{dz}{g(z)} > \psi(x_0+a) - \psi(x_0+). \tag{10.2.10}$$

Then, there exists $x_1 \in (x_0, x_0+a]$ such that the solution $y(x)$ is unique in $[x_0, x_1]$.

Proof. Assume that $y(x)$ and $\bar{y}(x)$ are two solutions of (10.1.3) satisfying $y(x_0) = \bar{y}(x_0) = y_0$ and both are defined in $[x_0, x_0+a]$. Then, in view of (10.1.6) and P5 for every $x \in (x_0, x_0+a]$ we have the inequality

$$\begin{aligned} |y(x)-\bar{y}(x)| &= \left|\int_{x_0}^x D[F(s,y(t)) - F(s,\bar{y}(t))]\right| \\ &\le \int_{x_0}^x g(|y(t)-\bar{y}(t)|)d\psi(t) \end{aligned} \tag{10.2.11}$$

$$= \int_{x_0}^{x_0+\delta} g(|y(t)-\bar{y}(t)|)d\psi(t) + \int_{x_0+\delta}^x g(|y(t)-\bar{y}(t)|)d\psi(t)$$

where $0 < \delta < x - x_0$. Further, from P5 and P6 we have

$$\int_{x_0}^{x_0+\delta} g(|y(t) - \bar{y}(t)|)d\psi(t) = g(|y(x_0) - \bar{y}(x_0)|)(\psi(x_0+) - \psi(x_0))$$

$$+ \lim_{\bar{x}\to x_0} \int_{\bar{x}}^{x_0+\delta} g(|y(t) - \bar{y}(t)|)d\psi(t)$$

$$\leq \sup_{t\in(x_0,x_0+\delta]} g(|y(t) - \bar{y}(t)|)(\psi(x_0+\delta) - \psi(x_0+)) = A(\delta)$$

because $g(|y(x_0) - \bar{y}(x_0)|) = g(0) = 0$. Since $\lim_{\delta\to 0^+}[\psi(x_0+\delta) - \psi(x_0)] = 0$ we also have $\lim_{\delta\to 0^+} A(\delta) = 0$. Hence, for every $\varepsilon > 0$ and $x \in [x_0+\delta, x_0+a]$, (10.2.11) and the increasing behavior of g implies that

$$|y(x) - \bar{y}(x)| + \varepsilon \leq A(\delta) + \varepsilon + \int_{x_0+\delta}^{x} g(|y(t) - \bar{y}(t)| + \varepsilon)d\psi(t).$$

Now an application of Lemma 10.2.3 to the above inequality gives that for $x \in [x_0+\delta, x_0+\lambda]$, $0 < \lambda \leq a$

$$|y(x) - \bar{y}(x)| + \varepsilon \leq G^{-1}(G(A(\delta)+\varepsilon) + \psi(x) - \psi(x_0+\delta)) \tag{10.2.12}$$

provided $G(A(\delta)+\varepsilon)+\psi(x_0+\lambda)-\psi(x_0+\delta) \leq G(A(\delta)+\varepsilon)+\psi(x_0+\lambda)-\psi(x_0+) < \beta = \lim_{\phi\to\infty} G(\phi)$. But, this can be achieved by the choice of sufficiently small $\varepsilon > 0$, $\delta > 0$ and $\lambda > 0$.

Since inequality (10.2.12) is the same as $G(|y(x) - \bar{y}(x)| + \varepsilon) \leq G(A(\delta) + \varepsilon) + \psi(x) - \psi(x_0+\delta)$, for $x \in [x_0+\delta, x_0+a]$, we have

$$G(|y(x) - \bar{y}(x)| + \varepsilon) - G(A(\delta)+\varepsilon) = \int_{A(\delta)+\varepsilon}^{|y(x)-\bar{y}(x)|+\varepsilon} \frac{dz}{g(z)} \tag{10.2.13}$$

$$\leq \psi(x) - \psi(x_0+\delta) \leq \psi(x_0+\lambda) - \psi(x_0+\delta) \leq \psi(x_0+\lambda) - \psi(x_0+).$$

If $|y(x^*) - \bar{y}(x^*)| = k > 0$ for some $x^* \in (x_0, x_0+\lambda]$ then we can choose $\varepsilon > 0$ and $\delta > 0$ such that $A(\delta) + \varepsilon < \Delta(k)$ and in this case from (10.2.13) it follows that

$$\int_{\Delta(k)}^{k} \frac{dz}{g(z)} \leq \int_{\varepsilon+A(\delta)}^{\varepsilon+k} \frac{dz}{g(z)} \leq \psi(x_0+\lambda) - \psi(x_0+) \leq \psi(x_0+a) - \psi(x_0+).$$

However, this inequality contradicts (10.2.10). Hence, $|y(x) - \bar{y}(x)| = 0$ for every $x \in [x_0, x_0+a]$. ∎

From (10.2.13) it follows that in Theorem 10.2.7 the inequality (10.2.10) can be replaced by the following : there exists no positive constant k such that the inequality

$$\int_{\varepsilon+A(\delta)}^{\varepsilon+k} \frac{dz}{g(z)} \leq \psi(x_0+a) - \psi(x_0+\delta) \tag{10.2.14}$$

holds for all ε, $\delta > 0$ and every function $A(\delta)$ such that $\lim_{\delta\to 0^+} A(\delta) = 0$. This restriction is similar to that of condition (iii) of Theorem 1.12.3.

It is also clear that in Theorem 10.2.7 the hypothesis (ii) can be replaced by the condition (1.4.1). Thus, Theorem 10.2.7 includes in particular Osgood and Lipschitz type uniqueness results.

Let $B \subset R$ be an open set and let the function $f(x,y)$ satisfy the Carathèodory conditions in $[c,d] \times B$. Further, let there exist a Lebesgue integrable function $\mu(x)$ in $[c,d]$ such that $\int_c^d \mu(t)dt < \infty$ and for all (x,y) $(x,\bar{y}) \in [c,d] \times \bar{B}$, $|f(x,y) - f(x,\bar{y})| \leq \mu(x)|y-\bar{y}|$. Then, for $y \in \bar{B}$, x, $x_0 \in [c,d]$ the Lebesgue integrable function $F(x,y) = \int_{x_0}^{x} f(t,y)dt$ satisfies the inequality

$$|F(x_2,y) - F(x_1,y) - F(x_2,\bar{y}) + F(x_1,\bar{y})| \leq \left|\int_{x_1}^{x_2} \mu(t)dt\right| \, |y-\bar{y}|.$$

Thus, if we define $\psi(x) = \int_c^x (M(t)+\mu(t))dt$ then this $F(x,y) \in \bar{F}(H,\psi,g)$ with $g(z) = z$. Hence, for the generalized differential equation (10.1.3) associated with the ordinary differential equation $y' = f(x,y)$ Lipschitz type uniqueness result holds. Theorem 10.1.1 then implies that the initial value problem (1.1.1) has at most one extended solution in $[x_0, x_1]$.

Theorem 10.2.7 provides only the local uniqueness of the solutions of the generalized differential equation (10.1.3). An immediate extension of this result to global uniqueness is the following :

Theorem 10.2.8. Assume that $F(x,y) \in \bar{F}(H,\psi,g)$, and $y(x)$ and $\bar{y}(x)$ are two solutions of the generalized differential equation (10.1.3) in $[c_1,d_1]$ and $[c_2,d_2]$, respectively. If the condition (10.2.10) holds for every $x_0 \in [c,d)$ and a sufficiently small $a > 0$ and if $y(\bar{x}) = \bar{y}(\bar{x})$ for some $\bar{x} \in [c_1,d_1]\bigcap[c_2,d_2]$, then $y(x) = \bar{y}(x)$ for all $x \in [c_1,d_1]\bigcap[c_2,d_2]\bigcap[\bar{x},d]$. ∎

84. **J.Kurzweil,** Generalized ordinary differential equations, Czech. Math. J. 8(1958), 360-389.

85. **J.Kurzweil,** Unicity of solutions of generalized ordinary differential equations, Czech. Math. J. 8(1958), 502-504.

86. **G.E.Ladas** and **V.Lakshmikantham,** Differential Equations in Abstract Spaces, Academic Press, New York, 1972.

87. **V.Lakshmikantham,** Uniqueness theorems for ordinary and hyperbolic differential equations, Michigan Math. J. 9 (1962), 161-166.

88. **V.Lakshmikantham,** On the Kamke's function in the uniqueness theorem of ordinary differential equations, Proc. Nat. Acad. Sci. India Ser.A. 34(1964), 11-14.

89. **V.Lakshmikantham,** Some results in functional differential equations, Proc. Nat. Acad. Sci. India Ser.A. 34(1964), 299-306.

90. **V.Lakshmikantham,** Properties of solutions of abstract differential equations, Proc. London Math. Soc. 14(1964), 74-82.

91. **V.Lakshmikantham** and **G.R.Shendge,** Functional differential Inequalities, Ann. Acad. Brasil Ci. 39(1967), 7-14.

92. **V.Lakshmikantham** and **S.Leela,** Differential and Integral Inequalities, Vols.I and II, Academic Press, New York, 1969.

93. **V.Lakshmikantham** and **S.Leela,** Nonlinear Differential Equations in Abstract Spaces, Pergamon Press, Oxford, 1981.

94. **V.Lakshmikantham, D.D.Bainov** and **P.S.Simeonov,** Theory of Impulsive Differential Equations, World Scientific, Singapore, 1989.

95. **J.LaSalle,** Uniqueness theorems and successive approximations, Annals of Math. 50(1949), 722-730.

96. **M.Lavrentieff,** Sur une èquation diffèrentielle du premier ordre, Math. Z. 23(1925), 197-209.

97. **R.Lemmert,** Über einen Salz von Witte, Math. Z. 145(1975), 289-291.

98. **R.Lemmert,** Zur Eindeutigkeitsbedingung von Nagumo, Comment. Math. Univ. Carolin. 29(1988), 293-294.

99. **P.Lèvy,** Processus Stochastiques et Mouvement Brownien, Gauthier-Villars, Paris, 1948.

100. **R.Lipschitz,** Sur la possibilitè d'intègrer complètement un système donnè d'èqations diffèrentielles, Bull. Sci. Math. Astro. 10(1876), 149-159.

NOTES

Chapter 1

In literature the initial value problem (1.1.1) is known as Cauchy problem. From the last one hundred years this problem has been a subject matter of numerous investigations. This includes qualitative as well as quantitative properties of its solutions. What we have discussed in this monograph is a cumulation of up-to-date uniqueness results, which is just one aspect of the qualitative study of (1.1.1). Lipschitz condition (1.2.1) appeared in the year 1876, [100]. Lemmas 1.2.1 and 1.2.2 are available in almost every book on differential equations, e.g., [2]. Lemma 1.2.3 is a particular case of the well-known Gronwall inequality. The first proof of Theorem 1.2.4 is classical, whereas its second proof is modeled after Diaz and Walter [40]. The third proof of Theorem 1.2.4 is based on Ziebur [181]. Theorem 1.2.5 can be found at many places, e.g., [2]. Theorem 1.2.6 is also well-known, e.g., see [2] or Rudin [141]. Theorem 1.2.7 is borrowed from Hales and Sells [57]. Theorem 1.2.8 can be found in Kamke [71]. Theorems 1.3.1 - 1.3.3 are also classical and are known since 1890 [129]. Lemma 1.4.1 is essentially due to LaSalle [95], however it is better known as Bihari's lemma [12]. Osgood's condition (1.4.3) appeared in the year 1898 [125]. Theorem 1.4.2 is based on Bihari [12]. Theorems 1.4.3 and 1.4.4 are available at several places, e.g., Petrovski [133]. The second proof of Theorem 1.4.4 is adapted from Chu and Diaz [34]. Theorem 1.5.1 appeared in the year 1925 [158], and also a year later in 1926 [111]. Condition (1.6.1) with the restriction $k < 1$ first appeared in the work of Rosenblatt [140], whereas in the year 1926 Nagumo [116] allowed $k \leq 1$. Lemma 1.6.1 and the first proof of Theorem 1.6.2 can be found at various places, e.g., [2]. The second proof of Theorem 1.6.2 is borrowed from Diaz and Walter [40]. Examples 1.6.1 - 1.6.2 are from Perron [132], and Diaz and Walter [40], respectively. Initial value problem (1.6.5) appeared in the work of Müller [113], but for a different purpose. Also, see Coddington and Levinson [36]. Examples 1.6.4 and 1.6.5 are due to Haviland [61]. Lemmas 1.2.3, 1.4.1 and 1.6.1 have also been proved in Hille [65]. Theorem 1.7.1 is taken from Krasnosel'skii and Krein [80]. Theorem 1.8.1 is taken from Kooi [79], whereas the proof of his second uniqueness criterion Theorem 1.8.2 is modeled after Luxemburg [103]. All results in Section 1.9 are adapted from Rogers [139]. Theorem 1.10.1 is due to Witte [175]. In [97] Lemmert showed that this result can be deduced from the Nagumo Uniqueness Theorem 1.6.2. Lemma 1.10.2, Theorem 1.10.3 and its following Corollary 1.10.4 are modeled after Banaŝ and Rivero [10]. Conditions (1.10.7) and (1.10.8) have appeared in Lemmert [98], and Doŝla and Doŝlŷ [45], respectively. Perron [131] established Theorem 1.11.4 in the year 1926, however with the additional assumption that $g(x,z)$ is monotone increasing in z Bompiani [13] proved the same result a year earlier in 1925. All results in Section 1.12 are taken from Györi [56]. Although, Corollary 1.12.6 is deduced from Theorem 1.12.3 it is essentially due to Yang [176]. Theorem 1.13.1 was proved in the year 1928 [69], however its proof we have modeled after Shimizu [152]. Theorem 1.14.1 is borrowed from Inaba [68]. Theorem 1.15.1 was established in the year 1930 [71]. Its proof which is based on Lemma 1.15.2 is from Olech [122]. Corollary 1.15.3 is due to Szarski [155]. Corollaries 1.15.4 and 1.15.5 are adapted from Wintner [173]. Corollary 1.15.6 was independently proved by Filippov [47] and LaSalle [95]. Its consequences (v) and (vi) are from van Kampen [160], and Iyanaga [69], respectively. Theorem 1.16.1 is taken from [151]. Section 1.17 is based on the work of Mikolajska [110]. Theorem 1.18.1 is due to Brauer [26]. Lemma 1.18.2 and the proof of Theorem 1.18.1 are taken from Walter [165]. Theorem 1.19.1 is a special case of the result of Wend [170]. Theorem 1.20.1 is due to van Kampen [159], however its proof is adapted from Hartman [60]. All results in Section 1.21 are borrowed from Yosie [179]. Section 1.22 is based on the work of Kitagawa [78]. Theorem 1.23.1 is the first known nonuniqueness criterion which appeared in the year 1922 [156]. Theorem 1.24.1 is from Lakshmikantham [88]. Samimi [143] contains Theorem 1.25.1.

Chapter 2

Theorem 2.2.1 is due to Diaz [39,42]. Theorem 2.2.2 is adapted from Wallach [163]. Theorem 2.3.1 and the following remarks are taken from Diaz and Walter [40]. Theorem 2.4.1 is based on the work of Chu and Diaz [34]. Theorem 2.5.1 is the contribution of Borùvka [14], however its proof is modeled after Chrastina [33]. Theorem 2.6.1 is a special case of the result of Wend [171]. Section 2.7 is based on the work of Zabreiko [180]. Lemmas 2.8.1 and 2.8.2 and Theorem 2.8.3 are taken from Lakshmikantham [87]. Theorems 2.9.1, 2.10.1 and 2.11.1 are adapted from Bownds [19], Bownds and Metcalf [21], and Gard [49], respectively. All results of Section 2.12 are borrowed from Majorana [104].

Chapter 3

In Section 3.2 the Lemma 3.2.1 is classical, and the second proof of the Lipschitz Uniqueness Theorem 1.2.4 for n - solutions is modeled after Diaz and Walter [40]. Theorems 3.3.1 - 3.3.3 are direct extension of the corresponding Theorems 1.3.1 - 1.3.3. Lemma 3.4.1 and Theorem 3.4.2 are borrowed from Heins [63]. Theorem 3.5.1 is adapted from Giuliano [51,52]. Theorem 3.6.1 is taken from Hartman [60]. Banaš and Rivero [10] form the bases of Section 3.7. Theorem 3.8.1 is from Coddington and Levinson [36], also see Sansone and Conti [145]. Theorem 3.9.1 is adapted from Wend [171]. Bownds [20] contains Theorem 3.10.1. Theorem 3.11.1 is due to Gard [49]. Theorem 3.12.1 is adapted from Norris and Driver [118]. Theorems 3.13.1 and 3.13.2 are based on Goebel and Zur [53]. Theorem 3.14.1 is taken from Bownds and Diaz [22]. Theorem 3.15.1 is due to Okamura [119-121], however its proof as well as other results in Section 3.15 are based on Yoshizawa [178]. Theorem 3.16.1 is borrowed from Chu and Diaz [34]. Satò [146] forms the bases of Section 3.17. Theorems 3.18.1 and 3.18.3 are from Lakshmikantham and Leela [92], whereas Theorem 3.18.2 is modeled after Brauer and Sternberg [23,24]. In Section 3.19 all the results are borrowed from Moyer [112], except Remark 3.19.1 and Theorem 3.19.6 which are due to George [50]. All the results in Section 3.20 are taken from Bernfeld, Driver and Lakshmikantham [11]. Section 3.21 is based on Kaper and Kwong [72,73]. Theorem 3.22.1 and its following Corollaries 3.22.2 - 3.22.7 are modeled after Tesařovà and Došlý [157]. Although, Corollary 3.22.7 is deduced from Theorem 3.22.1 it is a special case of the result of Perov [130]. Theorem 3.23.1 is due to Antosiewicz [3], and its generalization Theorem 3.23.2 is adapted from [157]. All the results in Section 3.24 are taken from Samimi and Lakshmikantham [144]. Theorem 3.25.1 and Corollaries 3.25.2 and 3.25.3 are borrowed from Kalas [70]. The last nonuniqueness criterion Theorem 3.26.1 is due to Samimi [143].

Lyapunov type functions have also been used to study uniqueness results in Chandra [30], Rao [134,135], Subramaniam and Lakshmikantham [154]. Stettner [153] seems to be the first to use Lyapunov functions to obtain nonuniqueness criterion, however the proof of his theorem is not correct.

Some closely related results of Chapters 1-3 have also appeared in Hartman [59], Krasnosel'skii and Krein [81], Lèvy [99], Markus [106], McShane [107], Scorza-Dragoni [149,150], Zwirner [182-184].

Most of the uniqueness criteria for (1.1.1) in Chapters 1-3 are directly related to the question of convergence of classical Picard's method of successive approximations to its solution. Although, the uniqueness of the solution of (1.1.1) does not imply such convergence, it has been shown that in addition to the sufficient conditions for uniqueness a certain requirement of monotonicity do imply convergence. For the results in this direction see Brauer and Sternberg [23,24], Brauer [25,26], Coddington and Levinson [35], Dieudonnè [43], Kooi [79], LaSalle [95], Luxemburg [101-103], Müller [113,114], Nikliborc [117], Olech and Plis [124], van Kampen [160], Viswanatham [161], Walter

[166], Wazewski [168,169], Wintner [172], Witte [175].

A classification of several known uniqueness criteria of Chapters 1-3 has been established in Walter [165,167]. Some of the extensions of this classification are available in Banaś, Hajnosz and Wedrychowicz [9].

For an excellent survey of uniqueness results up to the year 1948 see Cafiero [28,29].

Chapter 4

Lipschitz type uniqueness results for the second order differential equations are available in Kearns [76] and Rao [136]. Their proof is similar to that of the first proof of Theorem 1.2.4. All the three proofs of Theorem 4.2.2 are analogs of the corresponding proofs of Theorem 1.2.4. Theorem 4.3.1 is based on Merli [109], however its present proof is different. A particular case of Theorem 4.3.1 first appeared in Wintner [174]. Theorem 4.4.1 and its following Corollaries 4.4.2 and 4.4.3 are modeled after Kikodze [77]. Theorem 4.5.1 is adapted from Wend [170]. Lemma 4.6.1 as well as Theorem 4.6.2 is taken from Redheffer [138]. Theorems 4.7.1 and 4.7.3 are due to Heidel [62], whereas Theorem 4.7.2 is deduced from the work of Bernfeld, Driver and Lakshmikantham [11].

Chapter 5

For a detailed discussion of differential equations in abstract spaces see [86,93]. Theorem 5.2.1 is available in [86]. Theorems 5.2.2 and 5.2.3 are adapted from Aziz and Diaz [6]. Theorems 5.3.1 and 5.4.1 are borrowed from Medeiros [108]. Theorems 5.3.2, 5.4.2 and 5.4.3 are based on Diaz and Weinacht [41], whereas Theorems 5.4.4 and 5.4.5 are taken from [6]. Theorem 5.5.1 is due to Lakshmikantham and Leela [93]. Theorem 5.6.1 is from Olech [123]. Lemma 5.7.1 and Theorem 5.7.2 are modeled from Kato [74], and Goldstein [54], respectively. Theorem 5.8.1 is based on Arrate [4], whereas Corollary 5.8.3 is essentially due to Goldstein [55]. Lemma 5.9.1 and Theorem 5.9.2 are from Ramankutty [137]. Theorem 5.10.1 is a contribution of Murakami [115]. Theorem 5.11.1 is taken from Lakshmikantham [90]. For related results see Mamedov [105], Sadovskii [142].

Chapter 6

Theorem 6.2.1 is taken from Deo and Lakshmikantham [38].

Chapter 7

Lemma 7.2.1 and Theorem 7.2.2 are borrowed from Lakshmikantham [89]. Theorem 7.3.1 is due to Lakshmikantham and Leela [92]. Lemma 7.4.1 as well as Theorem 7.4.2 is taken from Lakshmikantham and Shendge [91]. Furumochi [48] forms the basis of Section 7.5.

Chapter 8

For a detailed discussion of impulsive and measure differential equations see Bainov and Simeonov [8], Lakshmikantham, Bainov and Simeonov [94], and Pandit and Deo [128]. Theorem 8.2.1 is due to Hu and Lakshmikantham [67]. Lemma 8.3.2 and Corollary 8.3.3 are from Pandit [126]. Theorem 8.3.4 is adapted from Pandit [127].

Chapter 9

Existence and Lipschitz type uniqueness results for the problem (9.1.1) have appeared in [16]. Lemma 9.2.1 and Theorems 9.2.2 and 9.3.1 are adapted from Borzdyko [17]. Lemma 9.4.1 and Theorem 9.4.2 are modeled after Borzdyko [18]. Theorem 9.5.1 is from Borzdyko [15] and can be applied to Prandtl's model of elastoplastic fiber.

Chapter 10

For a detailed discussion of Kurzweil - Henstock integral see Kurzweil [83] and Schwabik [148]. This integral is nonlinear. In [31] Chew has provided necessary and sufficient conditions so that a Kurzweil - Henstock integral can be expressed in terms of a Kurzweil - Henstock linear integral with a kernel function. Generalized Riemann integral and its limitations in the theory of differential equations has been discussed in Hu and Lakshmikantham [66]. For a detailed discussion of generalized differential equations see Chew and Flordeliza [32], Kurzweil [83,84], and Schwabik [148]. All the results in Section 10.2 are based on Schwabik [147]. Kurzweil [85] contains two more uniqueness criteria.

In an attempt to unify the theory of continuous and discrete dynamic systems Aulbach and Hilger [5] and Hilger [64] proposed the study of dynamic systems on time scale, i.e., on a closed subset of reals. This demands a fresh study, starting from the basic problems of existence and uniqueness of the solutions of dynamical systems on time scale. In fact, for such systems Lipschitz and Perron type uniqueness criteria have been established recently in [64], and Kaymakcalan [75], respectively.

Finally, we remark that while for the Cauchy problem uniqueness results have required a monograph of this size, for the boundary value problems for ordinary differential equations only few uniqueness criteria are known [1].

REFERENCES

1. **R.P.Agarwal,** Boundary Value Problems for Higher Order Differential Equations, World Scientific, Singapore, 1986.

2. **R.P.Agarwal** and **R.C.Gupta,** Essentials of Ordinary Differential Equations, McGraw-Hill, Singapore, New York, 1991.

3. **H.A.Antosiewicz,** An inequality for approximate solutions of ordinary differential equations, Math. Z. 78(1962), 44-52.

4. **M.Arrate,** A uniqueness criterion for ordinary differential equations in Banach spaces, Proc. Amer. Math. Soc. 81(1981), 421-424.

5. **B.Aulbach** and **S.Hilger,** Linear dynamic processes with inhomogeneous time scale, Nonlinear Dynamics and Quantum Dynamical Systems, Akademie-Verlag, Berlin, 1990, 9-20.

6. **A.K.Aziz** and **J.B.Diaz,** On a mean value theorem of the differential calculus of vector - valued functions, and uniqueness theorems for ordinary differential equations in a linear - normed space, Contrib. Diff. Eqs. 1(1963), 251-269.

7. **G.Bachman** and **L.Narici,** Functional Analysis, Academic Press, New York, 1966.

8. **D.D.Bainov** and **P.S.Simeonov,** Systems with Impulse Effect, Ellis Horwood Limited, Chichester, 1989.

9. **J.Banaś, A.Hajnosz** and **S.Wedrychowicz,** Relations among various criteria of uniqueness for ordinary differential equations, Comment. Math. Univ. Carolin. 22(1981), 59-70.

10. **J.Banaś** and **J.Rivero,** Remarks concerning J.Witte's theorem and its applications, Comment. Math. Univ. Carolina 28(1987), 23-31.

11. **S.R.Bernfeld, R.D.Driver** and **V.Lakshmikantham,** Uniqueness for ordinary differential equations, Math. Systems Theory 9(1976), 359-367.

12. **I.Bihari,** A generalization of a lemma of Bellman and its applications to uniqueness problems of differential equations, Acta Math. Acad. Sci. Hungar. 7(1956), 81-94.

13. **E.Bompiani,** Un teorema di confronto ed un teorema di unicitá per l'equazione differenziale $y' = f(x,y)$, Atti. Accad. Naz. Lincei Rend. Cl. Sci. Fis. Mat. Nat. 1(1925), 298-302.

14. **O.Borùvka,** Über eine Verallgemeinerung der Eindeutigkeitssätze für Integrale der Differentialgleichung $y' = f(x,y)$, Acta Facultatis Rerum Natur. Univ. Comen. 1(1956), 155-167.

15. **V.I.Borzdyko,** A uniqueness theorem of S.R.Bernfeld, R.D.Driver - V. Lakshmikantham theorem type for differential equations with hysteresis nonlinearities, Dokl. Akad. Nauk Tadzhik SSR 30(1987), 74-77.

16. **V.I.Borzdyko,** Conditions for the existence and uniqueness of a solution of the Cauchy problem for a system of ordinary differential equations with hysteresis nonlinearities, Dokl. Akad. Nauk Tadzhik SSR 30(1987), 766-770.

17. **V.I.Borzdyko,** Uniqueness theorem for differential equations with hysteresis nonlinearities, Diff. Equa. 23(1987), 623-626.

18. **V.I.Borzdyko,** Uniqueness conditions for differential - equation systems with hysteresis nonlinearities, Diff. Equa. 24(1988), 827-830.

19. **J.M.Bownds,** A uniqueness theorem for $y' = f(x,y)$ using a certain factorization of f, J. Diff. Equs. 7(1970), 227-231.

20. **J.M.Bownds,** A uniqueness theorem for non-Lipschitzian systems of ordinary differential equations, Funkcial. Ekvac. 13(1970), 61-65.

21. **J.M.Bownds** and **F.T.Metcalf,** An extension of the Nagumo uniqueness theorem, Proc. Amer. Math. Soc. 27(1971), 313-316.

22. **J.M.Bownds** and **J.B.Diaz,** On restricted uniqueness for systems of ordinary differential equations, Proc. Amer. Math. Soc. 37(1973), 100-104.

23. **F.Brauer** and **S.Sternberg,** Local uniqueness, existence in the large, and the convergence of successive approximations, Amer. J. Math. 80 (1958), 421-430.

24. **F.Brauer** and **S.Sternberg,** Errata to our paper "Local uniqueness, etc." Amer. J. Math. 81(1959), 797.

25. **F.Brauer,** A note on uniqueness and convergence of successive approximations, Canad. Math. Bull. 2(1959), 5-8.

26. **F.Brauer,** Some results on uniqueness and successive approximations, Canad. J. Math. 11(1959), 527-533.

27. **F.E.Browder,** Non-linear equations of evolution, Ann. Math. 80(1964), 485-523.

28. **F.Cafiero,** Sui teoremi di unicità relativi ad un'equazione differentiale ordinaria del primo ordine, Giorn. Mat. Battaglini 78(1948), 10-41.

29. **F.Cafiero,** Sui teoremi di unicità relativi ad un'equazione differentiale ordinaria del primo ordine, Giorn. Mat. Battaglini 78(1948), 193-215.

30. **J.Chandra,** On uniqueness of solutions of a non-linear system of differential equations, Proc. Nat. Acad. Sci. India Sec.A. 32(1962), 187-190.

31. **T.S.Chew,** On Kurzweil generalized ordinary differential equations, J. Diff. Equs. 76(1988), 286-293.

32. **T.S.Chew** and **F.Flordeliza,** On $x' = f(t, x)$ and Henstock-Kurzweil integrals, Diff. and Int. Equs. 4(1991), 861-868.

33. **J.Chrastina,** A note on the criteria of uniqueness of the solution of equation $y' = f(x, y)$, Arch. Math. (Brnö) 5(1969), 75-77.

34. **S.C.Chu** and **J.B.Diaz,** On uniqueness conditions for the initial value problem for the differential equation $y' = f(x, y)$, Revista de la Uniòn Matemàtica Argentina 25(1970), 109-121.

35. **E.A.Coddington** and **N.Levinson,** Uniqueness and the convergence of successive approximations, J. Indian Math. Soc. 16(1952), 75-81.

36. **E.A.Coddington** and **N.Levinson,** Theory of Ordinary Differential Equations, McGraw-Hill, New York, 1955.

37. **P.C.Das** and **R.R.Sharma,** Existence and stability of measure differential equations, Czech. Math. J., 22(1972), 145-158.

38. **S.G.Deo** and **V.Lakshmikantham,** On complex differential equations, Bol. Soc. Mat. Sao Paulo 18(1966), 3-8.

39. **J.B.Diaz,** On existence, uniqueness, and numerical evaluation of solutions of ordinary and hyperbolic differential equations, Ann. Mat. Pure Appl. 52(1960), 163-181.

40. **J.B.Diaz** and **W.Walter,** On uniqueness theorems for ordinary differential equations and for partial differential equations of hyperbolic type, Trans. Amer. Math. Soc. 96(1960), 90-100.

41. **J.B.Diaz** and **R.J.Weinacht,** On nonlinear differential equations in Hilbert spaces, Appl. Anal. 1(1971), 31-41.

42. **J.B.Diaz,** A comparison of two uniqueness theorems for the ordinary differential equation $y' = f(x,y)$, Inequalities III, Academic Press, New York (1972), 65-82.

43. **J.Dieudonnè,** Sur la convergence des approximations successives, Bull. Sci. Math. 69(1945), 62-72.

44. **J.Dieudonnè,** Deux exemples singuliers d'èquations diffèrentielles, Acta. Sci. Math. 12(1950), 38-40.

45. **Z.Došla** and **O.Došlý,** A remark on uniqueness criteria for initial value problem, Comment. Math. Univ. Carolin. 30(1989), 447-453.

46. **R.D.Driver** and **M.J.Norris,** Note on uniqueness for a one - dimensional two - body problem of classical electrodynamics, Ann. Physics 42(1967), 347-351.

47. **A.F.Filippov,** Sufficient conditions for uniqueness of solutions of differential equations, Dokl. Akad. Nauk SSSR (N.S.) 60(1948), 549-552.

48. **T.Furumochi,** On the uniqueness of solutions of a system of functional - differential equations, Funkcial. Ekvac. 15(1972), 149-159.

49. **T.C.Gard,** A generalization of the Nagumo uniqueness criterion, Proc. Amer. Math. Soc. 70(1978), 167-172.

50. **J.H.George,** On Okamura's uniqueness theorem, Proc. Amer. Math. Soc. 18(1967), 764-765.

51. **L.Giuliano,** Sull'unicità della soluzioni dei sistemi di equazioni differenziali ordinarie, Boll. Un. Mat. Ital. 2(1940), 221-227.

52. **L.Giuliano,** Su un notevole teorema di confronto e su teorema di unicità per i sistemi di equazioni differenziali ordinarie, Rend. R. Acc. Italia 1(1940), 330-336.

53. **K.Goebel** and **L.Zur,** Some remarks concerning of uniqueness conditions of Lipschitz type, Annales Uni. Mariae Curie - Sklodowska Lublin - Polonia, 25(1971), 27-32.

54. **J.A.Goldstein,** Uniqueness for nonlinear Cauchy problems in Banach spaces, Proc. Amer. Math. Soc. 53(1975), 91-95.

55. **J.A.Goldstein,** The exact amount of nonuniqueness for singular ordinary differential equations in Banach spaces with an application to the Euler-Poisson-Darboux equations, in Nonlinear Equations in Abstract Spaces, ed. V.Lakshmikantham, Academic Press, New York, 1978, 95-103.

56. **I.Györi,** A generalization of Bellman's inequality for Stieltjes integrals and a uniqueness theorem, Stud. Sci. Math. Hungar. 6(1971), 137-145.

57. **A.W.Hales** and **G.R.Sells,** Multiple solutions of a differential equation, Amer. Math. Monthly 73(1966), 672.

58. **P.Hartman,** A differential equation with non - unique solutions, Amer. Math. Monthly 70(1963), 255-259.

59. **P.Hartman,** On uniqueness and differentiability of solutions of ordinary differential equations, Proc. Symp. Non-Linear Problems, Madison, Wisconsin 1963, Univ. Wisconsin Press, Madison, Wisconsin (1963), 219-232.

60. **P.Hartman,** Ordinary Differential Equations, Wiley, New York, 1964.

61. **E.K.Haviland,** A note on the convergence of the successive approximations to the solution of an ordinary differential equation, Amer. J. Math. 54(1932),632-634.

62. **J.W.Heidel,** Uniqueness, continuation, and nonoscillation for a second order nonlinear differential equation, Pacific J. Math. 32(1970), 715-721.

63. **M.Heins,** Some remarks on unicity and continuity theorems for ordinary differential equations, Michigan Math. J. 10(1963), 85-89.

64. **S.Hilger,** Analysis on measure chains - a unified approach to continuous and discrete calculus, Res. in Mathematics 18(1990), 18-56.

65. **E.Hille,** Lectures on Ordinary Differential Equations, Addison -Wesley, Reading, Mass., 1969.

66. **S.Hu** and **V.Lakshmikantham,** Some remarks on generalized Riemann integral, J. Math. Anal. Appl. 137(1989), 515-527.

67. **S.Hu** and **V.Lakshmikantham,** Continuous dependence and differentiability of solutions with respect to initial values of impulsive differential equations, Preprint.

68. **M.Inaba,** Über eine hinreichende Bedingung für die Unität der Lösung der gewöhnlichen Differentialgleichung, Proc. Phys. Soc. Japan 11(1929), 69-72.

69. **S.Iyanaga,** Über die Unitätsbedingungen der Lösung der Differentialgleichung: $\frac{dy}{dx} = f(x,y)$, Jap. J. Math. 5(1928), 253-257.

70. **J.Kalas,** Nonuniqueness for the solutions of ordinary differential equations, Czech. Math. J. 29(1979), 105-112.

71. **E.Kamke,** Differentialgleichungen reeller Funktionen, Akademische Verlagsgesellschaft, Leipzig, 1930.

72. **H.G.Kaper** and **M.K.Kwong,** Uniqueness results for some nonlinear initial and boundary value problems, Arch. Rational Mech. Anal. 102(1988), 45-56.

73. **H.G.Kaper** and **M.K.Kwong,** Uniqueness for a class of nonlinear initial value problems, J. Math. Anal. Appl. 130(1988), 467-473.

74. **T.Kato,** Nonlinear semigroups and evolution equations, J. Math. Soc. Japan 19(1967), 508-520.

75. **B.Kaymakcalan,** Existence and comparison results for dynamic systems on time scale, J. Math. Anal. Appl. (to appear).

76. **D.A.Kearns,** Elementary uniqueness theorems for differential equations, Amer. Math. Monthly 68(1961), 275-277.

77. **M.S.Kikodze,** On the question of uniqueness of a solution of the Cauchy problem and the convergence of successive approximations, Differencial'nye Uravnenija 2(1966), 1553-1560.

78. **T.Kitagawa,** An application of Prof. Perron's existence - theorem for a differential equation $\frac{dy}{dx} = f(x,y)$, Jap. J. Math. 9(1932), 153-160.

79. **O.Kooi,** The method of successive approximations and a uniqueness theorem of Krasnoselskii and Krein in the theory of differential equations, Indag. Math. 20(1958), 322-327.

80. **M.A.Krasnosel'skii** and **S.G.Krein,** Nonlocal existence theorems and uniqueness theorems for systems of ordinary differential equations, Dokl. Akad. Nauk SSSR (N.S.) 102(1955),13-16.

81. **M.A.Krasnosel'skii** and **S.G.Krein,** On a class of uniqueness theorems for the equation $y' = f(x,y)$, Uspehi. Mat. Nauk 11(1956), 209-213.

82. **M.A.Krasnosel'skii** and **A.V.Pokrovskii,** Systems with Hysteresis, Springer-Verlag, Berlin, 1989.

83. **J.Kurzweil,** Generalized ordinary differential equations and continuous dependence on a parameter, Czech. Math. J. 7(1957), 418-449.

101. **W.A.J.Luxemburg,** On the convergence of successive approximations in the theory of ordinary differential equations, Canad. Math. Bull. 1(1958), 9-20.

102. **W.A.J.Luxemburg,** On the convergence of successive approximations in the theory of ordinary differential equations, II, Indag. Math. 20(1958), 540-546.

103. **W.A.J.Luxemburg,** On the convergence of successive approximations in the theory of ordinary differential equations, III, Nieuw Arch. Wisk. 6(1958), 93-98.

104. **A.Majorana,** A uniqueness theorem for $y' = f(x,y), y(x_0) = y_0$, Proc. Amer. Math. Soc. 111(1991), 215-220.

105. **Ja.D.Mamedov,** One-sided estimates in the conditions for existence and uniqueness of solutions of the limit Cauchy problem in a Banach space, Sibirsk. Math. Z. 6(1965), 1190-1196.

106. **L.Markus,** A uniqueness theorem for ordinary differential equations involving smooth functions, Proc. Amer. Math. Soc. 4(1953), 88.

107. **E.J.McShane,** On the uniqueness of the solution of differential equations, Bull. Amer. Math. Soc. 4(1939), 755-757.

108. **L.A.Medeiros,** On nonlinear differential equations in Hilbert spaces, Amer. Math. Monthly 76(1969), 1024-1027.

109. **L.Merli,** Sul teorema di unicità per il problema di Cauchy relativo alla equazione differenziale $x^{(n)} = f(t, x, x^{(1)}, ..., x^{(n-1)})$, Rend. Accad. Naz. Lincei, Cl. Sc. Fis., mat. nat. 22(1957), 580-584.

110. **Z.Mikolajska,** Une remarque sur le thèorème de Jürgen Witte sur l'unicitè de la solution de l'èquation diffèrentielle $y' = f(x,y)$, Ann. Polon. Math. 38(1980),51-64.

111. **P.Montel,** Sur l'intègrale supèrieure et l'intègrale infèrieure d'une èquation diffèrentielle, Bull. Sci. Math. 50(1926), 205-217.

112. **R.D.Moyer,** A general uniqueness theorem, Proc. Amer. Math. Soc. 17(1966), 602-607.

113. **M.Müller,** Über das fundamentaltheorem in der Theorie der gewöhnlichen Differentialgleichungen, Math. Z. 26(1927), 619-645.

114. **M.Müller,** Neuere Untersuchungen über den Fundamentalsatz in der Theorie der gewöhnlichen Differentialgleichungen, Jber. Deutsch. Math. Verein. 37(1928), 33-48.

115. **H.Murakami,** On non-linear ordinary and evolution equations, Funkcial. Ekavac. 9(1966), 151-162.

116. **M.Nagumo,** Eine hinreichende Bendingung für die Unität der Lösung von Differentialgleichungen erster Ordnung, Jap. J. Mat. 3(1926), 107-112.

117. **W.Nikliborc,** Sur l'application de la mèthod des approximations successive dans la thèorie des èquations diffèrentielles, Studia Math. 1(1929), 201-209.

118. **M.J.Norris** and **R.D.Driver,** A uniqueness theorem for ordinary differential equations, SIAM J. Math. Anal. 12(1981), 141-144.

119. **H. Okamura,** Sur l'unicitè de la solution de $\frac{dy}{dx} = f(x, y)$, Mem. Coll. Sci. Univ. Kyoto Ser. A 17(1934), 319-328.

120. **H.Okamura,** Sur l'unicitè des solutions d'un système d'èquations diffèrentielles ordinaires, Mem. Coll. Sci., Kyoto Ser. A. 23(1940), 225-231.

121. **H.Okamura,** Condition nècessaire et suffisante remplie par les èquations diffèrentielles ordinaires sans points de Peano, Mem. Coll. Sci. Kyoto Imp. Univ. Ser. A. 24(1942), 21-28.

122. **C.Olech,** Remarks concerning criteria for uniqueness of solutions of ordinary differential equations, Bull. Acad. Polon. Sci. Ser. sci. math. astr. et phys. 8(1960), 661-666.

123. **C.Olech,** On the existence and uniqueness of solutions of an ordinary differential equation in the case of Banach space, Bull. Acad. Polon. Sci. Ser. sci. math. astr. et phys. 8(1960), 667-673.

124. **C.Olech** and **A.Plis,** Monotonicity assumption in uniqueness criteria for differential equations, Colloq. Math. 18(1967), 43-58.

125. **W.Osgood,** Beweis der Existenz einer Lösung der Differentialgle ichung $\frac{dy}{dx} = f(x, y)$ ohne Hinzunahme der Cauchy - Lipschitzschen Bedingung, Monatsh. Math. Phys. 9(1898), 331-345.

126. **S.G.Pandit,** On the stability of impulsively perturbed differential systems, Bull. Austral. Math. Soc. 17(1977), 423-432.

127. **S.G.Pandit,** Systems described by differential equations containing impulses : existence and uniqueness, Rev. Roum. Math. Pure. Appl. 26(1981), 879-887.

128. **S.G.Pandit** and **S.G.Deo,** Differential Systems Involving Impulses, Lecture Notes in Mathematics, Springer-Verlag, Heidelberg, 1982.

129. **G.Peano,** Dèmonstration de l'intègrabilitè des èquations diffèrentielles ordinaires, Math. Ann. 37(1890), 182-228.

130. **A.I.Perov,** On uniqueness theorems for ordinary differential equations, Dokl. Akad. Nauk SSSR 120(1958), 704-707.

131. **O.Perron,** Über Ein-und Mehrdeutigkeit des Integrals eines Systems von Differentialgleichungen, Mat. Ann. 95(1926), 98-101.

132. **O.Perron,** Eine hinreichende Bendingung für die Unität der Lösung von Differentialgleichungen erster Ordnung, Math. Z. 28(1928), 216-219.

133. **I.G.Petrovski,** Ordinary Differential Equations, Prentice-Hall, Inc. Englewood Cliffs, N.J., 1966.

134. **M.Rama Mohana Rao,** The local uniqueness and successive approximations, Bull. Inst. Politehnic Iasi 2(1963), 13-18.

135. **M.Rama Mohana Rao,** Some problems on general uniqueness and successive approximations, Proc. Nat. Acad. Sci. India Sec.A. 33(1963), 205-212.

136. **M.Rama Mohana Rao,** On some uniqueness theorems for ordinary non-linear differential equations, Proc. Nat. Acad. Sci. India Sec.A. 36(1966) , 49-53.

137. **P.Ramankutty,** Kamke's uniqueness theorem, J. London Math. Soc. 22(1980), 110-116.

138. **R.Redheffer,** An elementary proof of sharp uniqueness theorems for second-order initial-value problems, Amer. Math. Monthly 81(1974), 377-379.

139. **T.Rogers,** On Nagumo's condition, Canad. Math. Bull. 15(1972), 609-611.

140. **A.Rosenblatt,** Über die Existenz von Integralen gewöhnlicher Differentialgleichungen, Arck. Mat. Astro. Fys. 5(1909), 1-4.

141. **W.Rudin,** Nonuniqueness and growth in first-order differential equations, Amer. Math. Monthly , April (1982), 241-244.

142. **B.N.Sadovskii,** On the question of uniqueness conditions for ordinary differential equations, Uspehi Mat. Nauk 21(1966), 263-265.

143. **M.Samimi,** Nonuniqueness criteria for ordinary differential equations, Appl. Anal. 13(1982), 291-296.

144. **M.Samimi** and **V.Lakshmikantham,** General uniqueness results for ordinary differential equations, Appl. Math. Comp. 12(1983), 77-88.

145. **G.Sansone** and **R.Conti,** Non - Linear Differential Equations, Pergamon Press, Oxford, 1964.

146. **T.Satò,** Contribution à unicitè de la solution d'une èquation differentielle ordinaire, Jap. J. Math. 13(1936), 1-6.

147. **S.Schwabik,** Uniqueness for generalized differential equations, Recent Trends in Mathematics, Reinhardsbrunn 1982, BSB B.G.Teubner, Leipzig 263-269.

148. **S.Schwabik,** Generalized Ordinary Differential Equations, World Scientific, Singapore, 1992.

149. **G.Scorza-Dragoni,** Sulle condizioni sufficienti per l'unicità degli integrali di un'equazione differenziale, Rend. Circ. Mat. Palermo 54(1930), 430-448.

150. **G.Scorza-Dragoni,** A proposito di un teorema di Rosenblatt, Atti. de la Akad. Naz. Lincei 14(1931), 7-11.

151. **Xin-yao Shen,** Remark on the uniqueness of solutions of differential equations, Shuxue Jinzhan 6(1963), 267-271.

152. **T.Shimizu,** On sufficient conditions for the uniqueness of the solution of $\frac{dy}{dx} = f(x, y)$, Proc. Imperial Acad. of Japan 4(1928), 326-329.

153. **H.Stettner,** Nichteindeutigkeit bei gewöhnlichen Differentialgleichungen, Math. Nachrichten 64 (1974), 233-237.

154. **M.M.Subramaniam** and **V.Lakshmikantham,** On the general uniqueness theorem in ordinary differential equations, Proc. Nat. Acad. Sci. India Sec.A. 29(1960), 350-353.

155. **J.Szarski,** Remarque sur un critére d'unicité des intégrales d'une équation différentielle ordinaire, Ann. Polon. Math. 12(1962), 203-205.

156. **J.Tamarkine,** Sur le thèorème d'unicitè des solutions des èquations diffèrentielles ordinaires, Math. Z. 16(1922), 207-213.

157. **Z.Tesařovà** and **O.Doślý,** General uniqueness theorems for ordinary differential equations, Arch. Math. (Brnö), 16(1980), 217-224.

158. **L.Tonelli,** Sull' unicità della soluzione di un'equazione differenziale ordinaria, Rend. Acc. Naz. Lincei (6), 1(1925), 272-277.

159. **E.R.van Kampen,** Remarks on systems of ordinary differential equations, Amer. J. Math. 59(1937), 144-152.

160. **E.R.van Kampen,** Notes on systems of ordinary differential equations, Amer. J. Math. 63(1941), 371-376.

161. **B.Viswanatham,** The general uniqueness theorem and successive approximations, J. Indian Math. Soc. 16(1952), 69-74.

162. **B.Viswanatham,** A generalization of Bellman's lemma, Proc. Amer. Math. Soc. 14(1963), 15-18.

163. **S.Wallach,** The differential equation $y' = f(y)$, Amer. J. Math. 70(1940), 345-350.

164. **W.Walter,** Eindeutigkeitssätze für gewöhnliche, parabolische und hyperbolische Differentialgleichungen, Math. Z. 74(1960), 191-208.

165. **W.Walter,** Bemerkungen zu verschiedenen Eindeutigkeitskriterien für gewöhnl iche Differentialgleichungen, Math. Z. 84(1964), 222-227.

166. **W.Walter,** Über sukzessive Approximation bei Volterra-Integral- gleichungen in mehreren Veränderlichen, Ann. Acad. Sci. Fennicae Ser. A. I. Mathematica No. 345, 32p (1965).

167. **W.Walter,** Differential and Integral Inequalities, Springer-Verlag, Heidelberg, 1970.

168. **T.Wazewski,** Sur une extension du procèdè de I. Jungermann pour ètablir la convergence des approximations successives au cas des èquations diffèrentielles ordinaires, Bull. Acad. Polon. Sci. Ser. sci. math. astr. et phys. 8(1960), 43-46.

169. **T.Wazewski,** Sur l'existence et l'unicitè des intègrales des èquations diffèrenti elles ordinaires au cas de l'espace de Banach, Bull. Acad. Polon. Sci. Ser. sci. math. astr. et phys. 8(1960), 301-305.

170. **D.V.V.Wend,** Uniqueness of solutions of ordinary differential equations, Amer. Math. Monthly 74(1967), 948-950.

171. **D.V.V.Wend,** Existence and uniqueness of solutions of ordinary differential equations, Proc. Amer. Math. Soc. 23(1969), 27-33.

172. **A.Wintner,** On the convergence of successive approximations, Amer. J. Math. 68(1946), 13-19.

173. **A.Wintner,** On non-constant Lipschitz factors in the uniqueness problem of ordinary differential equations, Arch. Math. 7(1956), 465-468.

174. **A.Wintner,** On the local uniqueness of the initial value problem of the differential equation $\frac{d^n x}{dt^n} = f(t,x)$, Boll. Un. Math. Ital.11(1956), 496-498.

175. **J.Witte,** Ein Eindeutigkeitssatz für die Differentialgleichung $y' = f(x,y)$, Math. Z. 140(1974), 281-287.

176. **En-Ho Yang,** Note on the uniqueness of solutions of differential equations, Shuxue Jinzhan 8(1965), 183-186.

177. **J.A.Yorke,** A continuous differential equation in Hilbert space without existence, Funkcial. Ekavac. 13(1970), 19-21.

178. **T.Yoshizawa,** Stability Theory by Liapunov's Second Method, The Math. Soc. Japan, Tokyo, 1966.

179. **T.Yosie,** Über die Unität der Lösung der gewöhnlichen Differentialgleichungen erster Ordnung, Jap. J. Math. 2(1926), 161-173.

180. **P.P.Zabreiko,** On uniqueness theorems for ordinary differential equations, Differential'nye Uravnenija 3(1967), 341-347.

181. **A.D.Ziebur,** Uniqueness theorems for ordinary differential equations, Amer. Math. Monthly 69(1962), 421-423.

182. **G.Zwirner,** Criteri d'unicità per un problema di valori al contorno per equazioni e sistemi di equazioni differenziali ordinari d'ordine qualunque, Mat. Uni. Padova 13(1942), 9-25.

183. **G.Zwirner,** A proposito di un teorema di unicità per gli integrali della equazioni differenziali del primo ordine, Ann. di Mate. et. Pura. Appl. 24(1945), 153-156.

184. **G.Zwirner,** Criteri di unicità per gli integrali della equazioni differenziali del primo ordine, Rend. Seminario Mat. Univ. Padova 19(1950), 273-293.